PRODUCTION TECHNOLOGY

PRODUCTION
TECHNOLOGY

STANLEY A. KOMACEK

Department of Industry and Technology
California University of Pennsylvania

 DELMAR PUBLISHERS INC.®

NOTICE TO THE READER

COVER PHOTO CREDITS: Cover photos by COMSTOCK Inc./Jim Pickerell (top photo) COMSTOCK Inc./Michael Stuckey (bottom photo)

DELMAR STAFF
Associate Editor: Christine E. Worden
Developmental Editor: Christine E. Worden
Senior Project Editor: Christopher Chien
Production Coordinator: Teresa Luterbach
Art Manager: Rita Stevens
Art Coordinator: Michael Nelson
Design Supervisor: Susan C. Mathews

For information address Delmar Publishers Inc.
2 Computer Drive West, Box 15-015
Albany, New York 12212-5015

Printed in the United States of America
Published simultaneously in Canada
by Nelson Canada,
a division of The Thomson Corporation

10 9 8 7 6 5 4 3 2 1

Library of Congress Cataloging-in-Publication Data
Komacek, Stanley A.
    Production technology/Stanley A. Komacek.
     p.   cm.
    Includes index.
    ISBN 0-8273-4837-1
    1. Production engineering.   I. Title.
TS176.K638 1993
670—dc20

91-32520
CIP

CONTENTS

(Courtesy of Toyota Motor Manufacturing, USA, Inc.)

(Courtesy of Amoco Corporation)

SECTION THREE _____
PRODUCTION TOOLS, MACHINES, AND PROCESSES   103

(Courtesy of FMC Gold Company)

SECTION FOUR _____
MANUFACTURING SYSTEMS   169

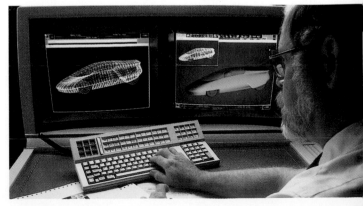

SECTION FIVE
CONSTRUCTION SYSTEMS   255

(Courtesy of Pat Dineen/photo by Shirley Nelson)

SECTION SIX
PRODUCTION IN THE FUTURE 383

(Courtesy of Lockheed Corporation/photo by Dick Luria)

PREFACE

Production technology is central to the way we live. Without the products made with manufacturing technology and the structures built with construction technology, our lifestyles would be very different. Modern production technology makes possible the buses, cars, and bicycles we use every day to go to school or to work. Telephones, televisions, radios, computers, even the paper that this book is printed on, would not exist without manufacturing technology. Homes, shopping malls, baseball stadiums, bridges, highways, even your school, would not exist without construction technology. Together, manufacturing and construction are the basic systems of production technology. Production technology affects almost every aspect of our lives and of people's lives around the world.

Organization

Production is a complete technology composed of many parts. To make it easier for you to learn about the many aspects of production, this textbook includes 28 chapters. The chapters are grouped into six sections.

Section One: Technology and Production. This section introduces you to production and its relationship to people and the other technologies. You will read about the history of production, problem solving methods, careers in production, the organization and management of production technologies, and why manufacturing and construction are the two subsystems of production.

Section Two: Production Materials. This section covers the basic materials used in production technology. You will read about where materials come from, how raw materials are changed into production materials, the science of materials and the various classes of materials, including metals, woods, plastics, ceramics, and composites.

Section Three: Production Tools and Machines. This section gives you the basic information you need to use tools and machines to manufacture products and construct structures. You will be introduced to the importance of using tools and machines safely and efficiently, see how six basic machines provide mechanical advantage to tool users, and learn about the various classes of production tools and machines, including measuring and layout tools, separating tools, forming tools, and combining tools. Also, you will learn how computers, robots, and lasers are being used to automate production tools and machines with processes like CAD, CAM, CNC, and other high-tech processes.

Sections One, Two, and Three give you the "basics" you need to work safely and efficiently in production technology. The next two sections are the heart of any production technology course; these deal with manufacturing and construction.

Section Four: Manufacturing Systems. This section covers the various types of manufacturing systems. It explains how several manufacturing departments work together to design, make, and sell products. You will learn about the entire process from product design and production engineering to mass production to advertising and selling products. Also, you will learn about the importance of money to manufacturing systems.

Section Five: Construction Systems. This section covers the various types of construction systems. It explains the process of designing, engineering, and constructing buildings and other structures. You will learn about the entire process, including architectural design and drawing; writing specifications and contracts; preparing the construction site; building foundations, floors, walls, and roofs; installing utilities; and finishing and landscaping. Also, you will learn about the heavy construction technologies used to build bridges, skyscrapers, highways, dams, and towers.

Section Six: The Future of Production. This section covers the environmental impacts of production, including scrap, waste, and pollution. You will read about the importance of recycling and pollution control. This section also covers the possibility that production may someday be conducted in space on a permanent basis. The final chapter looks at a number of trends that may affect the future of production technology and society.

Near the end of the text are two very important sections. The first provides a number of plans for products and structures you and your classmates can produce. The other section provides several ''design briefs'' that pose problems that you may get the chance to solve. The production plans and design briefs will require you to use all the knowledge and skills you have gained by reading the textbook.

Safety

Working in, and studying, production technology can be dangerous. You should not be afraid, but you should realize that you can be hurt when working with tools and machines. Near the beginning of the text is a special section on safety. Be sure to read that section carefully. Also, throughout the text there are many references to the importance of safety. Always follow all safety rules, and remember the ABC's of safety; Always Be Careful!

Special Features

Production Technology uses a number of special features, including:

Safety Guidelines. A special safety section introduces the basic safety rules that every student should follow when working in a technology laboratory.

Key Terms. Listed at the beginning of each chapter, these important terms and phrases are highlighted within the text.

Boxed Articles. These are short stories of interesting or unusual information related to the chapter's subject.

Photographs and Illustrations. There are hundreds of color photos, illustrations, and line drawings that will help you understand the important parts of manufacturing.

Summary. The key points of each chapter are summarized.

Discussion Questions. These questions stress critical thinking and problem-solving skills.

Chapter Activities. Hands-on and/or minds-on activities are included in most chapters. Math and science concepts relating to each activity are presented.

Product Plans. There are numerous plans for products that you and your classmates can manufacture.

Design Briefs. Also included in this text are design briefs. A design brief is an activity that provides you with the basic information needed to make a product or structure. The design briefs let you apply the skills and knowledge you have developed by reading the textbook to a technological problem.

Technology Student Association. A special section describes the Technology Student Association and its manufacturing-related activities.

Glossary. A complete glossary of terms with definitions is included as an appendix to help you study.

Acknowledgments

The author wishes to express his deep appreciation to the many people that contributed ideas, provided assistance, and helped make this textbook a usable resource for learners studying production technology. Three very important contributors were Mark Huth, Andy Horton, and Ann Lawson. The material they provided, through previously published textbooks, made this textbook a possibility. Thanks are also given to Richard T. Kreh, Sr., Larry Jeffus, and Paul E. Meyers for their work in supplying photographs for this textbook. Finally, thank you to Chris Worden, Chris Chien, and the other people at Delmar Publishers for transforming rough manuscript into a real textbook.

The following individuals reviewed the manuscript and offered suggestions for improvement. Their assistance is appreciated.

Bruce Barnes
Bloomington Hubert Olson
 Junior High School
Bloomington, Minnesota

Mark Means
Park Hill Junior High School
Dallas, Texas

Chuck Bridge
Chisholm Trail Middle School
Round Rock, Texas

Barbara Brock
Galena Park Middle School
Galena Park, Texas

Bernard Maas
Handley Middle School
Fort Worth, Texas

Ernest Savage
Bowling Green State
 University
Bowling Green, Ohio

Al Troyer
Forest Meadow School
Dallas, Texas

About the Author

Dr. Stanley Komacek is a veteran of eleven years teaching in industrial arts and technology education, including three as a technology education curriculum consultant, and five years as a teacher educator. He currently teaches in the Technology Education Program of the Department of Industry and Technology at California University of Pennsylvania. This textbook is the second for Dr. Komacek with Delmar Publishers. He has written more than thirty-five activity modules, articles, and curriculum guides in production and other technology areas. He has also given more than thirty presentations at local, state, and national conventions in technology education. Dr. Komacek holds an Ed.D. degree from West Virginia University, an M.Ed. from Miami University, and a B.S. degree from California University of Pennsylvania.

The following individuals provided activities, product plans, and other information for the text. Their assistance is also appreciated.

Neil Eshelman
Williamsport High School
Williamsport, Pennsylvania

Paper Tower Design Brief #1

Henry Harms
Great Hollow Junior
 High School
Nesconset, New York

Homemade Robotic Arm,
Package Design

Jim Jordan
Roosevelt Middle School
Erie, Pennsylvania

Recycling Center
 Design Brief

Albert Komacek
Brownsville High School
Brownsville, Pennsylvania

Separating, Forming,
 Combining

George McCartney
Westfield High School
Westfield, Wisconsin

Sheet Metal Planter

Ray Miller
Fairmont State College
Fairmont, West Virginia

Picture Frame Product —
 Wood, Metal, and Plastic

Jerry Murphy
Gardiner, Montana

Surveying the Market

Fred Postuma
Westfield High School
Westfield, Wisconsin

Re-Use It

A special thank you is expressed to the author's family; Cathie, Claire, and Anne, for the considerable and valuable patience, encouragement, love, and understanding.

SAFETY GUIDELINES

Safety is a very important concern in production and in school production laboratories. In the production industry safety is so important that people are given a special job to watch over the safety of other production workers. These people are called safety managers.

Safety Manager

Companies often have safety managers who run special safety programs. It is the job of the safety manager to teach other workers about the importance of safety. Learning about safety may be done through classes on the safe use of tools and machines, reading safety posters, and special programs to get workers thinking about safety. The safety manager also makes sure that tools, machines, and the workplace are free of safety hazards. The manager may inspect the workplace and find ways to remove any safety hazards that exist. In very large companies there may be several safety managers to find all the potential problems.

OSHA

In 1970, the U.S. government started the Occupational Safety and Health Administration (OSHA). OSHA (pronounced 'oh sha) has written rules and laws that require employers to meet certain safety standards. OSHA checks on the safety of workers in all areas of industry. Many of the laws of OSHA relate to production technology.

Personal Safety

Even though OSHA requires the employer to keep the workplace safe, the workers are responsible for their own personal safety. Production workers on the job must work safely to avoid accidents. Students in the school production laboratory must also work safely. School laboratories are as safe as a production job can be, but if students in that laboratory do not practice good safety habits, the chance of serious injury is high.

In this text you will find several safety cautions. There are many general safety rules that apply to working in production. There are other rules for specific tools and machines. Also, your teacher may have other safety rules for the tools and machines in your school production laboratory. Most of these rules may seem very natural to you; some may not. Read these rules carefully. Ask your teacher for help if you do not understand all the rules. Then, be sure to practice them. It is not enough to know all the safety rules; you should practice them so they become habits. Here are some general safety rules:

- Do not operate any machine or use any tool until you have been taught how to use it by your teacher.
- Use tools and machines only in the manner in which they were intended to be used.
- Do not distract others while they are working.
- Running and horseplay are not permitted in the laboratory, classroom, or other locations where the production class may meet.
- Wear proper safety clothes and equipment (e.g., safety glasses, hard hats, sturdy shoes, welding gloves) at all times when working in the production class.
- Use electrical equipment only with proper grounding and in dry areas.
- Operate a power tool only when all safety guards are in place.
- Do not work with dull or damaged tools.
- Know the locations of emergency devices (e.g., fire extinguishers, electrical disconnects, first aid kits) and how to use them.
- Do not attempt to lift or carry heavy loads—get help.
- Do not leave hazards for others (e.g., hot metal, exposed nail or screw points, a running machine).
- Report any injury, no matter how slight, to your teacher.
- Follow the ABC's of safety—**A**lways **B**e **C**areful!

SECTION ONE

TECHNOLOGY AND PRODUCTION

(Courtesy of Toyota Motor Manufacturing, U.S.A., Inc.)

CHAPTER 1

Introduction to Production Technology

OBJECTIVES

After completing this chapter, you will be able to:

- Define and explain the differences between technology, production technology, manufacturing technology, and construction technology.
- Explain the relationship between production technology and the technologies of communication, transportation, and energy and power.
- Describe the two basic types of manufacturing technology.
- Describe the two basic types of construction technology.
- Summarize the history of production technology from earliest times to the age of automation.
- Identify various positive and negative impacts of production technology.

KEY TERMS

Automation
Bartering
Communication technology
Computer
Conservation
Construction technology
Cottage industry
Custom manufacturing

Energy and power technology
Heavy construction
Impact
Industrial Revolution
Innovations
Inventions
Light construction
Manufacturing technology

Mass production
Production technology
Recycling
Resources
Structures
Technology
Transportation technology

What Is Technology?

Technology is all around us. It is part of our everyday lives. Without it our lives would be very different. Production is based on technology. What, then, is technology? Simply defined, **technology** is the use of tools, materials, and processes to meet human needs and wants. The food we eat has been grown, packaged, and shipped to the store through technology. Your home was planned and built by people using tools, materials, and processes. The clothes we wear and the furniture we use were also made using technology.

Technology is used by people to increase their power to make or do something. If you had a board you wanted to cut in half, you would not

FIGURE 1-1 Technology, like the computers shown here, needs humans to operate it. Technology includes tools, materials, and processes used to meet human needs and wants. *(Courtesy of LTV Aircraft Products Group, Dallas, TX)*

be able to do it with your bare hands. But with a tool of technology, a saw, your power to cut the board would be increased. If you wanted to get to the movies by a certain time and you had to walk, you would have to leave very early. With technology — a bicycle, automobile, or train — your power to get to the movies on time would be increased. Technology makes certain jobs and tasks easier to do.

Without people, technology would not exist. Technology requires people. People use their knowledge to apply tools, materials, and processes, Figure 1-1. Every day, we make decisions about how we will use the tools, materials, and processes of technology. These decisions include whether to write a letter or call on the telephone; whether to walk, ride a bike, or drive a car; and whether to buy a product or try to make it. Making decisions is a very important part of using technology.

Technology always causes change. Think of the changes in technology since your grandparents were children. Your parents have also seen many changes in technology during their lifetimes. Changes in technology can affect peo-

ple's lifestyles, their health, and how they work and learn, Figure 1-2. Technology also changes the natural environment. Water, air, and land are

FIGURE 1-2 Changes in technology affect the way we learn today. *(Courtesy of Eaton Corporation — Cleveland, OH)*

affected by changes in technology. Because humans make the decisions about the use of technology, they can control the changes caused by technology. To make the best choices, we must know how technology works. This book is about one part of technology: production. Reading this book will help you learn how to make better decisions about the use of technology.

Production Technology

Production technology is the study of how the products and structures (buildings) we use every day are made. Production technology is broken into two fields: manufacturing and construction, Figure 1–3. **Manufacturing technology** can be defined as the making of products in a factory.

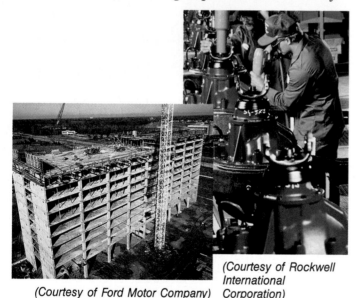

(Courtesy of Rockwell International Corporation)

(Courtesy of Ford Motor Company)

FIGURE 1–3 Manufacturing and construction are the two fields of production technology.

Furniture, automobiles, clothing, and airplanes are examples of products made with manufacturing technology. **Construction technology** can be defined as the building of **structures** that cannot be moved. Buildings, roads, bridges, and pipelines are examples of structures built with construction technology. Manufacturing and construction are both production technologies because they both produce (make) a product. The products of manufacturing and construction are made to meet human needs and wants. We need shelters to live in — construction technology provides homes and apartments. We need furniture and clothing — manufacturing technology provides them.

Production and the Other Technologies

Production technology is related to all the other technologies. The other major technologies are communication, transportation, and energy and power, Figure 1–4. **Communication technology** is the study of sending and receiving information. Radios, televisions, cameras, and computers are examples of communication technology. **Transportation technology** moves people and things using vehicles like buses, trucks, cars, airplanes, and even the space shuttle. **Energy and power technology** is the base of all the other technologies. Energy and power provides the muscle needed to communicate information, transport things, construct structures, and manufacture products.

The space shuttle is a good example of how production and the other technologies are related,

Energy and power *(Courtesy of New York Power Authority)* **Transportation** *(Courtesy of Aluminum Association, Inc.)* **Communication** *(Courtesy of Contel Corporation)*

FIGURE 1–4 Production technology is related to communication, transportation, and energy and power technologies.

FIGURE 1–5 To get the space shuttle off the ground, it takes all the technologies working together. *(Courtesy of NASA)*

FIGURE 1–6 Today, most products are mass produced; that is, large numbers of the same product are made. *(Photo by Kenneth A. Deitcher, M.D.)*

Figure 1-5. The shuttle itself is a large product that was manufactured in a factory. The tower and launch pad from which the space shuttle lifts off was constructed. The shuttle has very complex communication devices that the astronauts use to send information back to earth when they are in space. The space shuttle itself is an example of a transportation technology. Its job is to transport satellites and astronauts into space. Finally, energy and power is needed to get the heavy space shuttle off the launch pad and into space. Almost anything you can think of, from a package of chewing gum to the space shuttle, uses all of the major technologies.

Types of Manufacturing

There are two basic types of manufacturing: custom and mass. **Custom manufacturing** involves one person making one product by hand. All the parts on the product are made by the worker. Today, very few products are made by custom manufacturing. The products that are custom manufactured today include one-of-a-kind products like the space shuttle. Most products today are mass produced. **Mass production** involves making a large number (a mass) of the same product, Figure 1–6. A group of people work together to mass produce the product. Each person in the group is given one small job to do in mass production.

Types of Construction

There are also two basic types of construction: light and heavy. **Light construction** is the

FIGURE 1–7 Light construction refers to houses and other small buildings. *(Courtesy of the California Redwood Association)*

building of homes and other small buildings, Figure 1–7. These may be single-family homes, small apartment buildings, offices, stores, and so on. Larger structures are not included in this group because the tools and materials used are different from those used in light construction. **Heavy construction** includes the building of roads, bridges, tunnels, and factories. This type

The Tallest Building

The 110-story Sears Tower in Chicago is the world's tallest building. It contains 4½ million square feet of space, making it the largest private office building in the world. Designed by the architectural company of Skidmore, Owings & Merrill, the tower reaches 1,454 feet above the ground. Twin antenna towers atop the building bring its total height to 1,707 feet (nearly one third of a mile). Sears Tower has some spectacular statistics:

Courtesy of Sears, Roebuck & Company

■ Construction of Sears Tower took three years. During peak times, 1,600 people worked on the project.

■ The framework of the tower consists of 76,000 tons of steel. The building contains enough concrete to build an eight-lane highway, five miles long. It has 16,000 bronze-tinted windows and 28 acres of black aluminum skin.

■ 114 concrete piles support the 222,500-ton building and each is securely socketed into the bedrock.

■ The building's 104-cab elevator system, including 16 double-decker elevators, divides the building into three separate zones, with lobbies between. A special elevator for baby strollers and wheelchairs serves the public areas of the tower.

■ Six automatic window-washing machines clean the building exterior eight times a year.

■ 12,000 people work in Sears Tower, including 6,500 Sears, Roebuck and Co. national headquarters employees.

■ Sears occupies the lower half of the building: the remainder is leased to tenants.

■ The tower contains 30 stores featuring everything from souvenirs to women's and men's clothing. Seven restaurants offer everything from a quick bite to a full dinner.

of construction uses larger equipment, heavier materials, and more time than light construction. There are three different types of heavy construction: commercial, industrial, and civil. Commercial construction includes large buildings like skyscrapers, Figure 1–8. Industrial construction includes factories and other structures built for industry, Figure 1–9. Power plants, steel mills, and oil refineries are examples of industrial construction. Civil construction involves building roads, dams, tunnels, airport runways, and bridges, Figure 1–10. Civil construction is more involved with moving dirt and earth than the other types of construction.

The History of Production

People have been producing items to meet their needs for thousands of years. The first humans used simple tools and natural materials from their local area to make things. An early example is weapons for hunting and protection. These early people did not make anything more than

FIGURE 1–9 Industrial construction builds structures like this power plant. *(Courtesy of Niagara Mohawk Power Corporation)*

FIGURE 1–10 Highways are examples of civil construction. *(Courtesy of Conrail)*

FIGURE 1–8 Office buildings and skyscrapers are examples of commercial construction. *(Courtesy of PPG Industries)*

what they needed. Their tools were basic, simple, and very important to their survival, Figure 1–11. During this time, people did not build structures. They lived in caves and other shelters they could find.

Between 10,000 and 9000 B.C., people discovered how to grow food. It was then that the first small towns appeared. Farming tools, like the hand plow, greatly changed the people's lives, Figure 1–12. Once people could grow their own food, they had more time to produce other things. Some of the first things they produced included homes, clothes, and cooking and eating utensils. Each person or family had to make everything they needed to live. There were no shops or stores in which to buy things. The population steadily grew during this period and the size of villages increased.

FIGURE 1–11 Early people used simple technology to meet their basic needs.

FIGURE 1–12 When people began to grow their own food, they had more time to create new technology. *(Photo by Edith Raviola)*

Cottage Industries

The next important stage in the history of production was cottage industries. A **cottage industry** is a small production plant set up in the home (cottage), Figure 1–13. During this period, homes were made from logs, stones, and other available materials. Families began to specialize in one job or another. All the members of the family became experts in that job. In a village there would be a number of different families, each with a different job that they did well. One family would make clothes, another furniture, another candles, and so on. Still others would specialize in building homes or other structures. All of the products from the cottage industries were made by hand.

A system of **bartering** developed. Bartering is another word for trading. The families would barter (trade) the products they made for products made by another family. Bartering still exists today in many places. However, money quickly replaced it as the most common way to trade products.

The Industrial Revolution

The **Industrial Revolution** refers to the changes in production that happened around 1750 to 1850. These changes affected almost all parts of people's lives. The changes did not happen over night.

The main changes were in the area of production. Many new **inventions** (new technology) and **innovations** (improved technology) were created during this time. Many of the inventions and innovations made the job of production easier and less time consuming than the handmade cottage industry. Three types of inventions and innovations brought about the Industrial Revolution:

1. replacing hand tools with power tools and machines,
2. using the steam engine as a new source of energy and power, and
3. working in large factory systems.

Work once done with hand tools was replaced by power tools and machines. The main difference between hand tools and power tools and machines is the source of energy and power.

FIGURE 1–13 Here is an early drawing of a cottage industry. The entire family is working to make cloth. *(Courtesy of the Museum of American Textile History)*

Hand tools need a person to supply the power, Figure 1–14. Power tools and machines use a power source other than a person. A person is still needed to operate a machine, but another power source is used to actually do the work.

Up to this point in history, wind, water, and muscle power were the most common sources of energy and power. Because wind and water depended on the weather and the seasons, a more reliable source of power was needed. The invention of the steam engine by James Watt around 1770 provided the new power needed in the Industrial Revolution, Figure 1–15. Without the steam engine, the change from hand tools to power machines would not have been possible.

FIGURE 1–14 During the Industrial Revolution, hand-operated tools were replaced by power machines. *(© Wayne Michael Lottinville)*

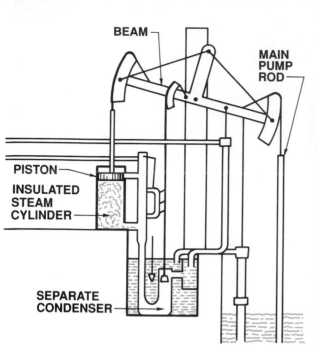

FIGURE 1–15 James Watt's steam engine gave the Industrial Revolution the new source of energy needed for power machines.

FIGURE 1–16 This farm tools plant was typical of early factories in the Industrial Revolution. *(Courtesy of Deere & Company.)*

The factory system was invented because the steam engine and the new machines were too costly for individuals to own and operate in the home. As a result, it became common for several people to combine their money and form large companies. These people built special buildings that would hold many machines and workers, Figure 1–16. This resulted in the movement of production from the home to the factory.

With most production work being done in large factories, cities developed. People wanted to live close to where they worked. Apartment buildings became a popular way to live in the city close to the factory.

The Age of Automation

The Industrial Revolution lasted for more than one hundred years. During these years, both workers and other people tried to improve on the factory system. Many methods were used to make the job of production workers easier and faster. As production improved, we entered the Age of Automation. We are still in the Age of

Automation today. **Automation** is when most or all of the machines and processes run with little or no human control. Automated (electronic) devices do many of the jobs once done by assembly workers. Automation is often better because electronic machines do not take breaks or vacations like people. Automated machines can work twenty-four hours a day without pay in working conditions that people would not be able to withstand. Some people are still needed to set up, operate, and repair the machines.

The most important electronic machine used in automated manufacturing today is the computer. A **computer** is a machine that can be programmed to accept, process, and display information. A common use for computers in production is to control machines, Figure 1–17. Companies also use computers for a number of other purposes. Three special uses of computers in manufacturing are computer-aided manufacturing (CAM), computer-integrated manufacturing (CIM) and computer numerical control (CNC). These will be explained in detail in later chapters.

FIGURE 1–17 Today, computers often control tools, machines, and processes in industry. *(Courtesy of Dow Chemical Company)*

FIGURE 1–18 A positive impact of production technology is the clothes and other products we have available to meet our needs and wants. *(Courtesy of Sony Corporation of America)*

The Impacts of Production

Production technology creates certain impacts. An **impact** is an effect on something or the result of something. For example, if you study for a test, your studying may have an impact on your grade. If you get a good grade, you could say your studying had a positive impact. If you studied the wrong chapters in the book and got a poor grade, you could say your studying had a negative impact. Production technology produces both positive and negative impacts. Although production technology tries to meet our needs and wants in positive ways, negative impacts sometimes occur.

Some of the most important types of impacts from production technology include meeting human needs and wants, jobs, pollution, and resource use.

Meeting Needs and Wants

As mentioned earlier, production technology meets our needs and wants by providing structures and products. Usually, the structures and products have positive impacts on our lives. Our homes and the buildings we work and live in protect us from the rain, wind, and snow. Our products, like clothes, furniture, and other devices, provide us with protection, comfort, and enjoyment, Figure 1–18. These are all examples of positive impacts. Sometimes, a structure or a product may not work as we hoped. Then, we have a negative impact. If our house or clothes started to fall apart, it could be dangerous, costly, and embarrassing. These are negative impacts.

Jobs

Production technology provides jobs for people who want to work. People work to make money so they can pay for a home, an automobile, clothes, and food. A production job also gives workers a good feeling of knowing they helped to make products or build structures that are used by other people. Production technology provides a wide variety of different types of jobs. People with a wide range of knowledge, skills, and talents are needed to work in production. Providing jobs is a positive impact of production.

When automated equipment replaces workers, that can be a negative impact for the human. Unemployment caused by job automation is just another example of the changes caused by technology.

Pollution

Manufacturing products and building structures can produce pollution. Pollution is a negative impact of production technology on water, air, or land. Some manufacturing plants dump their waste materials into local rivers or lakes.

This causes water pollution that kills fish and makes the water undrinkable. Other plants have chimneys that place smoke, dust, and other materials into the atmosphere, causing air pollution. Finally, some construction projects, like dams, roads, and parking lots, can make land unsuitable for other uses.

Resource Use

All technologies use resources. Our **resources** include raw materials like woods, metals and other minerals, energy sources (coal and oil), and water. As technology continues to grow, our use of resources also increases. Technology has grown so fast in recent decades, some people are afraid we may soon deplete (use up) our available supply of resources, Figure 1–19. In order to slow down the use of our resources, experts suggest we use conservation and recycling methods.**Conservation** means making wise use of available resources. **Recycling** means reusing what has already been used once. Depleting our supply of resources is a negative impact of production. By using conservation and recycling methods, we can lessen the negative impact of resource use.

Summary

Technology is the use of tools, materials, and processes to meet human needs and wants. Technology is used to increase people's power to do certain jobs and tasks. Production is one type of technology. The two types of production technology are manufacturing and construction. Manufacturing technology produces products in a factory. Construction technology produces (builds) structures that cannot be moved. The two basic types of manufacturing are custom manufacturing and mass production. The two basic types of construction are light and heavy construction. Production technology works together with the other technologies of communication, transportation, and energy and power.

Throughout the history of production, there have been many inventions and innovations that tried to make work easier. The history of production has gone from a time when people made

FIGURE 1–19 As production technology increases, our natural resources, like these trees, are being used more. This is a negative impact. *(Courtesy of Weyerhaeuser Company Inc.)*

tools for survival, to the cottage industries, to the Industrial Revolution, to the Age of Automation. During each stage, changes in technology caused changes in the way people lived and worked.

Production technology creates positive and negative impacts. Four examples of the impacts of production technology are meeting needs and wants, providing jobs, creating pollution, and using resources.

DISCUSSION QUESTIONS

1. How would you define the following terms: technology, production technology, manufacturing technology, and construction technology? What are the differences between these technologies?

2. How does production technology relate to other technologies like communication, transportation, and energy and power?

3. What are the two basic types of manufacturing technology? How are they different?

4. What are the two basic types of construction technology? How are they different?

5 How would you describe the history of production technology? What were some of the major events or periods?

6. Can you name four positive and four negative impacts of production technology?

CHAPTER ACTIVITIES

 ## Toolmaker

OBJECTIVE

This activity will help you understand how the earliest people made the simple tools they needed to survive. They used only easily found materials, human power, and creative thinking.

EQUIPMENT AND SUPPLIES

1. Handout "Biography of a Toolmaker"
2. Any materials found locally:

- rocks and stones
- mud
- shells
- sticks and branches
- clay
- vines (other rope-like materials)

3. Safety eyewear
4. Assorted tools and machines

PROCEDURE

1. As a class, discuss early toolmaking methods and uses for types of tools. Refer to the toolmaking handout.
2. Complete Part I of the handout "Biography of a Toolmaker."
3. Decide what type of tool you would like to make. Two ideas are provided in Figure 1.
4. Find the right toolmaking materials. Remember, you should make your tool from materials that can be found locally.
5. Carefully shape and put together (produce) your tool.
   NOTE: Be sure to wear safety glasses for this activity. Early people did not have safety devices, but we do and they should be used.
6. Report the information about your toolmaker and share your newly made tool with the class. If possible, demonstrate the use of your tool.

MATH AND SCIENCE CONNECTIONS

Scientists who dig up the tools and artifacts of early people are called archaeologists. To determine the age of the artifacts, they use a chemical method called carbon dating. Carbon is an element found in most materials. When the artifacts are buried in the ground, the amount of carbon in them decreases slowly and steadily. The half-life of carbon is about 5730 years. This means that one half of the carbon in an artifact will be lost in 5730 years. By measuring the amount of carbon, scientists can calculate the age of the artifacts.

RELATED QUESTIONS

1. Good toolmaking skills often made the difference between life and death for early humans. Now that you have tried to make a simple tool, what would you do differently next time, if your life depended on it?

2. The first tools were used for hunting and defense. What might have been produced next as tools, materials, and processes improved? Explain what materials and processes might have been used.

3. What impact do you think controlled fire-making had on early people, especially in terms of making tools?

4. How are the materials and processes used in this activity still used today? What new materials and processes have replaced them?

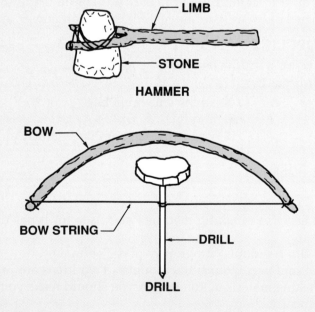

Figure 1. Some toolmaking ideas

Home Town Production Technology

OBJECTIVE

There are two types of production technology—manufacturing and construction. In this activity, you will research the types of production technology in your home town or local community.

EQUIPMENT AND SUPPLIES

1. Several telephone books or business directories
2. Pencil and paper

PROCEDURE

1. As a group, think of all the different examples of production technology in your home town or local community.
2. With the help of your teacher, use the telephone book or business directory to find examples of production technology businesses in your community.
3. Prepare a list of the manufacturing and the construction businesses.
4. Call the business owners and ask them what type of work they do. Ask them how many people work for that business.
5. Discuss the impact of these businesses on your local community.

MATH AND SCIENCE CONNECTIONS

If the companies are large ones, ask them for their annual reports. The annual report includes data on the money made by the company. Look at the figures to help you decide what impact the company has on your community.

RELATED QUESTIONS

1. How many businesses did you find in your local community related to production technology? How many were manufacturing or construction companies?
2. What types of businesses did you find? Were they large or small?
3. How many people are employed by the production businesses in your community?
4. What positive and/or negative impacts do these production businesses have on your local community?

Problem Solving and Technology Systems

KEY TERMS

Brainstorming	Outputs	System
CAD	Pictorial drawing	System model
Drafting	Problem solving	Technologists
Feedback	Processes	Trial and error
Inputs	Sketching	
Modeling	Subsystems	

As you read this book and study production, you will start to become a technologist. A **technologist** is a person who studies technology. Technologists have certain knowledge, skills, and attitudes about technology. They know how technology has changed through history. They have the skills to use tools and machines to perform processes. They can use tools and materials to build structures and make products. They also know how technology causes positive and negative impacts. A technologist is a special person.

One important set of skills that technologists need is problem solving. **Problem solving** is the ability to find solutions to problems or find better ways of doing something. As you read in Chapter 1, technology always causes changes and impacts. Along with changes and impacts come problems that must be solved. Throughout the history of technology, inventions and innovations have been created to help people solve problems. Just a few of the problems related to production technology include the following questions:

- how do we design safe products and structures that do not cost a lot of money?
- what is the best way to build a structure or make a product?
- which tools should be used to perform certain processes?
- how can we use technology so negative impacts are kept to a minimum?
- how can we best use technology to meet human needs and wants in the future?

Recent surveys of employers show that problem solving is one skill they want most in their workers. Some people say problem solving is a basic skill for living and working in our technology-based world.

How can you learn to be a good problem solver? One way is to read this chapter and learn more about the technology problem solving method. Another way is to use the problem solving method described in this chapter to solve problems. As with other skills, problem solving can be learned and improved through practice.

The Technology Problem Solving Method

There are a number of different ways that problems can be solved using the technology problem solving method. If you read other books or talk to other technologists, they may have a different set of steps to follow. In this book, we will describe the technology problem solving method with these seven steps:

1. Define the problem.
2. Consider the guidelines.
3. Create possible solutions.
4. Select the best solution.
5. Try the solution.
6. Evaluate the solution.
7. Make changes if necessary.

To give you an example of how the technology problem solving method works, consider the following sample problem:

"Design and build a model truss bridge that will support more weight without breaking than any other bridge built by your classmates."

Step 1: Define the Problem

The first step in solving any problem is answering the question, "What is the problem?" — or defining the problem. There is an old saying that a problem well defined is a problem half solved. It is very important to know exactly what the problem is. In our sample problem, it may seem obvious what the problem is — build a truss bridge that will support the most weight. But first, you have to know what a truss bridge is. You may have to do a little research of truss bridges to define the problem, Figure 2–1. Next, you may have questions about how you can solve the problem. These are the guidelines.

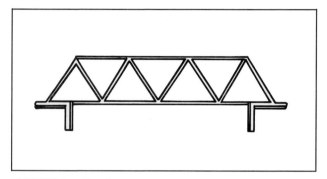

FIGURE 2–1 Design a strong truss bridge.

Step 2: Consider the Guidelines

Guidelines are the rules that must be followed in solving the problem. Considering the guidelines answers the question, What am I permitted to do or use to solve the problem? In the bridge-building problem you would need more information about the guidelines. Suppose you were given these guidelines:

1. You may use only thirty feet of $\frac{1}{8}$" x $\frac{1}{8}$" balsa wood.
2. Balsa wood pieces cannot be glued together to make pieces larger than $\frac{1}{8}$" square over 1" of length.
3. Balsa wood pieces will be connected with model cement.
4. The bridge must be free standing and no larger than:
   a. total length: 16"
   b. total height: 10"
   c. total width: 4"

5. Bridges must span a 14'' gap for testing.
6. A five gallon bucket will be hung from the bridge and filled with sand or other weight for strength testing.
7. You have one week to design and build your bridge.
8. Teams of three students will work together on the bridge.

These guidelines tell you what you can do and what you can use to solve the problem. They also tell you that you will be working with two other students, and how the bridges will be tested. This information makes the problem a little clearer. Now you can probably imagine the size and shape of the bridge in your mind. It's time to begin thinking about possible solutions.

Step 3: Create Possible Solutions

Usually, there is more than one possible way to solve a problem. But how do you create several possible solutions? One way is to use your past experience. Have you ever seen a truss bridge? What did it look like? Could you design your model bridge so it looked like a real truss bridge? Our past experiences can be used to help us create possible solutions.

A second way to create solutions is through research. For the bridge-building problem, you might read books about bridges, watch videos or television programs, or talk to bridge engineers or construction workers. Research can provide you with many good ideas for solving problems.

A third way to create solutions is called **brainstorming**. Brainstorming is a way for a group of people to create a list of possible solutions. In brainstorming, the members of a group discuss their ideas for solving the problem. All the ideas are written down. While the ideas are being offered and written down, no one says "that won't work" or "we can't do that." All the ideas are accepted and considered, no matter how "weird" they may seem at first. The point of brainstorming is to create a long list of ideas. The ideas can be judged later.

A fourth way to create solutions is **trial and error**. Trial and error means trying something and sometimes making mistakes. Trial and error is used quite often to solve problems. An example of using trial and error to solve problems

is when someone learns to ride a bicycle or skateboard. Most people make several attempts (trials) and falls (errors) before they are able to successfully ride a bicycle or skateboard. A motto for trial and error might be "let's try it and see if it will work." Trial and error can be used to try out some of the ideas generated with brainstorming.

Once possible solutions have been created, they should be sketched or drawn. Putting ideas down on paper as sketches or drawings allows people working on the problem to improve, refine, or eliminate certain ideas. Sketching and drawing are part of technical drawing. Technical drawing will be explained later in this chapter.

Step 4: Select the Best Solution

By using past experience, brainstorming, and trial and error, your team will probably come up with several possible solutions to the bridge-building problem. Now your task is to pick the one solution that you think will make the best possible bridge. This is a good time to check back over the guidelines. Compare your ideas with the guidelines. Do you have any ideas that do not meet the guidelines? These ideas can be removed from further consideration. This will limit the number of solutions you have to consider.

Another way to select the best solution is to use your past experience again. Compare your bridge ideas with actual truss bridges that you have seen before. Do you have any ideas that look like full-size truss bridges? These may work. Sometimes by just looking at sketches and drawings we get a feeling that one idea is best. This feeling may or may not be right. But people often act on these feelings. This is another example of using past experience. Your past experiences have prepared you to pick what you feel is the best solution.

Still another way to select the best solution is by combining the best features from several different ideas. Each team member may have a bridge design that might work. By selecting what seem to be the best parts from each bridge and combining them, you may find the best bridge of all.

Step 5: Try the Solution

Now it's time to try out your best solution. During this step, you get to find out how well you solved the problem. All the work in the first four steps was done to get to this step—trying the solution.

In our bridge-building problem, this is when you would actually build the bridge and test it to see how much weight it would hold, Figure 2–2. This may be the most important step. If you do not try your solution, you may never know if you could have solved the problem.

Technologists often use modeling during this step in the technology problem solving method. **Modeling** involves making a small version of the finished object. In manufacturing, two types of models are made; these are called mock-ups and prototypes. A mock-up is a model of a product that does not work. All the parts on a mock-up are there for appearance only. Prototypes are working models. Models are also used for large construction projects, like bridges and sky-scrapers. The bridge you are building is a model.

Models are usually built to scale and may be complete in every detail. They are usually made of paper, cardboard, wood, clay, or plastic. Models let the designer show what the finished product or structure would look like, Figure 2–3. Sometimes the models are tested to make sure they will work as planned. Making models is a quick and easy way to test out a design before making the real thing.

Step 6: Evaluate the Solution

Evaluate means to check and see if your solution works. During this step, the solution is once again compared to the guidelines. A successful solution will solve the problem and follow the guidelines. For the bridge-building problem, you would evaluate your solution by comparing the weight your bridge held with the weights held by other bridges in the class. Then, all the bridges would be compared with the guidelines. If the bridge that held the most weight did not follow the guidelines, it may not be considered as a successful solution.

Step 7: Make Changes if Necessary

Many times you will get several chances to solve a problem. When engineers design real bridges, they often build and test several models before the full-sized bridge is built. This gives the engineers a chance to evaluate their solution

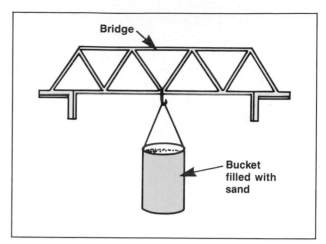

FIGURE 2–2 Test your bridge-building solution.

FIGURE 2–3 These models show what the finished buildings will look like. *(Courtesy of Northwestern Mutual Life Insurance Company)*

and make changes, if necessary. Making changes does not mean you failed. People who make products and build structures often make many changes to their designs as they move through the technology problem solving method.

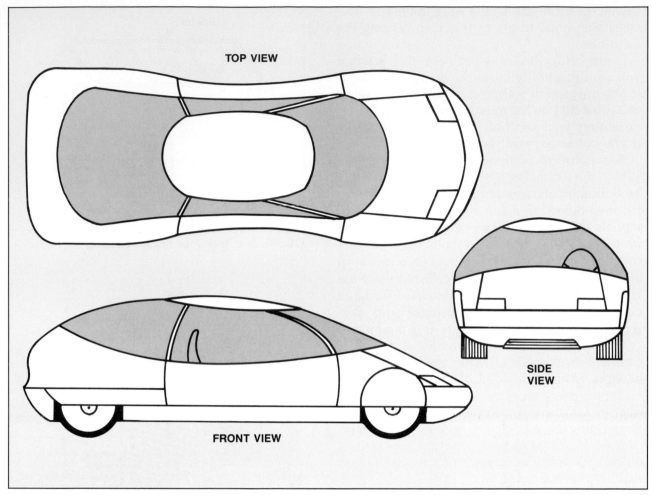

TOP VIEW

SIDE VIEW

FRONT VIEW

FIGURE 2–4 A basic technical drawing shows the front, top, and side view of an object.

Technical Drawing

Technical drawing is important to the technology problem solving method. Technical drawing involves putting ideas down on paper in the form of drawings. These drawings are done so other people can read and understand the information on the drawing. Consider the bridge-building problem described earlier. One way to communicate your ideas to the members of your group would be with technical drawings.

There are three ways to do technical drawings, sketching, drafting, and computer-aided drafting (CAD). No matter which method is used, a basic technical drawing will usually show three views of an object: the front view, the top view, and a side view (usually the right side view), Figure 2–4. These are called three-view drawings or multiview drawings. The front view always shows the most detail on an object. Since the most detailed view of the car is from the side, this is called the front view.

Another type of technical drawing commonly used in production is the **pictorial drawing**. Pictorial is another word for picture. A pictorial drawing shows what an object looks like to the eye. A pictorial drawing can be used in production to show what a new structure or product might look like, Figure 2–5.

Let's look at the three ways to do technical drawings—sketching, drafting, and CAD.

Sketching

Sketching involves drawing objects by hand. All that is needed to do a sketch is paper and a pencil. If you wanted to draw a floor plan for your home to tell a friend where your room

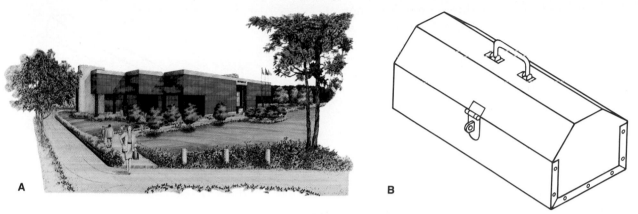

Figure 2–5 Pictorial drawings are like a picture of the final structure or product. *(Part A courtesy of NYNEX Properties Company; Einhorn, Yaffee, Prescott, Architecture and Engineering, P.C.)*

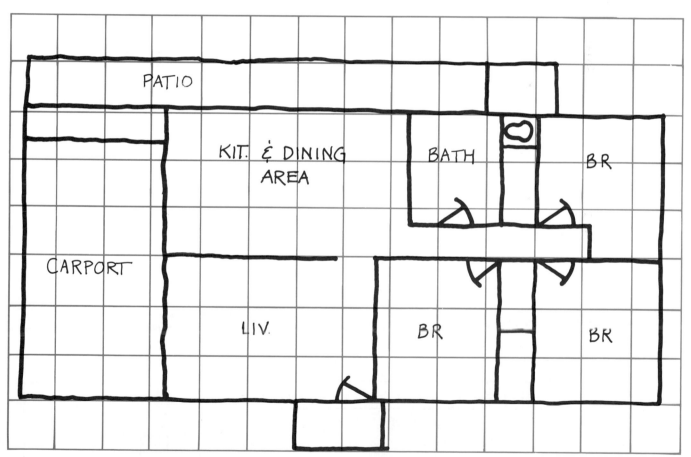

FIGURE 2–6 Sketches are a quick and easy way to do a technical drawing.

was, you might sketch a drawing on a piece of paper, Figure 2-6. The same method is used in production. When workers want to communicate their ideas quickly, they often use sketches. To make sketching more accurate and faster, grid paper can be used. The lines on the grid paper make it easy to measure sizes. They also help keep lines straight.

Sketches can be done as three-view drawings or pictorial drawings. Figure 2–7 shows a pictorial sketch for a spinning top toy. A special type of grid paper was used. Sketching is a fast way to do technical drawings.

Drafting

Drafting involves using tools to do technical drawings. Some of the basic drafting tools include a drawing board, T square, triangles, scales, compass, and templates. These tools make the job of drawing an object easier and more accurate. Learning how to use the various tools correctly and practicing with them is the best way to learn drafting skills. Drafting tools and methods can be used to make three-view drawings as well as pictorial drawings. Figure 2–8 shows a three-view drawing of the handle of the spinning top done with drafting tools.

CAD

CAD stands for computer-aided drafting. Like many other jobs in production, computers have been applied to drafting. The basic CAD system includes a microcomputer, input devices, output devices, and software, Figure 2–9.

Input devices are used by the operator to send information to the microcomputer. Three common input devices are the keyboard, mouse, and digitizer. The keyboard is like a typewriter with letters and numbers that can be input. A mouse is a hand-held device, Figure 2–10. When moved around a flat surface, the mouse can be used to indicate locations on drawings. A digitizer is used like a mouse, but it requires a special tablet, Figure 2–11.

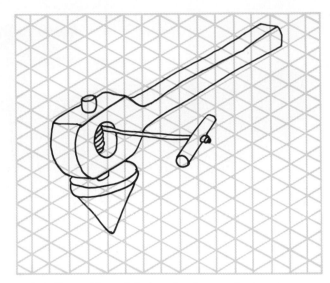

FIGURE 2–7 Pictorial sketch of a product.

Output devices are used to show the operator what the drawing looks like. The most used output device is the computer screen. Notice the drawing on the computer screen in Figure 2–11. The operator views the drawing on the screen while the drawing is being done. Once the drawing is complete, a copy of the drawing is needed on paper. Plotters are the most common way to output drawings onto paper, Figure 2–12.

Making and changing technical drawings with CAD is easier, faster, and produces better quality than drafting or sketching.

Measuring and Technical Drawing

Measuring is a basic skill in technical drawing. Being able to read a ruler and layout sizes is the first step in technical drawing. The English

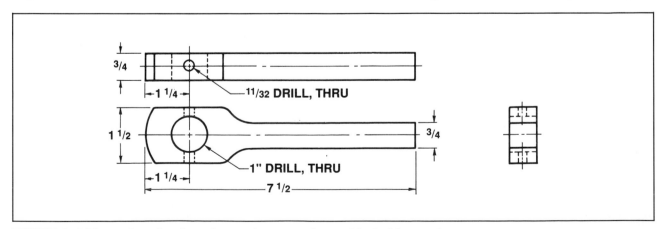

FIGURE 2–8 Three-view drawing of a product part done with drafting tools.

FIGURE 2-9 Microcomputer CAD system. *(Courtesy of Autodesk, Inc.)*

FIGURE 2-11 Digitizing tablet. *(Courtesy of Calcomp)*

FIGURE 2-12 This plotter prints out CAD drawings. *(Courtesy of Hewlett-Packard Company)*

system of measuring is commonly used in technical drawing for production. The English system uses the inch and fractions of an inch for basic measuring. The common fractions of an inch are $\frac{1}{16}''$, $\frac{1}{8}''$, $\frac{1}{4}''$ and $\frac{1}{2}''$. Figure 2-13 shows an enlarged view of one inch. Notice the division lines. The smallest division lines are for $\frac{1}{16}''$,

FIGURE 2-10 A mouse is a commonly used CAD input device. *(Courtesy of Mouse Systems)*

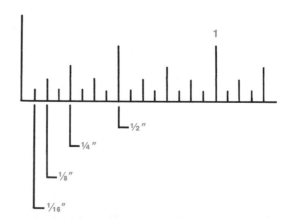

FIGURE 2-13 Fractions of an inch.

the next largest lines are for ⅛″, the next for ¼″, and so on. There are sixteen ¹⁄₁₆″ in one inch, eight ⅛″ per inch, four ¼″ per inch, and two ½″ per inch.

The fractional divisions are multiples of each other. For example, ²⁄₁₆″ equal ⅛″; ²⁄₈″ equal ¼″, ²⁄₄″ equal ½″, and ²⁄₂″ equal one inch. In every case, the fractions are reduced to their smallest multiple when measuring or laying out sizes. When sizes larger than one inch are measured, the whole inches and fractional parts of an inch are added together for the measurement. Being able to measure accurately is a skill that can be learned with practice.

Metrics in Production

Our nation is changing to the use of metric units of measure instead of inches and feet. As these changes take place, it will become necessary for production workers to be able to use metrics. Many parts of production are starting to use metrics at this time. However, many materials are made in standard sizes that use inches and feet. The basic unit of measure in the metric system is the millimeter. Some people say metric measurement is easier than the English system. When metric materials are made and used, metric scales will be used more in production.

Studying Systems of Technology

In order to study technology and solve problems, it is important to understand systems. A **system** is a group of parts that work together to do some task or job. There are many different types of systems. Our human bodies are an example of a system of biology. The combination of parts that make up our bodies includes the skeleton, muscles, nerves, organs, and so on. All of these parts must work together when we want to do a task. There are also systems of technology. A bicycle is an example, Figure 2–14. The bicycle parts include the frame, wheels, seat, handlebars, pedals, and chain. All of the parts of the bicycle must work together. There are two ways to study systems — the systems model and subsystems.

Systems Model

Using the **systems model** means breaking down a complex system into four basic parts — inputs, processes, outputs, and feedback, Figure 2–15.

FIGURE 2–14 A bicycle is an example of a system of technology. *(Courtesy of L.L. Bean, Inc.)*

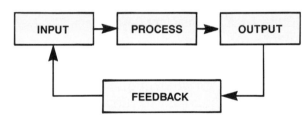

FIGURE 2–15 The systems model can be used to study production systems.

Inputs are everything that goes into starting the system. The inputs are needed to make the system work. The inputs for our human body include oxygen, food, water, and rest. The inputs for production include tools, materials, energy, people, and information, to name a few. All of these are needed to start a production system. Inputs are also called resources.

Processes are the actions that occur to make the system work safely and properly. A few of the processes of the human body include breathing oxygen, eating food, drinking water, walking, running, thinking, and sleeping. Two examples of processes that make production work are: designing products and structures, and using tools and machines to make products and build structures. Each of these processes is needed to make the production system work.

Outputs are the results of the processes. If our bodies are performing the processes needed to make us read this book, the result will be our learning something about production. In production, the results of performing processes include finished products and structures. Other outputs of production include the impacts discussed in Chapter 1, such as meeting needs and wants, and pollution.

Feedback is the final step in the systems model. Feedback happens throughout any system. It is the method of checking and changing the inputs, processes, and outputs. Our bodies give us feedback if we do not get enough sleep or food. When we feel hungry or sleepy, our bodies tell us to change the inputs. Production companies get feedback from the users of products or structures, Figure 2–16. Satisfied customers provide positive feedback to the company. Customers who are not satisfied will provide negative feedback.

The systems model can be used to study any system of technology. By identifying the inputs, processes, outputs, and feedback, a complex system can be made simpler and easier to solve.

Subsystems

Subsystems are smaller systems in larger systems. Once again, the human body system is a good example. The human body system is made up of a number of subsystems. For example, there is the skeleton (bone) system, muscle system, respiratory (breathing) system, circulatory (heart and blood) system, nervous system, hearing system, and so on. All of these subsystems must work together for our bodies to function correctly.

Technology systems also have subsystems. For example, remember that you learned in Chapter 1 that construction and manufacturing were two parts of production. They are considered subsystems of production. When you study manufacturing, you will learn that there are several subsystems called departments. Each department has a special job to do. There is a management department, an engineering department, a production department, and others. All the departments (subsystems) must work together in manufacturing. Similar examples can be found in the field of construction. Think about the subsystems found in a school building system.

FIGURE 2–16 Satisfied consumers can provide positive feedback. *(Reprinted courtesy of Eastman Kodak Company)*

Breaking complex systems down into their basic subsystems makes them easier to understand.

Summary

A technologist is a person who studies technology. Technologists must know how to use the technology problem solving method. The technology problem solving method involves defining the problem, considering the guidelines, creating possible solutions, selecting the best solution, trying the solution, evaluating the solution, and making changes if necessary. Technical drawings can be done through sketching, drafting, or CAD. Technology systems can be studied using the systems model (input, processes, output, feedback) and subsystems.

DISCUSSION QUESTIONS

1. What is a technologist? How are technologists different from scientists?
2. What are the seven steps in the technology problem solving method?
3. What are the differences between the systems model and subsystems?
4. What are the three types of technical drawings? What are the differences between them?
5. Why is measuring accurately important in technical drawing?
6. What is modeling? Why is it used in the technology problem solving method?

CHAPTER ACTIVITIES

 ## Solving Bridge-Building Problems

OBJECTIVE

In this chapter, you were introduced to the technology problem solving method with a sample bridge-building problem. In this activity, you will actually solve the bridge-building problem by using the technology problem solving method.

PROBLEM

Design and build a model truss bridge that will support more weight without breaking than any other bridge built by your classmates.

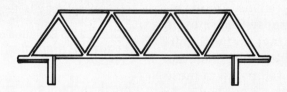

FIGURE 1. Design a strong bridge.

GUIDELINES

1. The bridge must be made from thirty feet of ⅛″ square balsa wood or other available wood.
2. Pieces of wood cannot be glued together to make pieces larger than ⅛″ square.
3. The joints on the bridge will be connected using model airplane-type cement, hot glue, or white glue.
4. The bridge must be free standing and have the following dimensions:
   a. total length: 16″
   b. total height: 10″
   c. total width: 4″

5. Bridges will be placed between two benches placed 14″ apart for testing.

6. A five-gallon bucket will be hung from the bridge and filled with sand or other weight for strength testing.

7. You have one week to design and build your bridge.

8. Teams of three students will work together to solve the problem.

EQUIPMENT AND SUPPLIES

1. 30 feet of ⅛″ square balsa wood per work team
2. X-Acto knives
3. Rulers or scales
4. Five-gallon bucket
5. Hook to hang bucket
6. Bathroom scale
7. Glue or adhesives
8. Pencils and ⅛″ grid paper
9. Wax paper

PROCEDURE

1. In your work team, define the problem. Make sure you know exactly what must be done.

2. Review the guidelines with the members of your work team.

3. Create possible solutions to the problem by drawing on your past experience, researching bridge designs, brainstorming, using trial and error, or sketching ideas.

4. Keep a notebook of the ideas your team brainstorms, the sketches you make, and the possible solutions you consider.

5. Remember, you have a limited supply of material. If you do any trial and error testing, be careful about how much material you use.

6. Select the best bridge design idea. Compare your ideas with the guidelines, combine the best parts of different ideas, or use your past experience to select the best idea.

7. To build the bridge, you will need a full-size drawing. Sketch the drawing on grid paper.

8. To build your bridge quickly, try this: Place the wax paper over your sketch. Then you can cut the balsa wood strips to the exact length using the X-Acto knife. Then, lay the pieces that make up the bridge on the wax paper to match the sketch and glue them in place. The wax paper will prevent the glue from sticking to the sketch.

9. Finish building your bridge and prepare for testing.

10. Try out your solution. Under the guidance of your teacher, place your bridge between two benches. Hang the bucket from your bridge. Add sand, water, or some other heavy material to the bucket until the bridge breaks.

11. Weigh the bucket and sand to find out which bridge held the most weight.

12. Evaluate your solution. If your bridge was the best in the class, write a short paragraph describing why you think it held the most weight. If your bridge was not the best, write a paragraph describing what improvements you would make if you built another bridge.

MATH AND SCIENCE CONNECTIONS

Scientists and engineers often want to know the efficiency of a design like your bridge. You can calculate the efficiency of your bridge by using the following formula:

$$\frac{\text{WEIGHT CAUSING BRIDGE TO FAIL}}{\text{WEIGHT OF BRIDGE}} \times 100 = \text{EFFICIENCY}$$

This formula will give you a percentage (%) for the efficiency. You can make your efficiency answer meaningful by saying "my bridge held x% of its own weight." It will not be uncommon to find model bridges in this activity with efficiencies of over 1000%. A bridge with an efficiency of 1000% can hold 1000% (or 100 times) its own weight.

Do you think the bridge that held the most weight will be the most efficient? Calculate the efficiency for all the bridges and find out.

RELATED QUESTIONS

1. What are the seven steps in the technology problem solving method?
2. Describe how you used the technology problem solving method to solve the bridge-building problem.
3. Which step in the technology problem solving method do you think is most important? Why?
4. What type of design was the best in the class? What do you think made that bridge hold more weight than any other bridge? Which bridge was most efficient?

(Popsicle) Stick To It!

OBJECTIVE

Technologists use the technology problem solving method to solve all kinds of problems. Two problems we face today are saving our resources and controlling our garbage. Every day, we use natural resources to make products, and then throw the products away. Garbage is a form of pollution, a negative impact of production systems.

Popsicle sticks are a good example. They are made from wood, a natural resource. After people eat popsicles, they throw away the wooden sticks. Millions of popsicle sticks are thrown in the garbage every year. The wooden resource is wasted. In this activity, you will solve the problem of the millions of wasted popsicle sticks.

PROBLEM

Design and produce a solution to the wasted popsicle stick problem.

GUIDELINES

1. Work in small groups of two to three students.
2. Your solution should:
   - Use the entire popsicle stick, but they can be cut, bent, or broken.
   - Not create more garbage than it saves.
   - Be safe and not hurt people or the environment.
   - Create a usable product or structure.
   - Be producible in your production lab.
   - Not require large amounts of money to make.
   - _____ (your teacher may have other guidelines) _____
   - _____

3. Consider some of these ideas:

- Popsicle sticks are made from wood: What are some uses for wood?
- Popsicle sticks are made a certain size: Can that size be used for anything?
- Popsicle sticks are often used to build models and crafts: Are there any models or crafts you could design?
- Popsicles are usually eaten by children: What new uses for the used popsicle sticks would children like? If you make a solution for use by children, make sure it is safe!
- _____ (your teacher may have other ideas) _____

EQUIPMENT AND SUPPLIES

1. A supply of popsicle sticks
2. Notepads and pencils
3. Assorted production tools and machines

PROCEDURE

1. Follow the technology problem solving method described in this chapter.
2. Define the Problem. Work with your teammates to define the problem. Write down your definition.
3. Consider the Guidelines. Review the guidelines give here. Your teacher may have other guidelines for you to follow.
4. Create Possible Solutions. Draw on your past experience, conduct research, brainstorm, and use trial and error to create solutions that might work. Get a supply of popsicle sticks to help your creative process.
5. Ask other teachers in your school how they might be able to use popsicle sticks in their class:

- art students could mix paint or do crafts
- math students could use them for measuring
- biology students could use them for potted plant labels

6. Put Ideas on Paper. Keep a list of the possible solutions created by your group. Use sketches and drawings to explain some of the ideas. Keep all your lists and drawings in a notebook.
7. Select the Best Solution. Review the guidelines, use your past experience again, or combine ideas to select the best solution.
8. Try the Solution. Use your supply of popsicle sticks to try out your ideas. Produce a model of your solution using the tools and machines in the production lab. Make sure you follow the safety rules!
9. Evaluate the Solution. Compare your solution with the guidelines. The different teams in your class could vote on the best solution.
10. Make Changes. If there is time, your teacher may ask you to make changes or suggestions for improving your solution.

MATH AND SCIENCE CONNECTIONS

Popsicle sticks are made in standard sizes. Measure the sizes of about a dozen popsicle sticks picked at random. Record all the sizes. You can then calculate the average size of a popsicle stick. You can also calculate how much the size of each stick varies from the average size. This could be important information as you solve this problem. Measuring the average size and variation is an important part of quality control in manufacturing.

RELATED QUESTIONS

1. What were some of the best solutions to the popsicle stick problem?
2. How did the group with the best solutions come up with their ideas?
3. Do you think the problem solving method you used in this activity could be used to solve other garbage problems?
4. What is the hardest part of problem solving?

CHAPTER 3

Careers in Production

OBJECTIVES

After completing this chapter, you will be able to:

- Identify career opportunities found in the production fields of manufacturing and construction.
- Describe why people with many skills and talents are needed in production.
- List the departments found in manufacturing companies and various careers in each department.
- List the four groups of workers in construction and various careers in each group.
- Identify and describe the various factors to consider when thinking about a career in production.
- Describe the impact of technology on the future of production careers.
- Describe the process of identifying and hiring employees.

KEY TERMS

Apprenticeship program
CEO
Department
Equal Employment Opportunity Act
Finance
Hierarchical order
Hire

Human resources
Job Application
Job description
Journeyman
Laborers
Management
Marketing

Salespeople
Semiskilled
Skilled
Supervisors
Technician
Unemployment
Unskilled

The main functions of production are to make products and build structures. Production also gives jobs to people. Production companies hire people with many types of skills and talents,

Figure 3-1. In this chapter, you will learn about the different types of jobs and careers in production. You will also read about the impact of new technologies on production jobs in the future.

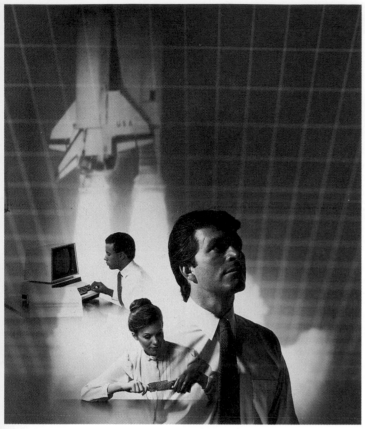

FIGURE 3–1 Production companies employ many people with different work interests. *(Courtesy of Dupont Corporation)*

Manufacturing — Several Departments, One Goal

Manufacturing companies often divide themselves into several departments. A **department** is a small part of the company. The departments in a manufacturing company can be thought of as subsystems. Just as in all systems, the subsystems must work together. In manufacturing, the departments work together to achieve the goal of making and selling products. Some of the more common departments found in large manufacturing companies include management, engineering, production, marketing, finance, and human resources.

Management Department

For a manufacturing company to be successful, all of the departments must work together toward one goal. Management's job is to make sure all the departments work together. Workers in **management**, called managers, make company rules and then make sure the other departments follow these rules. One type of management system is the **hierarchical order**. Figure 3–2 shows a simple hierarchical order for a manufacturing company. At the top of the hierarchy, a chief executive officer or **CEO** is in charge of the whole company.

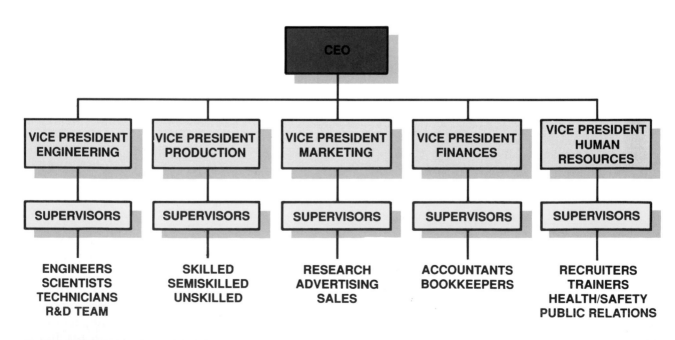

FIGURE 3–2 One form of manufacturing management structure is the hierarchical order.

Below the CEO are the vice presidents, who manage different departments in the company. The vice presidents communicate company rules to the supervisors, Figure 3–3. **Supervisors** make sure that workers in their departments follow rules made by the upper-level managers.

Engineering Department

Not only engineers work in the engineering department. Scientists conduct tests on new materials and processes in the engineering department, Figure 3–4. Technicians also work in the engineering department, Figure 3–5. A **technician** is a skilled worker. Only the most experienced and skilled technicians get to work in the engineering department as drafters, designers, and machinists. Finally, a number of different types of engineers work in this department, Figure 3–6. Some of the types of engineers found most often in manufacturing are listed in Figure 3–7.

Many engineering departments have a research and development team, Figure 3–8. This team is made up of the best engineers, technicians, and production workers. Their job is to create and test new mock-ups and prototypes.

FIGURE 3–3 These managers create company rules and make sure all the departments work together. *(Courtesy of Hexcel Corporation)*

FIGURE 3–4 Scientists, like this woman, work in the engineering department of manufacturing. *(Courtesy of Amoco Corporation)*

FIGURE 3–5 Technicians often work with computer-controlled tools, equipment, and machines. *(Courtesy Cincinnati Milacron Inc.)*

FIGURE 3–6 This engineer is testing a new automotive suspension system. Often, engineers and technicians work together on this type of job. *(Photo courtesy of EG&G, Wellesley, MA)*

FIGURE 3–8 This member of a research and development team is a highly skilled technician. He is developing a new toy design. *(Courtesy of Fisher-Price)*

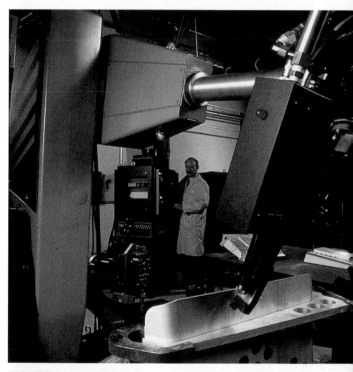

FIGURE 3–9 Skilled production workers get equipment, like this robotic welder, ready for other workers to use. *(Courtesy of FMC Corporation)*

COMMON ENGINEERING JOBS IN MANUFACTURING
aerospace engineer
ceramic engineer
chemical engineer
electrical engineer
electronic engineer
facilities design engineer
industrial engineer
manufacturing engineer
mechanical engineer
metallurgical engineer
product design engineer
product safety engineer
time-study engineer

FIGURE 3–7 These are some of the engineers who work in manufacturing.

Production Department

The production department sets up and uses tools and machines to make products. A supervisor is normally in charge of this department. The supervisor tells the production workers what to do and then makes sure they do it.

Production jobs use skilled, semiskilled, and unskilled labor. **Skilled** workers design and make any special tools that are needed. They also maintain tools and machines when they break down. Skilled workers include tool and die makers, pattern makers, millwrights, and mechanics, Figure 3-9. Because their jobs require precision skills, these workers usually have years of experience and training. **Semiskilled** workers run machines and use the special tools made by skilled workers, Figure 3-10. Semiskilled workers include machinists, machine tool operators, and welders. These workers get their training on the job. They do not have the experience of skilled workers. **Unskilled** workers do hard, routine jobs. They include material handlers, finishers, helpers, and assemblers. Unskilled workers do not have the skills to run machines.

Marketing Department

The **marketing** department advertises and sells the product. They do surveys to see what products consumers want and how much money they will pay, Figure 3-11. They also compare their product with others on the market. The "taste tests" shown on television by soda companies are examples of marketing in action.

Other career positions in marketing include advertisers, package designers, artists, and salespeople. **Salespeople** find ways to get people to buy the product. They point out the best features of a product and show consumers how to use the product, Figure 3-12.

FIGURE 3-11 Marketing workers determine consumer interest in new products. These workers are checking the child's interest in the new toy. (Courtesy of Fisher-Price)

Finance Department

Finance is another word for money. The finance department keeps track of all the company's money. This includes pay for workers, material costs, utility costs, and the costs of buying tools and machines. Accountants, bookkeepers, and other record keepers work in the finance department, Figure 3-13.

Human Resources Department

The **human resources** department does a number of jobs. Among their jobs are finding and hiring workers, working with unions, training new workers, making sure workers are safe, and making the public aware of recent events in the company. It is important that human resources workers get along with other people. Their job is working with people to make the company run smoothly.

FIGURE 3-10 Semiskilled production workers develop skills using tools to make products. (Courtesy of FMC Corporation)

FIGURE 3–12 This salesperson is telling a consumer about a new line of fishing equipment. *(Courtesy Zebco Division of Brunswick Corporation)*

FIGURE 3–13 This woman is using computers to keep track of finances. *(Courtesy of Fleetwood Enterprises)*

Construction Careers

The construction industry employs about one sixth of the working people in North America. Careers in construction can be divided into four groups — unskilled labor, skilled trades, technicians, and professionals.

Unskilled Labor

Unskilled construction workers are usually called **laborers**. Laborers do heavy work like moving materials, running errands, and helping skilled workers, Figure 3–14. Unskilled laborers must be strong for their heavy work.

Unskilled laborers are workers who have not reached a high level of skill in construction. They often work with skilled workers. A mason tender is an example. A mason tender does not have the skill of a bricklayer, but knows how to mix mortar, erect scaffolding, and care for the bricklayer's tools. Other laborers do whatever work needs to be done. Unskilled laborers are the lowest paid workers in construction.

FIGURE 3–14 Laborers usually work under the supervision of workers who are more skilled. *(Courtesy of the United Brotherhood of Carpenters and Joiners of America)*

Skilled Trades

There are over twenty skilled trades in construction, Figure 3–15. Skilled workers often learn their job in an **apprenticeship program**. Apprentices attend classes a few hours a week to learn their job. The rest of the week they work with a **journeyman**. A journeyman is a skilled worker who has finished an apprenticeship. The word journeyman has been used for centuries and probably will be for many more, even though many journeymen in construction today are women.

The skilled trades workers are among the highest paid workers in construction, Figure 3–16. However, work in construction depends on the weather. During the winter months in the North, many construction workers are unemployed. They must rely on money earned during the summer months.

STRUCTURAL TRADES	FINISH TRADES	MECHANICAL TRADES
Carpenter	Plasterer	Plumber
Mason	Terrazzo Worker	Electrician
Iron Worker	Tile Setter	Sheet Metal Worker
Operating Engineer	Painter	Millwright
	Floor Covering Installer	
	Cabinetmaker	

FIGURE 3–15 Some of the most common building trades.

FIGURE 3–16 Carpentry is a skilled trade. *(Courtesy of the United Brotherhood of Carpenters and Joiners of America)*

Technicians

Technicians provide a link between the skilled trades and the professionals. The work of a construction technician is very similar to the work of a professional. They both use math, computer skills, and knowledge of construction processes to solve problems, Figure 3–17.

Technicians spend much of their time in an office. They also visit the construction site. Figure 3–18 lists some technical careers in construction.

Technicians get their training in community colleges, technical schools, and colleges. The starting salary for technicians is about the same as the salary of the skilled worker. However, the technician is more certain of regular work.

FIGURE 3–17 Many construction technicians work with computers. *(Courtesy of Hewlett-Packard Company)*

TECHNICAL CAREER	SOME COMMON TASKS
Surveyor	Measures land, draws maps, lays out highways
Estimator	Calculates time and materials necessary for project
Expeditor	Ensures that labor and materials are properly scheduled
Drafter	Draws plans and designs some details
Inspector	Inspects project at various stages
Planner	Plans for best land use or community development

FIGURE 3–18 A few common categories of technicians.

FIGURE 3–19 Architects design new buildings like this one. *(Courtesy of Pella Windows and Doors)*

Professionals

Architects, engineers, and building code officials are examples of professionals in construction. Architects are the designers of new structures and buildings. It is their job to create useful, attractive buildings for the owners, Figure 3–19. Engineers in construction specialize in mechanical systems and electrical, structural, and civil engineering. Civil engineers design highways, dams, and bridges.

Construction professionals often use computers in their work. Computer-aided drawing and design (CADD) and computer-aided engineering (CAE) have completely changed the way construction professionals work today, Figure 3–20.

The construction professions require at least four years of college education. Some of the professions, such as architecture, require five years of college. The salary of the professionals is the highest in the construction industry.

Considering a Job in Production

There are many considerations when thinking about a job in production. One of the most important is the impact new technologies will have on jobs in the future. Other important factors to

FIGURE 3–20 Architects and engineers use CADD to design structures. *(Courtesy of Hewlett-Packard Company)*

consider include the education required, the nature of the work, and salary.

Impact of Technology

In the past, one person could have had the same job for thirty years before retiring. Today, the average worker could have twelve different jobs before retiring. One of the reasons for this is the rapid change caused by new technology. Does this mean that technology destroys jobs? Some new technologies, like robotics, are replacing thousands of workers today. Many of yesterday's jobs, such as the blacksmith, milkman, harness maker, and iceman, no longer exist because of changes in technology. But technology also creates new jobs that never existed before. Compare the robot to the backhoe. When the backhoe was first used, it did the work of ten workers, it could run without taking a break, and it needed only one operator, Figure 3–21. Many ditchdiggers were replaced by the backhoe, but new jobs were created in the fields of hydraulics, mechanics, control systems, and backhoe maintenance. Jobs were also created for workers who manufactured backhoes. Today, workers are afraid of losing their jobs because one robot can

FIGURE 3–21 The backhoe is a labor-saving device that gave jobs to semiskilled workers. *(Courtesy of Deere & Company, Moline, IL)*

FIGURE 3–22 The new jobs that will be created by technology, like these robot technicians, require a strong technology, math, and science education. *(Courtesy of NASA)*

replace seven workers. Robots can also work without taking breaks or vacations, they do not go on strike, and usually only one operator is required. But just as the backhoe created new jobs, new jobs will also grow around robot technology.

The Impact of Robots. Robots are a labor-saving technology. They will displace 100,000 to 200,000 jobs by the year 2000. Most of the resulting **unemployment** will affect unskilled and semiskilled workers.

On the other hand, robotics will create many jobs. The biggest growth is expected at the technician level, with up to 64,000 robot technicians needed by the year 2000. They will test, program, install, troubleshoot, and maintain the robots. More engineers will also be needed. Robots are made of electrical, electronic, mechanical, and fluid power systems. Companies will need more engineers in each of these fields, Figure 3–22.

Most of the jobs lost will be semiskilled or unskilled labor positions that require little education beyond high school. Many of the new jobs will require skills in technology, science, and math. Robot technicians must have two or more years of college training. Engineers need a four-year college degree or more.

Education Required

There are a wide variety of jobs and careers in production. Each job requires different skills and talents. The education required for these careers is as varied as the jobs themselves, Figure 3–23. Some of the different education training grounds follow.

College. Many jobs require a four-year college degree, called a bachelor's degree. Workers in manufacturing management, finance, and marketing, and construction professions often have college degrees. Some even have master's degrees or doctorates. Engineers and professionals have college degrees that require in-depth math and science training. Many technicians can receive a two-year college degree called an associate's degree. Because technicians work with engineers, their two years of education are filled with hands-on technical courses and the math and science courses needed to understand the basics of engineering.

Technical School. Many skilled and semiskilled workers receive their first educational training through a high school or postsecondary vocational or technical school. The program length varies with the school and the technical area. Programs may run from several months to several years.

FIGURE 3–23 Education is the key to the use of new technology in production. The ability to use computers may become a basic skill. *(Courtesy of NASA)*

Apprenticeship Training. Skilled workers usually must complete an apprenticeship training program. As apprentices, they work with skilled craftspeople on the job to learn specific skills. They also take classes that supply the knowledge base needed.

On-the-Job Training. Semiskilled workers get their training on the job. This can be done because their jobs do not require a high degree of skill that takes years to learn.

Future Education Requirements

With the increased use of new complex computer, robotic, and automation technologies, many more jobs in production will require some sort of education beyond high school. In the past, a high school education was required as a minimum for getting a good job. In the future, the minimum requirement will be some sort of postsecondary education.

Nature of the Work

All careers can be described by the amount of interaction with other people, data, or things. The *Dictionary of Occupational Titles* ranks jobs by these three factors. Jobs that require the most interaction with people include management, marketing, and professional positions, Figure

FIGURE 3–24 Managers work very closely with other people. *(Courtesy of The Stanley Works)*

3–24. Workers in these jobs must be able to get along with others, give and take instructions, and supervise. Workers in finance, marketing, and engineering spend most of their time working with data, Figure 3–25. They like to handle num-

FIGURE 3–25 This technician is checking the number of screwdriver handles being produced. He works with data. *(Courtesy of FMC Corporation)*

FIGURE 3–26 Production workers, like this machinist, enjoy working with things, such as tools and machines. *(Courtesy of Simpson Industries)*

Most people find they are more comfortable working with people, data, *or* things. Production today needs a new type of worker: one who can work easily with people, data, *and* things. These well-rounded workers are the technicians or technologists. They work with data by reading drawings and making calculations. They work with people by talking with engineers and other technicians. They work with things by turning ideas on paper into products and structures by using tools, materials, and processes of technology.

Everyone has unique talents and interests. When considering any career, first decide how much you like to work with people, data, and things. Then identify careers that seem to match your interests. To help you decide what might attract you, consider the terms in Figure 3–27, used in job descriptions.

bers, perform calculations, and compile information. Those people who spend most of their time working with things, such as tools, machines, and equipment, include the skilled, semiskilled, and unskilled production workers, Figure 3–26.

Salaries

Everyone wants to know how much money certain jobs pay. The *Occupational Outlook Handbook* provides some idea of the salary range for most jobs. In general, it seems that those who receive the highest wages are the same people who have the most education, work the longest hours, or have the most responsibility on the job.

TERM	DESCRIPTION
Competition	Compete with others for advancement
Confined space	Work in a small, crowded area
Creativity	Use your creative abilities
Details	Follow specifications, exact work
Hazardous	Involves danger
Help others	Train or teach other people
Influence others	Stimulate or inspire others to work
Initiative	See what needs to be done and do it
Outdoors	Work in all weather conditions
Physical stamina	Takes muscle power
Precision	Work to accurate standards
Public contact	Work with the general public
Repetitious	Do the same thing over and over again
Results seen	End result is a physical product
Solving problems	See problems, make carry out decisions
Teamwork	Work with others
Use tools machines	Use your hands to operate machines

FIGURE 3–27 These terms can be used to describe the nature of work in any job.

Equal Employment Opportunity Act

Employers must, by law, treat all job applicants and workers fairly. The **Equal Employment Opportunity Act** includes federal and state laws that require all employees and applicants to be treated fairly without regard to race, color, sex, age, marital status, religious belief, or national origin. This law protects workers and people looking for jobs from any type of discrimination.

Identifying and Hiring Workers

There is a need for thousands of good workers in production. Production companies need people with varied skills and talents. Companies **hire** employees they feel will help the company be successful. When a person with certain skills is needed, a **job description** is written, Figure 3–28. Job descriptions identify the experience, skills, education, and training needed for a specific job. When job descriptions are advertised, people can show their interest by filling out a **job application**, Figure 3–29. Company officials review job applications to see which people match the job description. The people who best match the description get an interview. Interviews are used to decide who to hire for a new job, Figure 3–30.

Reasons for Hiring

There are many reasons why certain people are hired and others are not. Three main reasons for hiring a person are knowledge, skills, and attitudes. Companies want people who know how to do a certain job. That is why people with work experience or training often have a good chance of getting the job. Figure 3–31 lists some general points companies want people to know. These are very important qualities in a worker.

Two types of skills important in production are communication skills and technical skills. Communication skills include reading, writing, listening, taking instruction, and learning new jobs. Technical skills relate to the use of tools and processes of the job.

Worker attitudes are very important to the safe operation of a production company. Without the proper attitudes, people with all the necessary knowledge and skills can have a negative impact on the company. Figure 3–32 lists some of the more important attitudes employers look for in employees.

TITLE: Manufacturing Engineer	
DUTIES:	**RELATIONSHIPS:**
■ Organize and manage manufacturing production department	■ Reports to vice president of engineering department
■ Identify most economical product production methods	■ Works closely with production department vice president and production supervisor
■ Implement and supervise production tools, equipment, and facilities	**EDUCATION/EXPERIENCE:**
■ Document parts lists, plant layout, process flow charts	■ College degree in manufacturing engineering
■ Determine production workers needed	■ Three or more years experience preferred

FIGURE 3–28 Job descriptions are created for each position in a company.

APPLICATION FOR EMPLOYMENT
(Please Print Plainly)

Name _____   Social Security No. _____
    Last   First   Middle Initial

Present Address _____   Telephone No. _____
    No.   Street   City   State

How long have you lived at above address? _____   Are you a citizen? _____

Position applied for _____   Rate of pay expected $_____ per week

Have you been previously employed by us? _____   If so, when? _____

Do you have any friends or relatives employed by us? _____

                                  Name   Address

_____   _____
    Name   Address   Name   Address

If application is considered favorably, when will you be available for work? _____

Any special skills or qualifications which you feel would especially fit you for work with the company? _____

EDUCATION RECORD

        Years Attended

High School _____ to _____   Did you graduate? _____   If not, highest grade _____

College _____ to _____   Did you graduate? _____   If not, highest grade _____

MILITARY SERVICE RECORD

Are you a veteran? _____   Any job-related experience? _____   If so, please state type _____

EMPLOYMENT RECORD

List your present and past employment for past five years

Name of Company	from/to	Type of Work	Salary

PERSONAL REFERENCES
(not relatives)

Name	Address	Telephone No.

Person to be notified in event of accident or emergency

Name _____ Address _____ Telephone No. _____

FIGURE 3–29 Companies gather information about potential employees with job applications.

FIGURE 3–30 This person is interviewing a potential employee. What questions do you think he will ask?
(Photo courtesy of UNISYS Corporation)

- The value of going to work regularly
- The value of getting to work on time
- How to take and follow directions
- How to work hard and get the job done
- How to communicate (read, write, speak, and compute)
- How to adapt to changes created by technology

FIGURE 3–31 Knowing these things can make a person a valuable worker for any company.

- **Ambition: a desire to be the best**
- **Cooperation: the ability to work with others**
- **Creativity: problem solving with good ideas**
- **Patience: remaining calm and in control**
- **Persistence: being determined; finishing a job**
- **Promptness: being on time to work or to do a job**
- **Quality awareness: concern for product quality**
- **Reliability: showing people they can count on you**
- **Safety awareness: concern for safety**

FIGURE 3–32 These attitudes are very important for production workers. Which of these do you think is most important?

Summary

Production companies employ large numbers of workers with varying types of education, training, and skills. Jobs in manufacturing are divided into six departments — management, engineering, production, marketing, finance, and human resources. Careers in construction are divided into four groups — unskilled labor, skilled trades, technicians, and professionals.

Four factors should be considered when studying careers in production—the impact of technology, the education required to do the job, the nature of the work, and the salary.

Companies identify and hire employees by writing job descriptions, reviewing job applications filled out by people, and interviewing promising applicants. When making a decision to hire an employee, companies consider the knowledge, skills, and attitudes of the new worker.

DISCUSSION QUESTIONS

1. What management structure was discussed for large companies? What would the management structure for smaller companies look like?
2. Several impacts of technology on careers were mentioned. Do you agree or disagree with the idea that new technology will create new jobs in the future? If so, will these jobs be better than the ones that will be replaced?
3. What are the departments commonly found in manufacturing?
4. What are the four groups of careers found in construction?
5. Compare the nature of the work, education required, and job outlook for skilled, unskilled, and professional production workers.
6. What three factors do companies consider when interviewing people for jobs? Which do you feel is most important? Why?

CHAPTER ACTIVITIES

 Considering a Career in Production

OBJECTIVE

The *Dictionary of Occupational Titles* (called the *DOT*) and the *Occupational Outlook Handbook* are fine resources for finding the educational requirements, working conditions, job descriptions, and future employment trends in various careers.

In this activity, you will use the *DOT* and the *Occupational Outlook Handbook* to learn more about some production jobs.

EQUIPMENT AND SUPPLIES

1. *Dictionary of Occupational Titles*
2. *Occupational Outlook Handbook*
3. List of Jobs in Production (handout from teacher)
4. Other references that describe production jobs

PROCEDURE

1. As a class, make a complete list of the jobs you would like to learn more about. Use the list from your teacher as a starting point.
2. Research the jobs using the *DOT* and the *Occupational Outlook Handbook*.
3. Write a short description for each job including the following:
   a. Nature of the Work — what kind of work is done?
   b. Involvement with Data, Things, and People — how much work is done with data, things, or people?
   c. Educational Requirements — what degree or training is required?
   d. Future Employment Trends — what is the outlook for jobs in this field in the future?

MATH AND SCIENCE CONNECTIONS

Many careers in production require workers to take math and science courses as part of their education. It is not uncommon for engineers, architects, and technicians to take several courses in physics, algebra, calculus, and chemistry.

RELATED QUESTIONS

1. Why do some production careers require four years of college, while others only require a high school or technical school education? What is the difference in these jobs?
2. Which production careers involve the most time working with people? With data? With things?
3. What are the differences among skilled, semiskilled, and unskilled production careers? Consider the kind of work, education, training, and pay.
4. What impact do you think changing technology will have on the job outlook for certain careers in production?

 ## Identifying and Hiring Workers

OBJECTIVE

Production companies must identify and hire the best workers for the jobs available. One important skill that companies look for is problem solving. In this activity, you will make a problem solving abilities tester and use it to identify and hire workers.

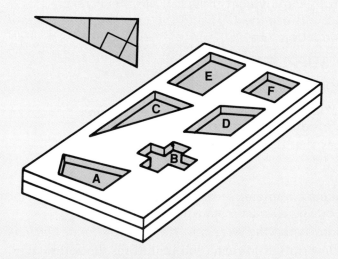

FIGURE 1. The problem solving abilities tester.

EQUIPMENT AND SUPPLIES

1. ⅜″ or ½″ thick plywood or plexiglass, ¼″ plywood
2. Jig or scroll saw
3. Disk sander
4. Sandpaper and wax
5. Stopwatch
6. Safety glasses

PROCEDURE

1. Make the problem solving abilities tester following all safety rules.
   - ▪ Lay out the puzzle pattern accurately on ⅜″ or ½″ thick plywood or plexiglass.

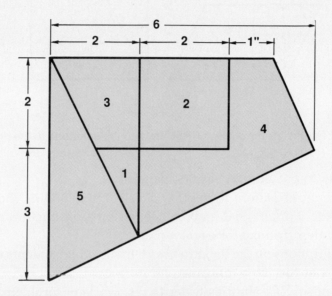

FIGURE 2. Lay out the pattern for these pieces accurately on a piece of plywood or plexiglass.

   - ▪ Cut out the pieces using the finest saw blade possible. Try to split the lines when cutting.
   - ▪ Sand the edges of the shapes using a disk sander. Try to leave just a bit of the line showing. Do not sand away the line.
   - ▪ Do not number the pieces; this will make it harder to complete the puzzle.
   - ▪ Sand the pieces to finish them and apply a coat of wax.
   - ▪ Lay out the pieces in the five patterns shown in the figure. Trace the patterns on a sheet of ¼″ plywood.

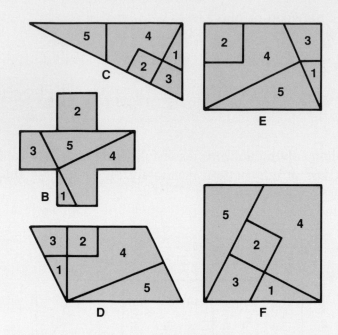

FIGURE 3. Five possible patterns for the pieces. These are also the solutions to the problem solving abilities tester.

- ■ Use a jigsaw to cut out the various patterns.
- ■ Attach the cut-out pattern to another sheet of plywood as shown in the figure.
2. Make a list of the workers that may be needed by a production company.
3. Write short job descriptions for each worker.
4. Post the job descriptions on the bulletin board. Invite other students to apply for the jobs.
5. Hold interviews of the job applicants.
6. Use the problem solving abilities tester to test workers' problem solving skills.
   - ■ Take the puzzle pieces and mix them up.
   - ■ Use a stopwatch to time students as they complete all five puzzles.
7. Use the results of the interviews and problem solving abilities test to make hiring decisions.

MATH AND SCIENCE CONNECTIONS

You can find the average problem solving abilities of your entire class. Add together the times for every student in your class. Calculate the average by dividing the total time by the number of students in the class. You can then compare your time with the class average.

RELATED QUESTIONS

1. What types of questions were asked in the interviews?
2. What qualities do you think companies look for in their workers?
3. During the interviews, what factors would make you not want to hire a person?
4. How important do you think problem solving skills are in production?

CHAPTER 4

Ownership, Organization, and Management

OBJECTIVES

After completing this chapter, you will be able to:

- Describe the importance of ownership, organization, and management to production companies.
- Identify the three major forms of ownership and list the advantages and disadvantages of each.
- List the steps required to organize a corporation.
- Describe the various functions of management.

KEY TERMS

Articles of incorporation	Management	Quality control
Board of directors	Organizing	Sole proprietorship
Bylaws	Partnership	Stockholders
Directing	Planning	
Estimating	QC circle	

Every production company must have a system of ownership, organization, and management. Deciding what type of ownership, organization, and management a company will have is the first step in starting any new enterprise or business.

Forms of Ownership

Production companies vary in size from small, single-person companies to very large international organizations. However, the size of the company does not tell you the form of ownership.

There are three major forms of ownership — sole proprietorship, partnership, and corporation. Each form of ownership has certain advantages and disadvantages for the owners.

Sole Proprietorship

The **sole proprietorship** is the easiest form of ownership to understand, Figure 4-1. The two words in the name of this form clearly describe it. "Sole" means only one, or single. The "proprietor" of a business is the owner and operator. So, a sole proprietorship is a business whose

FIGURE 4–1 Building trades workers sometimes start their own sole proprietorships.

FIGURE 4–2 The individual proprietorship is the oldest form of business ownership. *(Photo by David Tuttle)*

owner and operator are the same person. In a sole proprietorship production company, the owner is one of the main workers, Figure 4–2.

The advantages of the sole proprietorship are that the owner has complete control over the business. If the company is successful, the owner receives high profits. Another advantage of the sole proprietorship is that there is a minimum of government regulation. However, in the sole proprietorship, the owner is personally liable (legally responsible) for any debts of the company. The owner can be sued and possibly lose the company.

Partnership

A **partnership** is similar to a sole proprietorship, but with two or more owners. In a partnership, each partner shares the profits and losses of the business. The amount of profit or loss by each partner is equal to the amount of investment made by each partner. Partnerships are common among small engineering and architectural companies where each partner is an expert in a different field.

The main advantage of this form of ownership is that the partners share the expense of starting the business. Several partners can combine their money to start the company. Also, partnerships, like sole proprietorships, are not controlled by excess government regulations.

Corporation

In a corporation, a group of people own the company. The owners, called **stockholders**, buy shares of stock, Figure 4–3. The stock represents ownership in the corporation. Large corporations may have thousands of stockholders. The stock of many large corporations is bought and sold in public stock exchanges. Many other corporations are owned by private groups of investors.

Figure 4–4 illustrates the hierarchical order for a corporation. The stockholders are the owners of the company. At their annual meeting, they appoint a **board of directors** to set company policies and goals. The board of directors also hires a Chief Executive Officer (CEO). The day-to-day operation of the company is the responsibility of the CEO and the other managers.

The main advantage of a corporation is limited liability for the owners. The stockholders can only lose the money they invested in the shares of stock. They are not liable for the debts of the corporation. The corporation itself is liable for the debts.

FIGURE 4–3 Owners of corporations have shares of stock in the corporation.

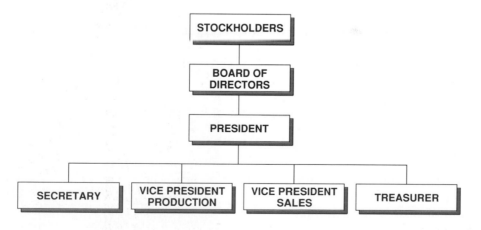

FIGURE 4–4 This is the basic hierarchical order for a corporation.

Because there is no individual who can be held liable for the actions of the company, the government has stricter regulations for corporations. Also, corporations are much more expensive to form than partnerships and sole proprietorships.

Organizing a Corporation

Corporations must be organized following certain laws and rules. Each state has specific laws that determine how a corporation may be organized. A company must file an application for incorporation with state officials. This application is called the **articles of incorporation**. The articles of incorporation require the following types of information:

- Name of the new corporation.
- Purpose(s) for forming the corporation.
- Location of the principal offices of the corporation.
- Description of the financing that will be used (loans, bonds, or stocks).
- Names of the CEO and other managers.
- Description of the type of business (construction, manufacturing, etc.) in which the corporation will participate.

A fee is paid to the state when the articles of incorporation are filed. Once the documents are approved by the state, a charter is issued to the new corporation. The charter makes the company a legal corporation. The next step is to create a list of **bylaws**. Bylaws are specific rules which are used to guide management in operating the company. Some of the items covered in the bylaws include:

- Date when the corporation begins work.
- Date when the corporation will cease work, if known.
- Description of the operation of the board of directors.
- Job descriptions for the CEO and other managers.
- Rules and regulations which must be followed by management.
- Description of how bylaws can be changed by stockholders.
- Date and times for stockholder's meetings.

The Functions of Management

Management can be defined as reaching goals through the work of others, Figure 4–5. It is concerned with achieving the goals of the company by making sure workers do their jobs. In production, there are managers at all levels of the organization. Their jobs, their authority, and the problems they solve are all very different.

Each aspect of a production company requires management. The functions of management include planning, organizing, directing, and controlling. This is true for whatever activity is being managed. In addition, the hiring and firing of employees is a function of management.

Planning

Planning is deciding what to do and how to do it. It demands the ability to analyze a situation and make decisions about it. Planning requires an understanding of the technical aspects of the operations being managed. This includes the tools, machines, materials, and processes.

One form of planning used in construction is **estimating**. Estimating means making a prediction of how much time, materials, money, or workers are needed to complete a job. One example of estimating is predicting the cost of a new building. Area or volume estimates are usually used. These estimates give "ball park figures" that are used for planning.

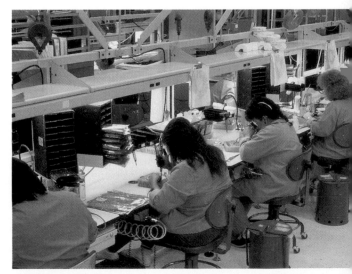

FIGURE 4–5 Management tries to achieve the goals of the company through the work of the employees. *(Courtesy Motorola Government Electronics Group)*

An area estimate is one in which a cost per square foot is multiplied by the number of square feet involved. For example, Figure 4–6 shows how the costs would be estimated for a two-story warehouse with a one-story office in front. The costs per square foot and total costs for each part of the building are estimated. When added together, they give an estimate of the total building costs.

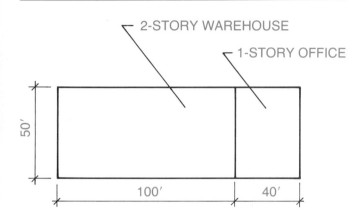

1st floor of warehouse at $20 per square foot = $100,000
2nd floor of warehouse at $25 per square foot = $125,000
Office space at $38 per square foot = $ 76,000

TOTAL = $301,000

FIGURE 4–6 Example of an area estimate.

120,000 cubic foot warehouse at $1.80 per cubic foot
16,000 cubic foot office space at $4.75 per cubic foot

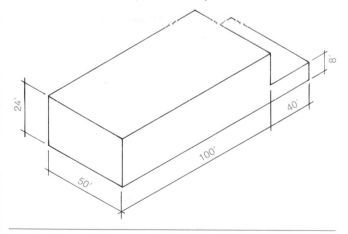

FIGURE 4–7 Volume estimate.

Volume estimates are sometimes slightly more precise than are area estimates. A volume estimate is based on the total volume of space enclosed by a building. Figure 4–7 illustrates a volume estimate. Can you calculate the total costs? In large companies, computers can also be used to improve the speed and accuracy of estimates, Figure 4–8.

Organizing

Organizing a group of people to work together is a difficult task. Management tries to get the workers to organize into an efficient team, Figure 4–9. Ideally, the organization of workers should prove that the whole is greater than the sum of its parts. This means a group of people working together can achieve much more than a group of individuals working alone.

In order for a sense of teamwork to develop, each worker must understand how he or she fits

FIGURE 4–8 Computers improve the speed and accuracy of estimates. *(Courtesy of Hewlett-Packard Company)*

in and contributes to the company. If the workers do not know why they are important to the company, they may not participate in the team. It is the job of management to make workers feel like an important part of the organization.

Directing

Directing is guiding workers. Another word for directing is supervising. Often, supervising

FIGURE 4–9 One aspect of organizing is getting workers to develop a sense of teamwork. *(Reprinted courtesy of Eastman Kodak Company)*

FIGURE 4–10 Teaching workers how to do their jobs is a part of directing. *(Courtesy of Polytech, Inc.)*

is thought of as telling workers what to do and how to do it. But it has been proven that good supervision and directing begin with motivating workers. Workers perform better when they know what is expected of them. They also perform better when they are taught how to do their jobs, Figure 4–10.

Workers have shown that they perform better when they have some decision-making power. Workers who feel they have some responsibility for directing their own work and the work of others often want to help the company make the highest quality product or structure possible, Figure 4–11.

Controlling

You have probably heard the term **quality control** before. Production companies have certain standards for the quality of the products they make and the structures they build. The steps they take to meet these standards are called quality control. Management sets the standards of quality and makes sure the standards are met.

Sometimes, special managers called quality control inspectors inspect products and structures and compare them with the standards, Figure 4–12. A recent development related to controlling quality is the quality control circle. The **QC circle** is composed entirely of workers, Figure 4–13. They meet weekly to discuss and solve quality problems. They study the causes

FIGURE 4–11 Workers who are given the power to make decisions on the job will often produce the best quality products and structures.

of poor quality, suggest changes to improve quality, and then put the changes to work. QC circles give the workers some responsibilities in making decisions about quality control. This type of responsibility motivates the workers to take pride in their work.

FIGURE 4–12 Quality control inspectors compare products with standards. *(Reprinted by permission of Dean Foods Company, © 1988. All rights reserved.)*

FIGURE 4–13 The members of this QC circle are discussing a quality control problem. *(Courtesy of International Public Affairs Division, Toyota Motor Corporation)*

Summary

Ownership, organization, and management are important factors in starting and operating a production company. There are three forms of ownership—sole proprietorship, partnership, and corporation. Each form of ownership has certain advantages and disadvantages for the owners.

In order for a corporation to organize, it must file an application with state officials. This application is called the articles of incorporation. After the articles of incorporation are filed, the state issues a charter to the corporation. Bylaws are written to set the rules by which management runs the company.

The four major functions of management are planning, organizing, directing, and controlling. Planning is deciding what to do and how to do it. Organizing is getting people to work as a team. Directing is guiding or supervising workers. Controlling is making sure standards are set and followed.

DISCUSSION QUESTIONS

1. What is the difference between ownership, organization, and management? Why is each important to production companies?
2. What are the three major forms of ownership?
3. What are the advantages and disadvantages of each form of ownership?
4. What is an article of incorporation? What is a charter? What are bylaws?
5. What are the four functions of management? Can you give one example of each?

CHAPTER ACTIVITIES

 ## Developing a Management Structure

OBJECTIVE

One of the most important decisions that must be made early in the life of a production company is what type of management structure to use. Managers will make most of the decisions during the life of a company. Choosing the right managers is important.

In this activity, you will make a chart that shows the management structure and managers for a company.

EQUIPMENT AND SUPPLIES

1. Poster board
2. Drafting tools or computer graphics program

PROCEDURE

1. Using drafting tools or a computer graphics program, draw the hierarchical management structure on a large piece of poster board.
2. Make the blocks on the chart large enough to include the job title and the name of a student.
3. Write the name of your teacher in as CEO. (Your teacher may decide to let someone else act as CEO.)
4. Hold an election to identify those people who would make good vice presidents and supervisors for the company.
5. Place the names of the elected managers on the poster board in the proper location.
6. Write the names of the workers in for each department.
7. Hang the management structure chart on the wall in your lab.

MATH AND SCIENCE CONNECTIONS

In science, families of birds and animals are organized on charts similar to the hierarchical structure used in this activity. Placing groups of people, animals, or birds on a structure makes it easy to see the relationship between any of the individuals.

RELATED QUESTIONS

1. What personal qualities do you think a good manager needs in production?
2. What are the disadvantages of the hierarchical management structure? What are some other structures?
3. Why is the work in production divided among labor and management?
4. What are the major jobs and tasks of management?

Corporate Charter and Bylaws

OBJECTIVE

The activities of a corporation are controlled by three things: the laws in the state where the corporation was formed, the corporation's charter (or articles of incorporation), and the corporation's bylaws. In this activity, you will draft the necessary documents, as listed above, to form a corporation.

EQUIPMENT AND SUPPLIES

1. Paper and pencils
2. Sample articles of incorporation

FIGURE 1.

PROCEDURE

1. Work in small groups to divide the work load.
2. Research the laws of incorporation in your state by contacting your local chamber of commerce.
3. Based on the information you find, draft a charter (or articles of incorporation) for a classroom production corporation.
4. The charter should outline certain facts about the new corporation, such as the following:
   - name of the corporation
   - purpose(s) for forming the corporation
   - names of people forming the corporation
   - location of principal offices
   - type of financing that will be used (loans, bonds, or stocks)
   - list of the management workers and officers

5. Outline the bylaws for the corporation. These should include:
   - date when the corporation begins work
   - date when the corporation will cease work
   - job descriptions for corporate officers
   - description of how bylaws can be changed by stockholders
   - date and times for stockholders' meetings
6. Once the articles of incorporation (charter) and bylaws have been developed, discuss them as a class. Make any changes in the charter or bylaws before beginning work.

RELATED QUESTIONS

1. Why do states have laws regarding incorporation?
2. What is the difference between a corporation's articles of incorporation and its bylaws?
3. List three specific areas usually covered in corporate bylaws that might differ between corporations.
4. How did your group make decisions about the articles of incorporation and bylaws? Was the democratic process used?

SECTION TWO

PRODUCTION MATERIALS

(Courtesy of Amoco Corporation)

Introduction to Production Materials

KEY TERMS

Ceramic	Fatigue strength	Recycling
Composites	Impact strength	Renewable
Compression strength	Malleability	Secondary manufacturing
Ductility	Plasticity	Standard stock
Elasticity	Polymers	Tensile strength
Exhaustible	Primary manufacturing	

The materials that make finished products and structures are produced themselves. For example, before a bicycle can be manufactured, the metal for the frame, rubber for the tires, plastics for the seat, and even the paint must be produced first. Then these materials can be combined to make a bicycle. The same is true for the concrete blocks, wall studs, floor boards, and other materials that are used to build a house.

The process of converting raw materials into standard forms of materials for production is called **primary manufacturing**. These standard forms of materials made by primary manufacturing companies can be sold to consumers or to **secondary manufacturing** companies. For example, you can go to a lumber yard and buy standard sizes of lumber and plywood. A secondary manufacturing company can also buy the lumber and plywood to make finished products such as furniture, cabinets, or homes to sell to consumers.

There are three classes of secondary manufacturing processes: forming, separating, and combining. Chapters 8, 9, and 10 look at the processes and tools secondary manufacturing companies use to make products from standard forms of materials.

This chapter discusses the primary manufacturing processes used to convert raw materials into usable materials for production. Also discussed are materials science, properties of materials, and the classification of manufacturing materials.

Renewable and Exhaustible Raw Material Resources

The source of all materials is the earth. The natural resources in the land, water, and air provide the raw materials that are the basis for all production materials. Raw material resources are either renewable or exhaustible, Figure 5–1.

Renewable Raw Material Resources

Renewable raw material resources are living, growing things. Examples include trees, plants, and animals. Trees are a resource for wood. Some plants provide materials like cotton and rubber. An example of a material that comes from an animal is the wool provided by sheep.

Renewable material resources can be replaced. When one tree, plant, or animal is harvested for raw materials, a new one can begin growing to renew the supply.

Exhaustible Raw Material Resources

Exhaustible raw material resources are substances found in the earth that are not living. Examples include mineral ores, petroleum, and natural gas. There are limited amounts of these materials available on earth. Once people have removed all the available resources, these raw materials will no longer exist. In other words, exhaustible material resources cannot be renewed.

Recycling: Additional Raw Material Resources

Many materials that come from renewable and exhaustible resources can be recycled. **Recycling** is the process of reusing waste materials. Every year, we throw away hundreds of millions of tons of scrap and waste. Many of these materials can be recycled. They include metals, plastics, glass, and paper. Figure 5–2 lists some yearly averages for recycled materials. Despite these numbers, the Institute of Scrap Recycling Industries says that more than 800 million tons of scrap iron and steel in the United States alone is waiting to be recycled. Recycling provides additional resources for raw materials. Recycling has other advantages like saving energy, reducing waste in landfills, and saving natural resources, Figure 5–3.

RAW MATERIALS RESOURCES	
RENEWABLE	**EXHAUSTIBLE**
TREES, PLANTS, ANIMALS	PETROLEUM, MINERAL ORES, NATURAL GAS

FIGURE 5–1 Raw material resources can be renewable or exhaustible.

RECYCLED SCRAP	TONS
Iron and Steel	56,000,000
Paper	24,000,000
Aluminum	2,500,000
Copper	1,500,000
Lead	1,300,000
Stainless Steel	800,000
Zinc	300,000

FIGURE 5–2 These figures represent only a small fraction of the scrap materials that can be recycled.

FIGURE 5–3 Recycling waste and scrap is an important way to conserve our valuable raw material resources. *(Courtesy of Aluminum Association, Inc.)*

Primary Manufacturing Processes

All raw materials must be processed: That is, they must be changed to make them more useful. Most materials cannot be used in their raw, natural state. The steps in primary manufacturing include obtaining raw materials, refining raw materials, and making standard materials.

Obtaining Raw Materials

Renewable raw materials are obtained from the earth by harvesting. Examples of harvesting include cutting down trees for wood, picking cotton from plants, and shearing wool from sheep.

Exhaustible raw materials are obtained from the earth by mining and drilling, Figure 5–4.

A

(Courtesy of Weyerhaeuser Company Inc.)

B

(Courtesy of the Office of Surface Mining Reclamation, United States Department of the Interior)

C

(Courtesy of Conoco)

FIGURE 5–4 Raw materials are obtained from the earth by (A) harvesting, (B) mining, and (C) drilling.

Mineral ores such as iron and coal must be mined. Miners dig shafts or large open pits in the ground and remove the raw ore with heavy equipment. Drilling is used to obtain petroleum and natural gas. Often, workers must drill holes several miles deep to locate petroleum and gas deposits.

Refining Raw Materials

Once obtained, raw materials must be refined. Refining separates the useful part of the raw material from unwanted and less useful materials. Refining is a way of purifying the raw material and making it more useful.

There are three major processes used to refine raw materials: mechanical processes, chemical processes, and thermal processes, Figure 5–5.

Mechanical refining processes use a mechanical force to change the raw materials. Cutting logs into lumber or boards at a sawmill is an example of a mechanical refining process. The saw blade applies the mechanical force. Mechanical forces are applied to mineral ores for crushing. Crushing is used to separate the wanted mineral ore from dirt and rocks.

Chemical refining processes use chemicals to change the raw material into a more useful form. Plastics are created by chemical reactions using petroleum or natural gas as raw materials. Lumber cut from a log can be treated with chemicals to make it resist decay in outside construction projects.

Thermal refining processes use heat. Ores, like iron, are often melted in a furnace. When melted, the iron ore separates from unwanted minerals. Wood is also treated with thermal refining. Excess moisture content in lumber is removed by heating and drying wood in kilns.

Making Standard Materials

First raw materials are obtained from the earth and refined. Then they are made into standard forms of materials called **standard stock**. Metals and plastics are produced in many of the same forms, such as plates, sheets, rolls, tubes, pipes, bars, rods, and other structural shapes. Wood is prepared in standard sizes, such as two-inch by four-inch lumber and four-foot by eight-foot plywood sheets. Making materials into standard stock makes them easier to use by production companies.

Thermal
(Courtesy of U.S. Iron and Steel Institute)

Mechanical
(Courtesy of Weyerhaeuser Company Inc.)

FIGURE 5–5 Thermal and mechanical processes are two of the ways that primary manufacturing refines raw materials.

Classification of Production Materials

Most production materials can be classified into four broad groups: metals, polymers, ceramics, and composites. These groups do not include every material used in production, but they do include a vast majority of the most important materials.

Metals are materials such as iron, steel, and aluminum. There are two main types of **polymers**: woods and plastics. Woods are natural polymers; plastics are synthetic or human-made polymers. Pottery, china, porcelain, and glass are examples of **ceramic** materials. **Composites** are combinations of two or more materials. Plywood is an example of a wood-based composite material. Fiberglass, a composite also, is a combination of glass fibers in plastic.

Each of the classifications of materials will be described in detail in later chapters. For now, let's look at materials science and the properties of materials.

Materials Science

Understanding the basic science of materials and the properties of materials will help you understand why certain materials are used in certain products or applications.

To a materials scientist, materials are composed of atoms and molecules. Atoms are the basic building blocks of all materials. They cannot be broken down into smaller parts. Molecules are combinations of two or more atoms, Figure 5–6.

Materials: Gas, Liquid, or Solid

Materials can be in the form of a gas, liquid, or solid. The form of a material is determined by the connections between molecules, Figure 5–7. Gases have no connections between molecules; liquids have only slight connections between molecules; solid materials have varying degrees of connection between molecules.

The molecules in certain solids, such as wood and plastic, link together to form long chains. These long chains make wood easy to split and certain plastics easy to melt.

The molecules in ceramic materials link together in flat grids called platelets. There is very little attraction between platelets. In fact, platelets slip against one another. This permits clay to be formed and shaped easily when wet. However, once clay materials are heated in an oven, the platelets melt together and fuse into solid masses. This process cannot be reversed.

Molecules in metals form together into three-dimensional shapes called crystals. The cross linking between molecules makes metals strong materials.

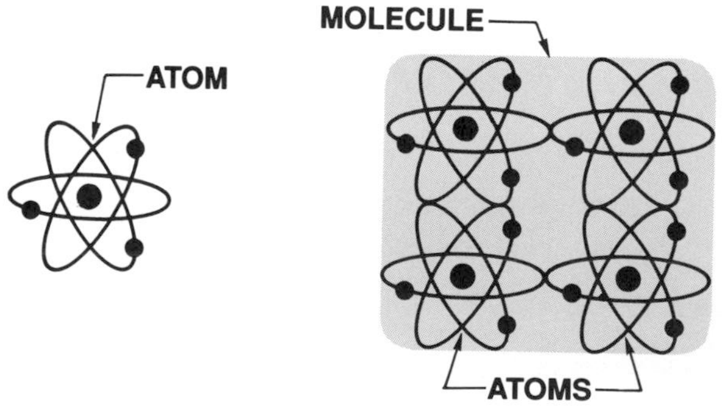

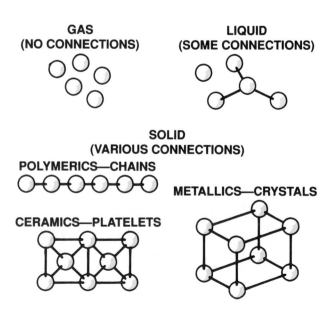

FIGURE 5–6 Atoms are the building blocks of all materials. Molecules are combinations of two or more atoms.

FIGURE 5–7 The connections between molecules are different for gases, liquids, and solids.

Changing the Form of Materials

The form of many materials can often be changed. Water (a liquid) can be frozen to form ice (a solid). Water can also be heated to create steam (a gas). The form of production materials can also be changed. Metals and certain plastics can be heated and made into liquid or semiliquid forms.

Being able to change the form of a material is useful in making products. For example, metals, plastics, and ceramics can be shaped much more easily when they are in a liquid form. When liquid materials are forced into a container, they take on the shape of that container.

Properties of Materials

Properties of materials relate to the characteristics of the material. The most common material properties are mechanical, chemical, thermal, and electrical. The properties of a material are based on the structure of its molecules. The connection between molecules and atoms determines how a material reacts when it is used.

Mechanical Properties

Mechanical properties relate to the reaction of a material to mechanical forces, such as pressing, pulling, and pounding. Mechanical properties described here include several measures of strength, elasticity and plasticity, malleability, ductility, and hardness.

Strength. There are several measures of strength. The most common are tensile, compression, fatigue, and impact, Figure 5–8. **Tensile strength** allows a material to resist being pulled apart. A fishing line being pulled by a fish is an example of tensile strength. **Compression strength** allows a material to resist crushing or compression. It is the opposite of tensile strength. Concrete blocks in a foundation resisting the crushing forces of the house are an example of compression strength. **Fatigue strength** is the ability of a material to resist failure after repeated forces in opposite directions. You can demonstrate fatigue by bending a paper clip back and forth several times until it breaks. Fatigue is the most common cause of failure in metals. **Impact strength** is the ability of a material to absorb energy during impacts. Another name for impact strength is toughness. Tough materials absorb energy during impact by changing their shape. Metals are tough materials. When a piece of sheet metal is hit with a hammer, the metal indents. The opposite of toughness is brittleness. Brittle materials do not absorb energy well during impact and they usually shatter. Glass and ceramic materials are brittle materials.

Elasticity and Plasticity. Elasticity allows a material to return to its original shape after being changed. Springs and rubber bands are examples of elastic materials.

Clay and putty are the opposite of elastic materials. They display plasticity. **Plasticity**

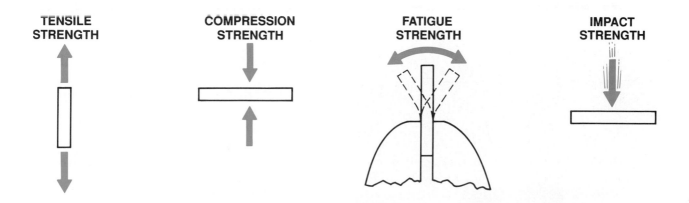

FIGURE 5–8 Four important measures of strength are tensile, compression, fatigue, and impact.

allows a material to be easily formed into various shapes that the material retains, Figure 5–9.

Malleability. Malleability lets a material be pounded, rolled, and formed into sheets. Many metals, such as copper and aluminum, are malleable. Aluminum foil is an example of a malleable material.

Ductility. Ductility allows a material to be drawn into a wire shape. Materials used in wiring, such as copper, are ductile.

Hardness. Hardness is the ability of a material to resist scratching. Diamonds are the hardest of all materials. Carborundum™, a commonly used abrasive material, is almost as hard as diamonds. Usually, hard materials are also strong. In general, metal and ceramic materials are harder and stronger than plastic and wood materials.

Chemical Properties

Chemical properties relate to the reaction of materials to certain chemicals. This reaction is called corrosion. One of the most common forms of corrosion is oxidation, or rusting. The iron in metals reacts with oxygen in the air to form rust. Tin is one metal that resists corrosion, Figure 5–10. In general, glass, ceramic, and plastic resist corrosion best.

FIGURE 5–10 Tin is used to plate steel cans because it is resistant to corrosion from food. *(Courtesy of the Canned Food Information Council)*

Thermal Properties

The thermal properties of materials determine how the materials react to heat. The important thermal properties relate to conductivity, fire, melting, and expansion. Conductivity allows a material to conduct heat. If you hold the end of a metal bar that is being heated on the other end, the heat will be conducted through the metal to your hand. Wood, on the other hand, is not a good conductor of heat. Certain materials, especially wood and plastics, will catch fire when exposed to high temperatures. Metals, glass, and certain plastics can melt when exposed to heat. Generally, plastics have a much lower melting point than glass and metals. Metals also expand when heated.

FIGURE 5–9 The plasticity of clay makes it useful for making models. *(Courtesy of Ford Motor Company)*

Can You Say "Xanthan"?

Xanthan is a microbial polysaccharide (my-'crow-bee-ull pah-lee-'sak-kah-ride). Microbial polysaccharides are large groups of synthetic chemicals that are produced by the work of microbes. You may have studied microbes in science or another class. Microbes are like little bugs that are too small to be seen without the help of a microscope. When placed into a solution, they go to work in changing the solution from one substance to another.

Xanthan is a gum, or slippery substance that has been produced by the microbial process. It is similar to car wax, and cannot be easily removed once it is in place. Xanthan has been used to recover oil in a process called Enhanced Oil Recovery.

Oil is usually found under pressure because of the squeezing effect the surrounding rock formations place on the oil. This pressure makes it easy to pump the oil out of the ground. When the oil is found in rock pores (holes in rock formations), there is not enough pressure to pump the oil out of the ground. It would be like sipping a milkshake through a straw that had holes in it! All of the force applied in getting the milkshake up the straw would be lost through the holes.

In Enhanced Oil Recovery, Xanthan is sent down into the oil deposit, covering the rock pores. This places pressure on the oil deposit, allowing the oil to be pumped up and recovered. The Xanthan "enhances" or helps get oil out of the ground that normally would be left in the ground.

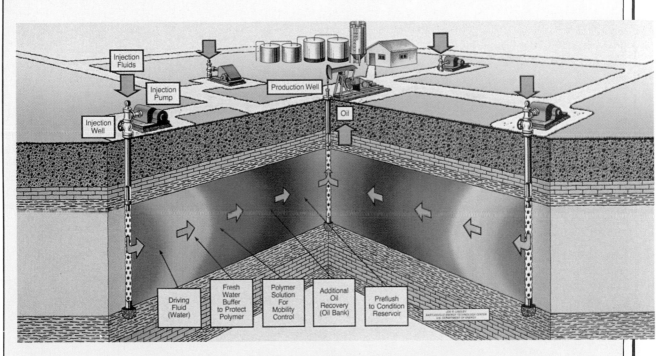

FIGURE 1. In this picture of Enhanced Oil Recovery, a polymer solution is used to trap the oil by sealing the holes in underground rock formations. Then water is sent down to force the trapped oil toward the well, where it can be pumped up. *(Courtesy of National Institute for Petroleum and Energy Research)*

Worth Their Weight In Gold

How would you like to obtain your weight value in gold? That is what's in store for some pretty hungry microbes at a company called U.S. Gold Corporation.

U.S. Gold Corporation has built an operating plant in Tonkin Springs, Nevada. The plans are to obtain huge amounts of ore that contains pyrite (fool's gold) and gold. The gold in this ore is in such small amounts that it usually cannot be extracted. it is too expensive to do so with common removal techniques. Bio-related technology has come up with a less expensive process that uses microbes to extract these small amounts of gold.

The ore taken from the earth will be crushed to a powder as fine as flour. Then, the crushed ore will be mixed with water, acid, and compressed air in very tall stainless steel tanks. These tanks are about forty feet in height. The temperature inside the tank will slowly rise to 100 degrees Fahrenheit. This makes a perfect environment for the microbes to go to work. The microbes are like little bugs that eat away impurities or unwanted substances in the mixture. They will not eat the gold found inside the mixture. Once the microbes have completed their task, the result is a heap of gold and byproducts. The byproducts are washed off the gold with cyanide. This process results in about 120 ounces of gold a day!

FIGURE 1. A gold processing plant that uses microbes to separate the gold from unwanted materials. *(Courtesy of U.S. Gold Corporation)*

The Importance of Strength-to-Weight Ratios

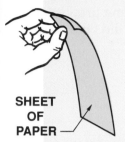 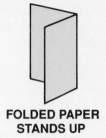 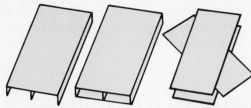

SHEET OF PAPER

FOLDED PAPER STANDS UP

OTHER FORMS ARE EVEN STRONGER

REINFORCED PAPER COMPOSITES

LAYERED PAPER COMPOSITES

FIGURE 1. Improving the strength-to-weight ratio of a material can be demonstrated by folding a sheet of paper into various forms or by making composites.

Strength is an important factor in seeing how well a material will work. In some products, especially vehicles, weight is also important. Airplanes, boats, bicycles, and automobiles must be lightweight. They use energy to move. The heavier they are, the more energy they use.

Designers are always looking for materials that are lightweight and strong. They conduct tests to find materials with high strength-to-weight ratios. These materials are both strong and light.

Strength-to-weight ratios can be improved by changing the shape of a material, reinforcing, making composites, or changing the molecular structure. You can demonstrate these improvements with a sheet of paper, Figure 1.

One reason why plastics are being used in so many products is their strength-to-weight ratios. Some plastics have strength-to-weight ratios greater than steel.

FIGURE 2. The plastic composite materials used in this kayak have excellent strength-to-weight ratios. *(Photo by Rob Lesser. Courtesy of Perception, Inc.)*

When choosing materials, consider the strength-to-weight ratio and not just strength alone, Figure 2.

Electrical Properties

Electrical properties relate to the reaction of materials to the flow of electricity. Materials with a low resistance to the flow of electricity are conductors. Materials that have a high resistance to the flow of electricity are insulators. Generally, metals are conductors, while plastics, ceramics, and woods are insulators.

Summary

Primary manufacturing obtains and refines raw material resources and changes them into standard forms of materials. Secondary manufacturing makes finished products from these standard forms of materials. All production materials come from either renewable or exhaustible raw material resources in the earth.

Recycling is one technique that can be used to help preserve our limited resources. Recycling also helps reduce energy use and keeps our environment free from wastes.

The major types of production materials include metals, polymers, ceramics, and composites. Understanding the science and properties of materials helps designers find the best materials for each application.

DISCUSSION QUESTIONS

1. What is the difference between primary manufacturing and secondary manufacturing?
2. Name at least four products that are produced by primary manufacturing processes.
3. What are the two types of raw material resources? What are the differences between these two types?
4. What production materials are recycled most? What are three reasons why recycling is a good idea?
5. What properties are important in a material? Pick a product or structure application and describe why you think a designer chose the material used.
6. What is the difference between an atom and a molecule? How are these important to the science of materials?

CHAPTER ACTIVITIES

 Reuse It

OBJECTIVE

Some people say we live in a wasteful "throw away" society. Many of the things we throw away could be recycled, Figure 1.

In this activity, you will make fire starters from a waste (sawdust) and discarded wax. You will see how easy it is to recycle waste products into a usable product.

EQUIPMENT AND SUPPLIES

1. 2 sheets of 8½″ x 11″ paper
2. 1 dowel, 1″ dia. x 8½″ long
3. 1 dowel, ½″ dia. x 10″ long
4. 2 sheets of tissue paper
5. 6″ pieces of lightweight string
6. Masking tape
7. 2 pints sawdust
8. Discarded wax candles
9. Tablespoon
10. Paint stir stick
11. Safety glasses/gloves
12. Double boiler/stove

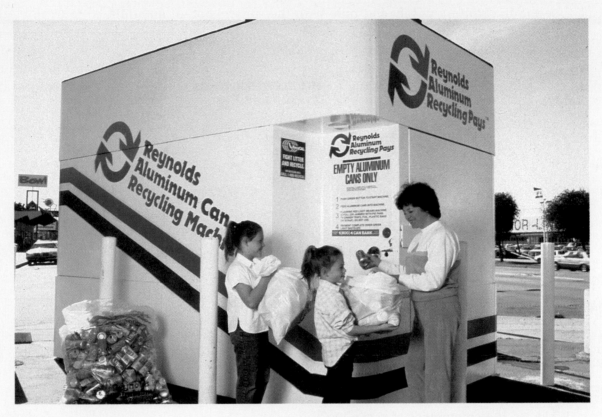

FIGURE 1. An aluminum can recycling center. *(Courtesy of Reynolds Aluminum Recycling Company)*

PROCEDURE

1. Roll one sheet of 8½″ x 11″ paper around the 1″ dowel to make a tube 8½″ long. Tape the tube along its length. Then slide the paper off the dowel. Make two tubes like this (see Figure 2).
2. Carefully melt the old wax using a double boiler. Do this only with the supervision of your teacher. Do not do this in an ordinary pan from home — the wax can ignite. This could be dangerous so be careful.
3. Place the sawdust in a metal container. Then, carefully pour in 1 pint of melted wax. Mix the wax-sawdust combination with the stir stick. Do this quickly so the mixture remains warm (see Figure 3).
4. Place several paper tubes in an upright position. Spoon the wax-sawdust mixture into the tubes. As you are doing this, gently ram the mixture into the tube with the ½″ dowel. Fill the tubes as full as possible.
5. Allow the wax-sawdust mixture to cool. Remove the paper from the outside. Handle carefully to keep from breaking the mixture.
6. Cut the cooled wax-sawdust rods into four equal-length pieces.

7. Wrap each of the pieces in tissue paper. Twist each end and then tie off the twists with the string.

8. Use the fire starters wisely and only with the supervision of an adult.

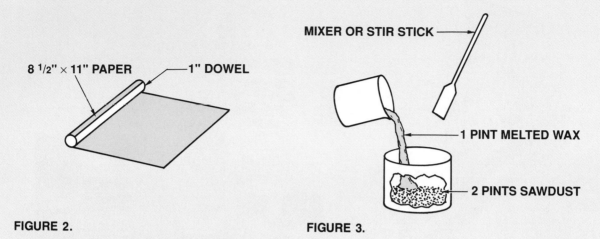

FIGURE 2.

FIGURE 3.

MATH AND SCIENCE CONNECTIONS

Wax will burn at 140 degrees Fahrenheit. The process that occurs when a fuel burns is called combustion. Combustion is a chemical reaction between the fuel, oxygen, and heat. If one of these three elements is not present, combustion cannot take place.

RELATED QUESTIONS

1. What are the advantages of recycling?
2. Why are exhaustible raw materials recycled most?
3. What specific materials are recycled most?
4. Why is it important for society to recycle?

 Testing Material Properties

OBJECTIVE

Before product designers choose materials for a product, they must know something about the properties of materials. In this activity, you will learn about the properties of various materials by conducting a few simple tests.

EQUIPMENT AND SUPPLIES

1. Materials for testing. Materials of different types (metals, polymers, ceramics, and composites) or materials of only one type (steel, aluminum, copper, lead) can be tested and compared.

2. Material-testing devices. Many tests can be done with simple tools; others will require testing devices. Figure 1 gives some ideas for various material-testing methods. Your teacher may have you make the devices.

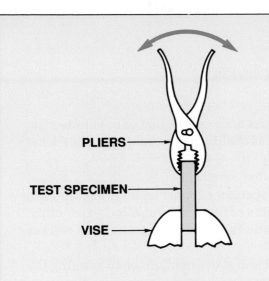

PLIERS

TEST SPECIMEN

VISE

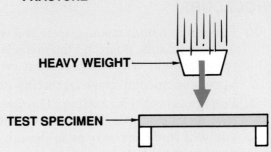

- A TOUGH MATERIAL WILL BEND AND NOT REACT
- A BRITTLE MATERIAL WILL CRACK OR FRACTURE

HEAVY WEIGHT

TEST SPECIMEN

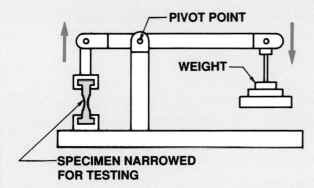

PIVOT POINT

WEIGHT

SPECIMEN NARROWED FOR TESTING

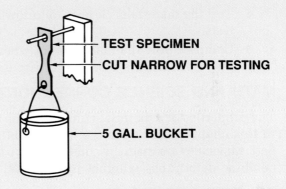

TEST SPECIMEN

CUT NARROW FOR TESTING

5 GAL. BUCKET

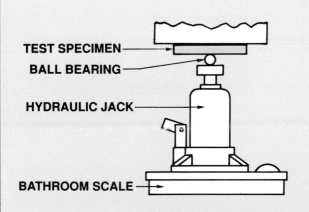

TEST SPECIMEN

BALL BEARING

HYDRAULIC JACK

BATHROOM SCALE

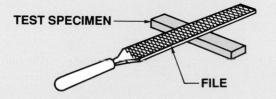

TEST SPECIMEN

FILE

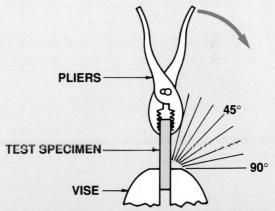

PLIERS

45°

TEST SPECIMEN

90°

VISE

- **BEND MATERIALS TO DIFFERENT DEGREES AND SEE IF THEY RETURN TO ORIGINAL SHAPE (ELASTICITY)**
- **IF THEY DO NOT RETURN TO ORIGINAL SHAPE, THEY DISPLAY PLASTICITY**

G. Elasticity/plasticity

FIGURE 1. Ideas for material-testing devices.

3. Material Testing Data Sheet (handout)
4. Safety glasses
5. Production tools and machines to make testing devices

PROCEDURE

1. Determine which material tests you will do. Discuss materials testing with your teacher and classmates. Based on the materials and tools available in your school, decide what materials you will test and what properties you will study.
2. Make any special material testing devices you may need.
3. Prepare samples for testing. The size of the test specimens must be identical for each test. As an example, if you test the fatigue strengths of two different-sized paper clips, you will find they vary even though both are made from the same material. Keep the cross-sectional area constant.
4. Test the materials. Be sure to follow all safety rules and use caution when running the tests.
5. Compare how the materials act when tested. Write down your findings. Your teacher may hold a class meeting to discuss properties of materials.

MATH AND SCIENCE CONNECTIONS

The properties of a material come from the structure of its atoms and molecules. As you read in this chapter, atoms are the basic building blocks of materials. The atom is the smallest part. Molecules are made by combining two or more atoms. The way that two or more molecules combine determines whether the material will be a gas, liquid, or solid.

RELATED QUESTIONS

1. What are the four different measures of strength?
2. What is the difference between plasticity and elasticity?
3. What is the difference between malleability and ductility?
4. Pick any product and describe why you think the company chose the particular materials used.

CHAPTER 6

Metal and Polymer Materials

OBJECTIVES

After completing this chapter, you will be able to:

- Name various ferrous and nonferrous metals.
- Identify and describe properties and uses of metals.
- Describe the difference between natural and synthetic polymers.
- Name various hardwoods and softwoods.
- Identify and describe properties and uses of woods.
- Describe the difference between thermosets and thermoplastics.
- Identify and describe properties and uses of plastics.

KEY TERMS

Alloy	Hardwoods	Softwoods
Blooming mill	Ingots	Synthetic
Board foot	Nonferrous	Thermoplastic
Ferrous	Polymerization	Thermoset

In the previous chapter, you learned about the primary manufacturing processes used to convert raw material resources into standard stock materials for production. You also read about basic materials science and the properties of materials. In this chapter, you will read about two of the basic types of production materials—metals and polymers.

Metal Materials

Metals are one of the most important of all production materials. Think of all the things pro-

duced from the many metals available today. Automobiles, airplanes, bicycles, piping, eating utensils, and tools are just a few examples of products manufactured from metals. Skyscrapers, bridges, and towers are a few examples of structures constructed with metals.

In general, metals are good conductors of electricity and heat, they are strong, and they can be formed and machined without breaking. They can also be melted for casting. There is a wide variety of metal types. More than seventy of the chemical elements found in the earth are metals. Also, there are over 70,000 metal alloys. (An

alloy is a combination of two or more metals.) As a result, the properties of metals can cover a very wide range. One example of this is shown by the difference between the melting points of mercury (-38 degrees Fahrenheit) and tungsten (6170 degrees Fahrenheit). Metals are also produced in a variety of standard shapes. Figure 6–1 shows the basic shapes available today.

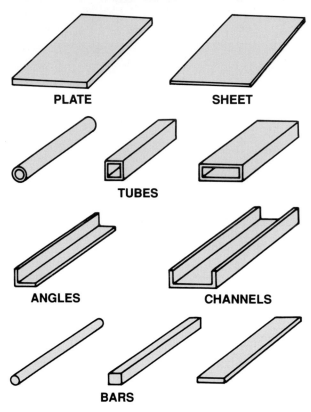

PLATE   SHEET

TUBES

ANGLES   CHANNELS

BARS

FIGURE 6–1 Metals are made in these standard stock forms.

There are two main types of metals—ferrous and nonferrous. The basis for these groups is the amount of iron in the metal. **Ferrous** metals contain iron as the primary element, while **nonferrous** metals contain little or no iron.

Ferrous Metals

The two major groups of ferrous metals are cast iron and steel. Cast iron is an alloy of iron, carbon, silicon, and other materials. It is a heavy, brittle metal that is often used to make engine blocks and the bases for machine tools.

Steel is by far the most important ferrous metal. Like cast iron, steel is an alloy of iron, carbon, and other materials. The 25,000 different steels can be grouped into two categories—carbon steels and alloy steels.

Carbon steels are rated by the amount of carbon they contain as shown in Figure 6–2. Very mild and mild structural steel are often combined and called low-carbon steel. Low-carbon steels contain less than 0.25 percent carbon and are relatively soft; they cannot be hardened by heat treating. Low-carbon steels are used for screws, rivets, pipe, bridges, and skyscrapers, Figure 6–3. The two medium-carbon steels (medium and medium-hard) contain between 0.25 percent and 0.65 percent carbon; these can be hardened by heat treating. Medium-carbon steels are commonly used in machines for shafts and axles. The two high-carbon steels contain more than 0.85 percent carbon. These steels are hard, and they are difficult to machine. High-carbon steels are heat treated and used to make springs and cutting tools, such as drill bits and taps.

TYPE	CHARACTERISTICS
Very Mild	0.05 to 0.15 percent carbon. Soft, tough steel used for sheets, wire, and rivets
Mild Structural	0.15 to 0.25 percent carbon. Ductile, machinable steel used for buildings and bridges
Medium	0.25 to 0.35 percent carbon. Stronger and harder than mild structural, used for machinery and construction
Medium-hard	0.35 to 0.65 percent carbon. Used where it is subject to wear and abrasion
Spring	0.85 to 1.05 percent carbon. Used in springs
Tool	1.05 to 1.20 percent carbon. Very hard and strong. Used for making cutting tools

FIGURE 6–2 The Classifications of carbon steel.

FIGURE 6–3 One important use for steel is for skyscrapers and bridges. *(Courtesy of Turner Construction Company)*

Steel can be alloyed with a number of different materials to give it special qualities. When chromium is the main alloy ingredient, the result is stainless steel or chromium alloy steel. Chromium makes steel more resistant to corrosion. Some steel alloys with carbon contents greater than 1.00 percent are called tool steels. The additional carbon gives these steels more strength and resistance to wear. Tool steels are used to make special high-quality cutting tools.

Shaping Steel

When steel is made, it is formed into huge slabs called **ingots**. Ingots of steel are further processed into a variety of shapes. The shaping process used and the ingredients of the steel determine the properties of the steel.

Blooming Mill. Steel ingots are heated and rolled in a blooming mill. A **blooming mill** is a set of rollers that rolls the ingots into a large slab called a bloom. Rolling in the blooming mill also tends to make the steel tougher.

Hot Rolling. The bloom may be reheated or it may go directly to the rolling mill. In the rolling mill, it is passed through a series of rollers to produce the desired shape, Figure 6–4. The

red-hot steel is passed through the rollers several times. The size is reduced on each pass. When the desired size and shape are reached, the piece is cooled and cut to length.

Cold Rolling. Hot rolling can cause small variations in the size of the steel parts. Cold rolling produces parts with very little variation in size. The steel is cleaned in a chemical solution, then rolled without heat. As it is cold rolled, it increases in strength but becomes less ductile.

Cold Drawing. Wire is formed by pulling cold steel through openings called dies, Figure 6–5. The wire is reduced in size each time it passes through one of a series of dies. This produces high-strength wire of an accurate size.

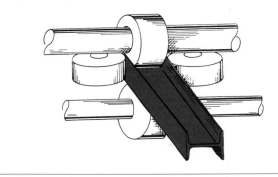

FIGURE 6–4 Structural steel shapes are formed by hot rolling. The space between the rollers is reduced on each pass.

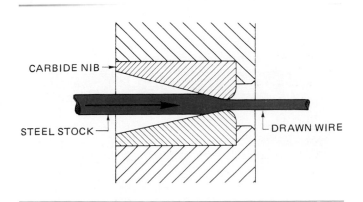

CARBIDE NIB

STEEL STOCK

DRAWN WIRE

FIGURE 6–5 Wire is formed by pulling steel through a drawing die.

Nonferrous Metals

Nonferrous metals contain little or no iron. Those most commonly used include aluminum, copper, zinc, and lead.

Aluminum is the most widely used nonferrous metal. It is found in a mineral ore called bauxite. Aluminum is lightweight and strong, is an excellent conductor, and resists corrosion. Automobile manufacturers use aluminum as a replacement for cast iron in engine blocks. This maintains strength but reduces weight. The outer skin on airplanes is also made from aluminum. Because aluminum resists corrosion and conducts heat and electricity, it is useful for beverage cans, metal cooking utensils, and electrical wiring, Figure 6–6. Aluminum can be rolled into structural shapes or sheets like steel for construction, Figure 6–7. It can also be extruded into complex shapes, Figure 6–8.

Silver is the best conductor of electricity. Copper is second to silver. The price of silver is too high for electrical applications. Copper, therefore, is the number one material for wiring, motor windings, and other electrical parts. Copper is also highly resistant to corrosion. This makes it an excellent material for pipes and fixtures in plumbing systems.

Two popular copper alloys are brass (copper and zinc alloy) and bronze (copper and tin alloy). These metals are valued for their color and beauty, Figure 6–9.

FIGURE 6–7 Aluminum roofing panels and siding are formed from sheets. *(Courtesy of Reynolds Metals Company)*

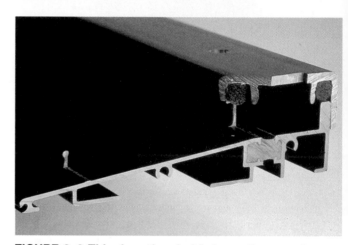

FIGURE 6–8 This door threshold shows the complex shapes possible with aluminum.

FIGURE 6–6 Aluminum cans keep soft drinks safe from corrosion. *(Used with permission. Copyright © PepsiCo, Inc., 1980.)*

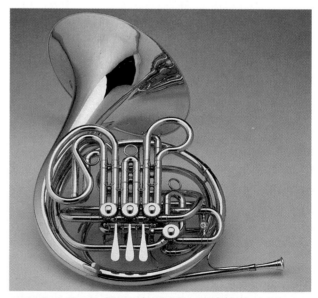

FIGURE 6–9 This French horn is made from brass, an alloy of copper and zinc. *(Photo courtesy of the G. Leblanc Corporation, Kenosha, WI)*

Lead is a very dense and heavy metal. Its density makes it valuable as a protective barrier when x-rays are taken. Lead is also highly resistant to corrosion by acids. This makes it useful in storage batteries. The natural lubricating characteristics of lead make it useful as a material in bearings. It is also used in solder because of its low melting point.

Zinc is a metal that is used to coat other structural metals such as iron and steel. A process called galvanizing is used to bond zinc to other metals. Since zinc is highly resistant to corrosion, galvanized metals are used in outdoor construction.

There are dozens of other important metals. Some of the more common metals are listed in Figure 6–10, along with their important mineral ores, properties, and uses.

Polymer Materials

The two categories of polymers are natural polymers, such as wood, and **synthetic** (human-made) polymers, such as plastics. Natural polymers are a renewable raw material resource. Synthetic polymers come from exhaustible raw materials.

Wood

Trees have been a valuable resource for thousands of years. They will continue to be important in the future. Wood isn't the only material to come from trees. Trees support the manufacture of thousands of products. These include paper, plywood, turpentine, and even a few resins that are used to make plastics. Most of the lumber from trees is used to build homes and similar structures, Figure 6–11. The lumber used in manufacturing goes primarily into furniture

FIGURE 6–11 Wood is used as the framing material in many buildings and homes.

METAL	IMPORTANT ORES	IMPORTANT PROPERTIES	IMPORTANT USES
Aluminum	Bauxite	Lightweight	Planes, foil, pots & pans
Chromium	Chromite	Corrosion resistant	Chrome plating, stainless
Copper	Chalcocite	Second best conductor	Wiring, brass, bronze
Gold	Gold	Most malleable metal	Jewelry
Iron	Hematite	Magnetic	Cast iron, steel
Lead	Galena	High density	Batteries, solder
Magnesium	Magnesite	Lightweight	Aircraft, tools
Nickel	Pentlandite	Corrosion resistant	Coins, nickel plating
Platinum	Platinum	Catalyst	Catalytic Converters
Silver	Argentite	Best conductor	Coins, jewelry, photography
Tin	Cassiterite	Corrosion resistant	Coating tin cans
Titanium	Ilmenite	Lightweight, strong	Gas turbine engines
Tungsten	Wolframite	High melting point	Light bulb filaments
Uranium	Pitchblend	Radioactive	Nuclear fuel
Zinc	Sphalerite	Corrosion resistant	Galvanized metals

FIGURE 6–10 This chart lists just a few of the more important metals.

Gateway Arch

Jefferson National Expansion Memorial, on the old St. Louis riverfront, is a national monument. The most famous part of the memorial is the Gateway Arch.

Steel is one of the main materials used in the Gateway Arch. The stainless steel-faced arch spans 630 feet between the outer faces of its legs at ground level. Its top soars 630 feet into the sky. Each leg is a triangle. The three sides of each leg are 54 feet long at ground level. They taper to 17 feet at the top. The legs have double walls of steel 3 feet apart at ground level and 7¾ inches apart above the 400-foot level. Up to the 300-foot mark, the space between the walls is filled with concrete. Beyond that point, steel stiffeners are used.

The double-walled triangular sections were placed one on top of another. They were then welded inside and out to build up the legs of the arch. Sections range in depth from 12 feet at the base to 8 feet for the two top sections. The complex construction is hidden from view. All that can be seen is its sparkling stainless steel outside skin.

Concrete foundations, sunk 60 feet into the ground, extend 30 feet into bedrock. To prepare the site for the arch foundations, the contractors had to move 300,000 cubic feet of earth and rock. Engineers who designed the arch report that a wind of 150 miles per hour will cause the top of the arch to bend only 18 inches from side to side.

Each leg has a hollow core 48 feet wide at the base, tapering to 15½ feet at the top. The inner skin is of ⅜-inch thick carbon steel. The outside surface contains 900 tons of polished stainless steel, ¼ inch thick. Inside the hollow core, a 40-passenger train operates to take visitors to the top of the arch.

The Gateway Arch.

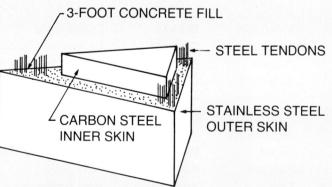

3-FOOT CONCRETE FILL

STEEL TENDONS

CARBON STEEL INNER SKIN

STAINLESS STEEL OUTER SKIN

Structure of the Gateway Arch.

(Courtesy of Jefferson National Expansion Memorial NHS/National Park Service)

Wood can be classified into softwoods and hardwoods. These classes have nothing to do with the hardness or softness of the wood. Rather, they relate to the type of tree. **Softwoods** come from conifers, or cone- bearing trees with needles. **Hardwoods** come from deciduous trees that have broad leaves. Examples of softwoods and hardwoods are listed in Figure 6-12.

Properties and Uses of Wood. There are hundreds of different species of trees. Each produces wood with unique properties. Some woods, like balsa, are extremely light and soft; others, like oak, are very heavy and hard. Because of the wide variety available, there are woods that can match almost any needs. In general, wood has been admired for years because it is a material that is easy to work, it is strong and durable, and it shows beautiful grain patterns, Figure 6–13.

One property of wood that is of special concern is moisture content. Wood is made up of tiny cellulose fibers held together by a substance called lignin. In wood, there are many pores and gaps between the wood fibers. When a tree is growing, moisture is contained within the cell structure of the wood. Before wood can be used, it must be dried to remove this moisture. As the moisture content decreases, wood shrinks. This shrinking would be disastrous for homes or products made from wood with a high moisture content.

Acceptable moisture content for construction lumber is approximately 19 percent. For furniture manufacturing, the moisture content must be held to 6 to 10 percent. Furniture made with a higher moisture content will shrink, warp, and crack when the wood dries in a home.

Two methods of drying wood are air and kiln drying. In air drying, wood is stacked outdoors under shelter and left to dry naturally. In kiln

HARDWOODS	SOFTWOODS
ash	cedar
basswood	fir
birch	pine
cherry	redwood
mahogany	spruce
maple	
oak	
poplar	
walnut	

FIGURE 6–12 This list shows the common hardwoods and softwoods.

FIGURE 6–13 In construction, wood is used for (A) siding, (B) stair railings, and (C) cabinets. *(A and C courtesy of Silz, Seabold & Associates)*

drying, the wood is forced dry by treating it in a kiln (oven), Figure 6–14. Kiln drying is a much faster method.

Specifying Wood. Lumber is sold in standard sizes and grades. Figure 6–15 lists sizes and grades for both hardwood and softwood lumber. Sizes are given in a nominal size and an actual size. The nominal size is the named size given; the actual size gives the exact measurements of the lumber. As an example, a common pine (soft-wood) two-by-four (nominal size) has an actual size of 1½ inches by 3½ inches.

Wood is also graded to uniform standards that indicate the presence or absence of defects in the wood, such as knots, checks (splits), stains, or pitch. The best lumber grades have the fewest defects.

Lumber is sold by a standard measure called a **board foot**. One board foot is equal to one inch thick, twelve inches wide, and twelve inches (one foot) long, Figure 6–16. To determine the number of board feet in a piece of lumber, use the formula in Figure 6–16.

Plastics

The primary raw material used to make plastics is the carbon found in petroleum and natural gas. Plastic polymers are formed by a chemical reaction called **polymerization**. In polymerization, individual molecules react together to form long chains.

There are two basic types of plastics, **thermoplastic** and **thermoset**. Thermoplastics are made of many long polymer chains with no links between them, Figure 6–17. Thermoplastic materials can be reheated and reshaped repeatedly. When heated, the bonds between the polymer chains weaken. The material can then be reshaped by

FIGURE 6–14 Lumber on its way to steam-heated drying kilns. *(Courtesy of Weyerhaeuser Company, Inc.)*

SOFTWOODS		HARDWOODS		SOFTWOODS	HARDWOODS
THICKNESS AND WIDTH		THICKNESS		GRADES	GRADES
NOMINAL	ACTUAL	GREEN Rgh*	DRIED S2S**	*Select*—good appearance and finish quality	*FAS*—first and seconds, highest quality, clear
1"	3/4"	3/8"	3/16"	Grade A—clear, no knots	
2"	1 1/2"	1/2"	5/16"	Grade B—high quality	
3"	2 1/2"	5/8"	7/16"	Grade C—for painting	*No. 1 Common* and Select, some defects
4"	3 1/2"	3/4"	9/16"	Grade D—lowest select	
5"	4 1/2"	(4/4) 1"	13/16"	*Common*—general use, not finish quality	
6"	5 1/2"	(5/4) 1 1/4"	1 1/16"	Construction/No. 1—best	*No. 2 Common*, for smaller cuttings
7"	6 1/2"	(6/4) 1 1/2"	1 1/4"	Standard/No. 2—good	
8"	7 1/2"	(8/4) 2"	1 3/4"	Utility/No. 3—fair	
9"	8 1/2"	(12/4) 3"	2 1/2"	Economy/No. 4—poor	
10"	9 1/2"				

*Rgh = Rough (directly from sawmill)
**S2S = Surfaced on two sides

FIGURE 6–15 Lumber is sold in the standard sizes and grades listed here.

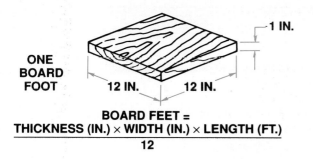

ONE BOARD FOOT

12 IN.   12 IN.   1 IN.

$$\text{BOARD FEET} = \frac{\text{THICKNESS (IN.)} \times \text{WIDTH (IN.)} \times \text{LENGTH (FT.)}}{12}$$

FIGURE 6–16 One board-foot is one inch thick, twelve inches wide, and twelve inches long. You can calculate the amount of board feet in any piece of wood with this formula.

pressure. When the reshaped thermoplastic cools, it retains the new shape. Thermoplastics are like wax in a candle that can be remelted and reshaped over and over again.

Thermoset plastics are also composed of many long polymer chains, but they have links between the chains. Once formed into shape, thermosets cannot be reheated and reshaped. This

process is like hard boiling an egg; once it is hardened, its shape cannot be changed with heat.

Generally, thermosets are harder and more stable at high temperatures than thermoplastics. Thermoplastics, on the other hand, are easier to work with and cost less. They are used more often because they are less costly and the wastes are easier to recycle. Figure 6–18 lists a number of the more common thermoset and thermoplastic materials along with their uses.

Properties and Uses of Plastics. In general, plastics are excellent thermal and electrical insulators. They are lightweight and easily formed with heat and pressure. Also, they need very little finishing such as sanding and painting. Some plastics can be corroded by certain solvents; for the most part, however, plastics have excellent corrosion resistance.

Plastics are being used in many different ways. New plastics are always being developed. In recent years, plastics have been applied where woods and metals were traditionally used. In some cases, certain plastics are stronger than steel. Some will not burn, and others are as transparent as glass. Some of the most common uses of plastics include electrical wiring insula-

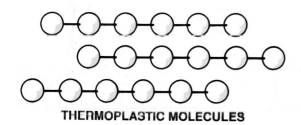

THERMOPLASTIC MOLECULES

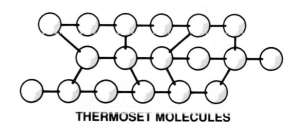

THERMOSET MOLECULES

THE SHAPE OF A THERMOPLASTIC CAN BE CHANGED WHEN THE MATERIAL IS REHEATED

ONCE A THERMOSET IS HARDENED BY HEAT, ITS SHAPE IS PERMANENT

FIGURE 6–17 A comparison of thermoplastics and thermosets.

MATERIAL	USES
THERMOPLASTICS	
ABS (Acrylonitrile-butadiene-styrene)	Telephones, car grilles, sports helmets
Acetal	Small gears in toys
Acrylic	Plexiglass, skylights
Cellulose acetate	Photographic film base, eyeglass frames
Nylon	Clothing, bearings, gears
Polycarbonate	Safety eyeglasses, space helmets
Polyethylene	Packing film, plastic bottles, garbage bags
Polypropylene	Battery cases, washing machine agitators
Polystyrene	Disposable tableware, meat trays, cups
Polyurethane	Packing foam, surface finish (varnish)
Polyvinyl chloride	Plastic pipe (PVC)
Vinyl acetate	Vinyl floor tiles, phonograph records
THERMOSETS	
Amino plastics	Plastic laminates (formica)
Epoxy	Adhesives, fiberglass
Polyester	Clothes, fiberglass
Polytetrafluoroethylene	Nonstick surfaces (Teflon)
Silicone	Printed circuit boards, sealants

FIGURE 6–18 Here are some of the more common thermoplastics and thermosets and their uses.

tion, foamed plastic board wall insulation (Figure 6–19A), vinyl flooring and house siding, polyvinyl chloride (PVC) piping (Figure 6–19B), plastic sheeting (Figure 6–19C), fiberglass showers (Figure 6–19D), plexiglass windows (Figure 6–19E), plastic bottles (Figure 6–19F), and plastic eating utensils (Figure 6–19G). Many paints, fabrics, caulking materials, and hundreds of other products are also made with plastics. They are very useful materials.

FIGURE 6–19B PVC plastic pipe. *(Courtesy of Larry Jeffus)*

FIGURE 6–19A Foamed plastic insulation board. *(Courtesy of Dow Chemical Company)*

FIGURE 6–19C Plastic sheet is used to prevent moisture from passing through the floor.

FIGURE 6–19D Fiberglass plastic is a popular choice for bathtubs and showers. (Courtesy of Larry Jeffus)

FIGURE 6–19E Plexiglass windows are used to prevent safety hazards. (Courtesy of Rohm & Hass Co.)

FIGURE 6–19F Plastic bottles and containers. (Courtesy of Dow Chemical Company)

FIGURE 6–19G Plastic eating utensils. (Photo by Brent Miller and Sonya Stang)

Summary

Metals and polymers are two common types of materials used in production. The two types of metals are ferrous and nonferrous. Ferrous metals, such as steel, contain iron. Nonferrous metals, such as aluminum, contain little or no iron. There is a wide variety of metals. Each metal has special properties and uses in production.

The two types of polymers are natural and synthetic. Wood is a common example of a natural polymer. Plastic is an example of a synthetic polymer. The two types of wood are hardwood and softwood. Hardwoods come from trees with leaves. Softwoods come from trees with needles and cones. The names hardwood and softwood do not indicate the hardness of the wood.

The two types of plastics are thermosets and thermoplastics. Thermoset plastics remain set or fixed once molded into shape. They cannot be reshaped. Thermoplastics can be reheated and reshaped again and again. Plastics have many special properties that make them desirable replacements for metals and woods in many production applications.

DISCUSSION QUESTIONS

1. What are the two types of metals? Can you name two or three examples of each type?
2. How are metals used in production? What special properties do metals have that make them useful?
3. What are the two types of polymers? Can you name two or three examples of each type?
4. How are polymers used in production? What special properties do polymers have that make them useful?
5. What is the difference between thermoset and thermoplastic materials?

CHAPTER ACTIVITIES

 ## Sheet Metal Planter

OBJECTIVE

Metal materials are used to make many products and structures. In this activity, you will produce a planter for flowers or other plants. The planter will be made of 28-gauge sheet metal. The planter should be leakproof and painted a bright color.

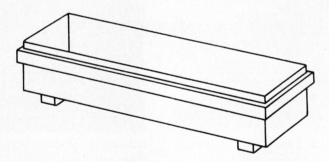

FIGURE 1. Sheet metal planter

EQUIPMENT AND SUPPLIES

1. 28-gauge galvanized sheet metal (11″ x 24″ and 1¾″ x 10″)
2. 28-gauge decorative sheet metal (1½″ x 50″)
3. Sheet metal box and pan brake
4. Solder and soldering gun
5. Tin snips or hand shears
6. Spray primer and paint
7. Combination square
8. Drill and drill bits
9. Marking gauge
10. Rivets
11. Ruler
12. Files
13. Scribe
14. Safety glasses

PROCEDURE

1. Prepare the pieces of sheet metal needed.
    CAUTION: Sheet metal edges and corners may have sharp burrs that can cause serious cuts. Be careful.
2. Lay out the lines for the planter plans using the square and scribe.
3. Wearing your safety glasses, cut the sheet metal pieces to the proper sizes using shears or snips.
4. File all edges to remove any burrs.
5. Bend the ¼ " hem on the planter body and feet at 180 degrees using the box and pan brake.

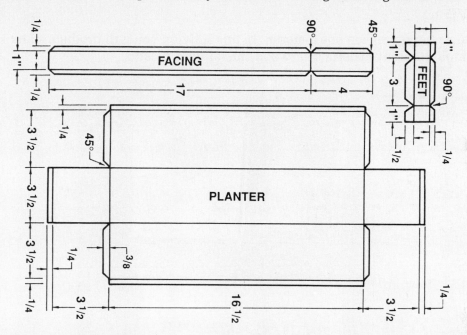

FIGURE 2. Sheet metal planter plans

6. Bend the sides of the planter facings at 90 degrees.
7. Bend the ⅜ " tabs and the ends of the planter at 90 degrees.
8. Bend the sides of the planter at 90 degrees.
9. Fix the box together so that the ⅜ " tabs are flush with the ends of the planter. Check the planter to make sure all angles are square.
10. Bend the ½ " line on the planter feet to 90 degrees.
11. Bend the corners of both the facing and the feet to 90 degrees.
12. Align the feet on the bottom of the planter, drill the necessary holes, and fasten the feet with rivets. Also, fasten the facing pieces.
13. Solder the inside joints of the planter.
14. Use spray primer and paint to finish the planter.

MATH AND SCIENCE CONNECTIONS

Galvanized sheet metal is steel coated with zinc. Electroplating processes are used to make the zinc stick to the steel. Zinc is resistant to corrosion. It protects the steel in the sheet metal from rusting. This makes galvanized sheet metal very useful in wet conditions. That is why it is used in planters where plants and flowers must be watered.

RELATED QUESTIONS

1. What does the term ''gauge'' mean in this activity?
2. What other products can you name that are made with sheet metal?
3. What type of metal is used in sheet metal?
4. Is the sheet metal you used a ferrous or nonferrous metal?

 ## Picture Frame Product — Wood, Metal, and Plastic

OBJECTIVE

Many products use more than one material. In this activity, you will produce a picture holder that uses three different materials — wood, metal, and plastic.

FIGURE 1. Picture holder

EQUIPMENT AND SUPPLIES

1. Plans for the Picture Frame (handout from teacher)
2. Base: Wood ($\frac{5}{8}$" x 2" x 7")
3. Frame: Plexiglass ($\frac{1}{16}$" x 3½" x 8")
4. Brace: Aluminum or steel ($\frac{1}{8}$" x 1" x 3¾")
5. Felt pad: ½" diameter
6. Miter saw, band saw, table saw
7. Buffer
8. Strip heater
9. Router
10. Files, sandpaper, and steel wool
11. Safety glasses
12. Safety gloves for hot materials

PROCEDURE

NOTE: Some of the steps in the procedure may require special tools and machines. These tools and machines should be set up by your teacher for the best results. Also, be sure to wear your safety glasses and use the safety gloves.

Preparing the Wood Base

1. Prepare the wood base to the sizes given on the plans using the miter saw.
2. Route a bead along one edge of the wood base.
3. Route the ⅛″ groove along the edge opposite the bead you just routed.
4. Route the ⅜″ groove on the top of the base.
5. Sand smooth and finish with varnish or lacquer.
6. Install four felt pads under base for feet.

Preparing the Plexiglass Frame

1. Cut the plexiglass to size. Your teacher may perform this process.
2. Polish the edges of the plexiglass on the buffer.
3. Wearing your gloves, heat one end of the frame at 1½″ from the end using the strip heater.
4. Bend the end of the frame over along the heated line.
5. Repeat steps 2 through 4 for the other end of the frame.

Preparing the Metal Brace

1. Cut the brace to size.
2. File and sand the edges of the brace smooth.
3. Bend one end of the brace at 90 degrees.
4. Bend the other end of the brace at 45 degrees.
5. Install the felt pad on the 45 degree bend.

Final Assembly

1. Press the brace into the small slot on the back of the wood base.
2. Insert the frame in the long slot on the top of the base.
3. Insert two of your favorite photographs.

MATH AND SCIENCE CONNECTIONS

Plexiglas is a trademark name. The scientific name for the plastic (popularly called plexiglass) is polymethyl methacrylate. It is also called Lucite. In Britain it is called Perspex. Plexiglass is a stable material. It also has good resistance to weathering. One unusual property of plexiglass is its light-transmitting ability. Light projected into the edge of a piece of bent plexiglass will be transmitted around the bend. Can you think of a use for this unusual property?

RELATED QUESTIONS

1. Was the wood used for the base a hardwood or softwood?
2. What is the difference between natural and synthetic polymers?
3. What is the difference between ferrous and nonferrous metals?
4. What are the two categories of plastics?

CHAPTER 7

Ceramic and Composite Materials

OBJECTIVES

After completing this chapter, you will be able to:

- Identify examples of clay-based, refractory, glass, and abrasive ceramic materials.
- Describe properties and uses of ceramic materials.
- Identify examples of layered, fiber-reinforced, and particle composite materials.
- Describe properties and uses of composite materials.

KEY TERMS

Ceramics	Mortar	Silicates
Fiber-reinforced composite	Particle composites	Slump test
Gypsum	Plaster	Tempered glass
Lath	Portland cement	Veneer
Layered composite	Rebar	Wallboard

In this chapter, you will read about the other two categories of production materials — ceramics and composites. Ceramics are made from clay and sand. Pottery, china, porcelain, and glass are just a few examples of ceramic materials. Composites are combinations of two or more materials. Plywood and concrete are two common examples of composite materials.

Ceramic Materials

Ceramics are one of the oldest production materials. Some of the first signs of technology were clay pots found when prehistoric living places were uncovered. Today, the field of ceramic materials is expanding. Many new types and applications are being found.

One reason for the growth in ceramics is the abundance of raw materials for ceramics. **Ceramics** are compounds of oxygen and another element, usually silicates. **Silicates** are one of the most common minerals on earth. They are found in sands and clays.

Properties and Uses of Ceramics

Ceramic materials are plastic and moldable when wet. This makes them easy to shape into finished products by a number of processes. Once the material dries, it becomes rigid and maintains its shape. After forming, ceramics are fired (heated) in a kiln or oven. Once fired, the shape is permanent and the material is hard and strong.

Ceramics are very good insulators. Two important applications are as furnace lining for glass and metalworking, and the heat shield tiles on the space shuttle. Ceramics are also hard, stiff, strong, and unaffected by weather. Common building bricks are one of the most common examples of a product that takes advantage of these properties.

Many different products are made from ceramic materials, Figure 7–1. Four of the more common categories of ceramics are clay-based, refractories, glasses, and abrasives. Clay-based ceramics are used to manufacture dinnerware, including plates and fine china, ceramic floor tiles, structural bricks, and porcelains used to coat metals. Refractories are made for high-temperature applications. Examples include furnace-lining bricks, spark plug insulators, and the tiles on the space shuttle, Figure 7–2. Ceramic glasses are manufactured into common window glass, fiberglass fibers, and fiber optics. Two common ceramic abrasives are silicon carbide and aluminum oxide. The trade name for silicon carbide is Carborundum™. Both these abrasives are used in grinding wheels and abrasive sheets.

Three important ceramic materials for construction are bricks and blocks, gypsum products, and glass.

FIGURE 7–2 Heat-resistant tiles on the space shuttle are an example of refractories. *(Courtesy of NASA)*

Bricks and Blocks

Bricks and blocks are two common ceramic materials used in construction. Bricks are small rectangular units made of fired clay or shale. There are several kinds of bricks. Most are either common brick, face brick, firebrick, or special brick. Common bricks are the red bricks used where special shapes and colors are not needed, Figure 7–3. Face bricks are similar to common bricks but have a smoother surface finish. Firebrick is used in areas that come in contact with fire or heat. Special bricks include such types as paving bricks and odd shapes.

Blocks are usually made from concrete. Concrete block is popular where heavy loads must be supported, such as in a house foundation, Figure 7–4. Concrete blocks come in assorted

FIGURE 7–1 Floor tiles, toilets, and porcelain-lined tubs are common ceramic products. *(Courtesy of Kohler Company)*

FIGURE 7–3 The bricks on the outside of this fireplace are common brick. The inside lining is firebrick.

FIGURE 7–4 Concrete blocks are frequently used for building foundations because of their strength. *(Courtesy of Richard T. Kreh, Sr.)*

sizes and shapes, Figure 7–5. The named size of a block may be different from its actual size. For example, an 8″ x 8″ x 16″ block is actually ⅜ inch smaller in all dimensions. This allows for a ⅜-inch joint between the blocks.

Many different styles of blocks are available to give a pleasing appearance or for special use. Figure 7–6 shows a few of the many styles of blocks available today.

Gypsum Products

Gypsum is a natural rock taken from mines. To make a usable product, the gypsum rock is crushed, then heated in kilns. The heat dries out the water in the rock. This leaves a fine powder called plaster. **Plaster** is widely used for finishing interior walls, Figure 7–7. It produces a durable surface that resists fire. Plaster is also used in plaster of paris, which is mixed with water and used for patching cracks on plastered walls. Gypsum is also made into **wallboard** and **lath**. Wallboard is a fireproof wall covering. It is made up of a gypsum plaster core covered with strong paper. Wallboard sheets are made in a standard width of 4 feet and in lengths from 8 feet to 16 feet. Lath is similar to wallboard. It is made in sheets 16 inches wide by 48 inches long, Figure 7–8.

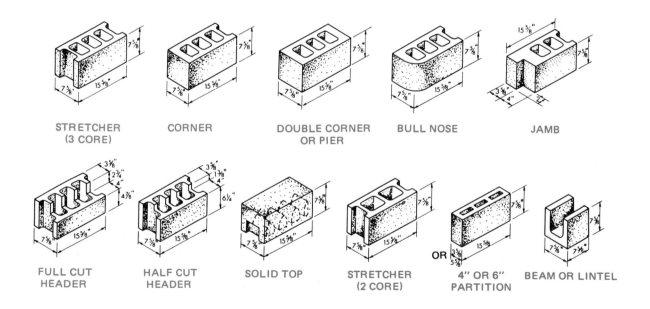

FIGURE 7–5 Common sizes and shapes of concrete blocks.

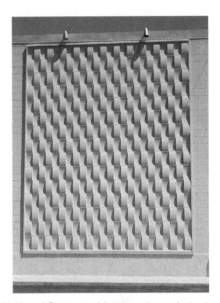

FIGURE 7–6 Several styles and uses of decorative concrete blocks. *(Courtesy of National Concrete Masonry Association)*

FIGURE 7–7 This plasterer is finishing an interior wall with gypsum plaster. *(Courtesy of United States Gypsum Company)*

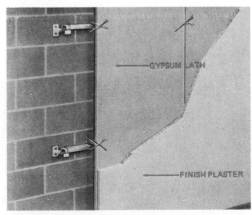

FIGURE 7–8 Gypsum lath is similar to wallboard, only made in narrower sheets. *(Courtesy of Gold Bond Building Products)*

Glass

Glass has been used for windows for hundreds of years. Although some additions have been made, the basic ingredients are still the same.

Glass can be made with special properties. **Tempered glass** is reheated, then cooled very rapidly. This produces a glass that is three to five times stronger than regular glass. It is a good safety glass. Reflective glass has metal flakes mixed in when it is made. Reflective glass is used to reflect unwanted heat and light, Figure 7–9. Insulating glass is made of two sheets sealed around the edges. The air space between the two sheets acts as a barrier to heat/cold and sound, Figure 7–10.

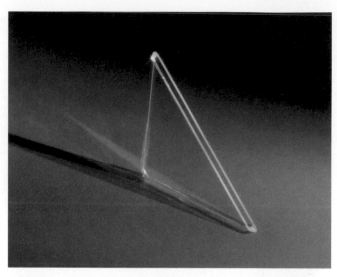

FIGURE 7–10 Insulating glass consists of two pieces of glass with an air space between them. *(Courtesy of PPG Industries)*

FIGURE 7–9 Reflective glass is attractive and reduces the heat gain inside from the sun. *(Courtesy of PPG Industries)*

Composite Materials

Composite materials are made of two or more materials bonded together. In a composite, each material used retains its original properties. When composited (combined), a superior material results. There are three basic types of composites — layered, fiber-reinforced, and particle.

Layered Composites

In a **layered composite**, layers of materials are sandwiched together by an adhesive or other binder. Plywood is an example of a layered composite, Figure 7–11. Thin layers of wood are glued together to create a composite board. The individual wood layers are fairly strong and flexible, but they can swell and shrink because of moisture. They therefore split easily. When composited, a material with superior strength results. It does not split easily and is more stable in different moisture conditions. Other examples of layered composites include plastic laminate counter tops, wax coated paper on milk cartons, cardboard, and the covering on textbooks.

Plywood. The thin layers of wood used to make plywood are called **veneer**. Most veneer for common plywood is peeled off a log. After peeling, the veneer is cut to size. Then several layers are glued together to make the plywood sheet.

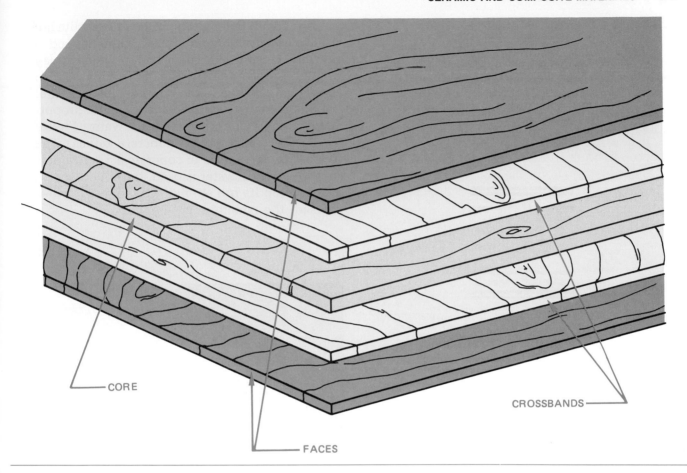

CORE

CROSSBANDS

FACES

FIGURE 7–11 Plywood consists of layers of veneer glued together. *(Courtesy of American Plywood Association)*

Plywood is available in several types and grades for a variety of uses. It may be bonded with waterproof or nonwaterproof glue. It may also be graded by the quality of its face and core veneers. The American Plywood Association sets standards for most plywood, Figure 7–12.

Fiber-Reinforced Composites

The most common **fiber-reinforced composite** is fiberglass. Glass fibers are used to reinforce a base of plastic. One common base material is epoxy plastic. Fibers of carbon, graphite, and boron are also common. Fibers are available in the form of loose fibers, ropes, and woven mats.

Fiber-reinforced composites have properties similar to most plastics (lightweight, insulating, corrosion resistant) with strength comparable to that of many metals.

Particle Composites

In **particle composites**, particles or small bits of reinforcement are used instead of fibers. Concrete is an example. A water and cement base is reinforced by rock and sand particles. Several common wood-based particle composites include hardboard, waferboard, and particle board. In each of these, wood chips or particles are suspended in an adhesive base and pressurized to form a board similar to plywood.

Concrete. Concrete has been used for centuries as a basic production material. The basic ingredients in concrete are aggregates, cement, and water. All concrete contains fine and coarse aggregates. The fine aggregate is normally sand. Coarse aggregates are crushed stone and gravel. When weight is a concern, lightweight aggre-

APA
RATED STURD-I-FLOOR
24 OC 23/32 INCH
SIZED FOR SPACING
T&G NET WIDTH 47-1/2
EXPOSURE 1
000 UNDERLAYMENT
PS 1-83 UNDERLAYMENT
NER-QA397 PRP-108

Panel grade → RATED STURD-I-FLOOR
Span Rating → 24 OC
Thickness ← 23/32 INCH
Tongue-and-groove → T&G NET WIDTH 47-1/2
Mill number ← 000
Exposure durability classification → EXPOSURE 1
Product Standard → PS 1-83 UNDERLAYMENT

APA
RATED SHEATHING
32/16 15/32 INCH
SIZED FOR SPACING
EXPOSURE 1
000
NER-QA397 PRP-108

Panel grade → RATED SHEATHING
Span Rating → 32/16
Thickness ← 15/32 INCH
Exposure durability classification → EXPOSURE 1
Mill number ← 000
Code recognition of APA as a quality assurance agency ← NER-QA397 PRP-108 → APA's Performance-Rated Panel Standard

APA
RATED SIDING
303-18-S/W
16 OC 11/32 INCH
GROUP 1
SIZED FOR SPACING
EXTERIOR
000
PS 1-83 FHA-UM-64
NER-QA397 PRP-108

Panel grade → RATED SIDING
Siding face grade ← 303-18-S/W
Span Rating → 16 OC
Thickness ← 11/32 INCH
Species group number ← GROUP 1
Exposure durability classification → EXTERIOR
Mill number ← 000
FHA recognition ← FHA-UM-64
Product Standard → PS 1-83

A-B·G-1·EXPOSURE1·APA·000·PS1-83

Face veneer
Exposure durability classification
Back veneer
Species group number
Mill number
Product Standard

FIGURE 7–12 Typical American Plywood Association (APA) trademarks, indicating the grade and use of plywood.

gates such as shale, vermiculite, and perlite are used. The most common type of cement used is **portland cement**. Portland cement is mainly clay and limestone or chalk. The cement reacts with the water to cure or harden the concrete mixture.

Concrete must be mixed with specific proportions of aggregates, cement, and water. Figure 7–13 lists some typical concrete mixes. The amount of water used in concrete is very important. If too little water is used, the concrete will not develop its full strength. If too much water is used, the concrete will be weak.

FIGURE 7–14 In a concrete slump test, the slump is the number of inches the concrete sags after being released from the cone. *(Courtesy of Portland Cement Association)*

PROPORTIONS FOR VARIOUS MIXES OF CONCRETE	CEMENT (cu. ft.)	COARSE AGGREGATE (cu. ft.)	FINE AGGREGATE (cu. ft.)	GAL. WATER	
				WET SAND	DRY SAND
3/4″ max. aggregate	1	2¼	2	5	6
1″ max. aggregate	1	3	2¼	5	6
1 1/2″ max. aggregate	1	3½	2½	5	6

FIGURE 7–13 Typical concrete mixes.

To check the quality of a concrete mixture, a **slump test** is performed. Freshly mixed concrete is placed in a slump cone. The concrete is then emptied onto a flat surface. The number of inches that the concrete settles is called the slump, Figure 7–14. A cement mason, construction crew supervisor, or construction inspector usually conducts a slump test.

Concrete is often reinforced with wire mesh or steel rods called **rebar**, Figure 7–15. Reinforcing concrete improves its strength. Ironworkers, or rodsetters, place the reinforcing steel in position in the concrete form.

Mortar. Mortar is a construction composite made of portland cement, hydrated lime, and sand, mixed with water. The proportions of a mortar mix are stated in that order and by volume. For example, 1:1½:4 mortar is made up of 1 part portland cement, 1½ parts lime, and 4 parts sand. The amount of water varies depending on how the mortar will be used. Mortar is used to bond bricks or blocks to make walls, foundations, and other parts of structures.

Wood-Based Particle Composites. Hardboard, particle board, and waferboard are three common wood-based particle composites used in production. Each of these materials is made of wood chips of various sizes mixed with glues and pressed into sheets. Figure 7–16 shows several common wood-based composite materials.

Summary

Ceramics are compounds of oxygen and silicates. They are plastic and moldable when wet, but become hard and brittle when fired. The four types of ceramics are clay-based, refractories, glasses, and abrasives. Examples of clay-based ceramics include bricks, blocks, gypsum mater-

A

B

FIGURE 7–15 Concrete is reinforced with wire mesh (A) and rebar (B). *(Photo courtesy of Niagara Mohawk Power Corporation)*

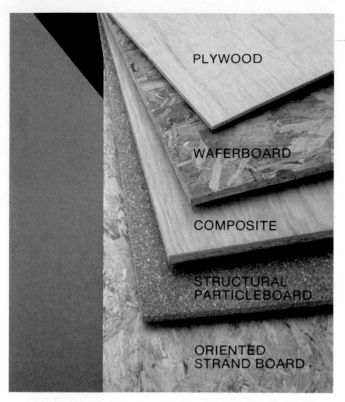

FIGURE 7–16 Common wood-based composite materials. *(Courtesy of American Plywood Association)*

ials, and dinnerware. Refractories include firebrick which is used in high-temperature applications. Glasses include window glass and the fibers in fiberglass. Abrasive ceramics are used to make grinding wheels and sanding sheets.

Composites are combinations of two or more materials. The three types of composites are layered, fiber-reinforced, and particle. Plywood is an example of a layered composite. Layers of thin veneer wood are bonded together to make plywood. Fiberglass is a common fiber-reinforced composite. Concrete, mortar, particle board, hardboard, and waferboard are examples of particle composite materials. Particle composites are made of a number of particles mixed together in a base material. Composite materials are superior to the individual materials used to produce the composite.

DISCUSSION QUESTIONS

1. What are the four types of ceramic materials? Can you name two examples of each type?

2. What are a few of the important properties and uses of ceramic materials?

3. What are the three types of composite materials? Can you name two examples of each type?

4. What are a few of the important properties and uses of composite materials?

CHAPTER ACTIVITIES

 ## Slump Test

OBJECTIVE

Construction workers often perform a slump test on a concrete mixture to test its stiffness. In this activity, you will prepare a concrete mixture. Then you will perform a slump test. The concrete mixture may be used in the next activity.

EQUIPMENT AND SUPPLIES

1. ¼ cubic foot portland cement
2. ½ cubic foot clean sand
3. ¾ cubic foot of ¾-inch aggregate
4. Slump cone
5. Folding rule or tape measure
6. Safety glasses and gloves

PROCEDURE

CAUTION: Safety glasses should be worn whenever you work with mortar or concrete. This is to prevent it from splashing in your eyes. Gloves should also be worn to protect against chemical burns, which can result from long contact with portland cement products.

1. Mix the sand and the aggregate with the portland cement.
2. Mix in five quarts of clean water. If the sand is very dry, it may be necessary to mix in an additional ½ quart (1 pint) of water.
3. Fill the slump cone with freshly mixed concrete and work out the air bubbles with a rod.

Measure the distance from the straightedge laying on top of the slump cone to the top of the concrete.
(Courtesy of Portland Cement Association)

4. Carefully empty the slump cone onto a flat surface.

5. Measure the vertical distance from a rod laid across the top of the empty cone to the top of the concrete sample. This is the slump of the batch. If the slump is between one and four inches, use this batch for the next activity.

MATH AND SCIENCE CONNECTIONS

The ingredients in portland cement are mainly clay and limestone or chalk. Portland cement is used to make mortar and concrete. It is a chemical reaction between the portland cement and water that causes mortar and concrete to harden.

Concrete and mortar are also tested for compressive strength. Samples of the hardened materials are squeezed with a hydraulic press until they shatter. Gauges on the press are used to determine the compressive strength.

RELATED QUESTIONS

1. What are the ingredients in concrete?
2. Why are slump tests conducted on concrete mixtures?
3. Who conducts slump tests?
4. What happens to the slump if too much water is used?

 ## Reinforced Concrete

OBJECTIVE

In this activity you will play the role of an ironworker. Follow the procedure outlined and add rebar to the forms provided. This activity will teach you about reinforcing concrete.

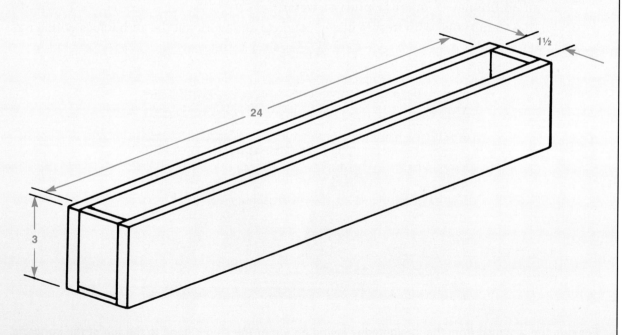

Form for concrete beam.

EQUIPMENT AND SUPPLIES:

1. 2 forms, 1½″W x 3″D x 23″L
2. Waxed paper or plastic wrap
3. Two pieces of #2 rebar, 24 inches long
4. Two chairs to support 2 bars at a height of ¾ inch
5. Concrete to fill forms
6. 4-pound hammer
7. Safety glasses, face shield, and gloves

PROCEDURE

1. Line the forms with waxed paper or plastic wrap to prevent cured concrete from sticking to the forms.
2. Place two pieces of rebar in one of the forms. The rebar should be positioned ¾ inch from the bottom of the form on chairs made of stiff wire.

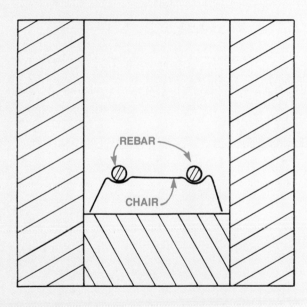

Cross-sectional view of form for concrete beam, with reinforcement in place. Chairs may be made of stiff wire.

3. Fill both forms with concrete and work out the air bubbles. Be careful not to displace the rebar.
4. Write your name in the top of the fresh concrete to help locate the top surface after the concrete is removed from the form.
5. Allow seven days for the concrete to cure.
6. After seven days, position each concrete beam between two supports. The reinforced beam should be positioned with the rebar on the bottom.
7. Wearing a face shield, strike each beam at the midpoint with a 4-pound hammer. Notice the difference between the beams as they are broken.

MATH AND SCIENCE CONNECTIONS

Hitting the beam with a hammer is an example of testing impact strength. Impact strength is a measure of the ability of the material to absorb the shock of the hammer. Impact strength is also called toughness. Materials that do not have high impact strength are brittle. They tend to shatter when hit with an impact load.

RELATED QUESTIONS

1. What happened when you hit the two beams with the hammer?
2. Why is concrete reinforced with steel bars?
3. What do you think would happen if the beams were made with a mixture that did not pass the slump test?
4. What do you think would happen if chairs were not used to support the rebar in the forms?

SECTION THREE

PRODUCTION TOOLS, MACHINES, AND PROCESSES

(Courtesy of FMC Gold Company)

CHAPTER 8

Introduction to Tools and Machines

OBJECTIVES

After completing this chapter, you will be able to:

■ Define tools, hand tools, power hand tools, and machines.

■ Describe the importance of safety in using tools and machines.

■ Explain how tools use the six basic machines to provide mechanical advantage.

■ List the various classes of tools for production.

■ Identify and describe various measuring and layout tools.

KEY TERMS

Builder's level	Go–no-go gage	Plumb bob
Chalk line	Hand tool	Power hand tool
Clamps	Jigs	Separating tools
Combining tools	Machines	Spirit level
Dividers	Mechanical advantage	Squares
Fixture	Micrometer	Tape measure
Forming tools	Pliers	Tools

Production workers use tools and machines to do their jobs. In general, tools and machines process (change) materials or information. In manufacturing and construction, workers use tools and machines to change materials into finished products and buildings. Finance workers use calculators and computers to keep track of the company's money. Marketing workers send product information to consumers in advertisements made with video and audio recording machines. Every production worker must know how to use the tools and machines of his or her trade.

Defining Tools and Machines

Tools extend human abilities in doing the work of processing (changing) materials or information. So, strictly speaking, machines are also tools. Tools extend human abilities by increasing the power, speed, efficiency, accuracy, and productivity of work. We cannot drive nails in boards with our bare hands, but we can drive nails with a tool — the hammer. We can do math problems in our head, but an electronic calculator is faster and more accurate. Both the hammer

Using Tools and Machines Safely

Using tools and machines safely is a big concern in production. Every year, thousands of workers are hurt using tools. Injuries happen not only with big, powerful machines, but also with hand tools. More people are hurt every year by screwdrivers and hammers than by large industrial machines.

Accidents and injuries happen for two reasons: (1) working conditions are unsafe, or (2) worker actions are unsafe. Using dull or broken tools, working in cluttered conditions, or taking poor care of machines are examples of unsafe working conditions. Not knowing how to use a tool, not knowing or following the proper safety rules, or using a tool or machine the wrong way are examples of unsafe worker actions.

Always follow the safety guidelines listed in Table 1 when using tools. Also follow other safety rules and guidelines set by your teacher. Using tools and machines safely should be your number one concern.

TABLE 1. SAFETY GUIDELINES FOR USING TOOLS AND MACHINES

- **Get permission from your teacher before using any tool or machine.**
- **Learn the safety rules for every tool and machine you will use; learn the dangers involved.**
- **Select the right tool for the job, match the tool to the work to be done.**
- **Use tools and machines in a safe and proper manner.**
- **Use tools and machines only for the uses for which they were designed.**
- **Keep tools and machines properly maintained and adjusted.**
- **Use tool and machine safety devices such as guards.**
- **Tools are for work: Never play with tools and machines.**
- **Respect the power of tools and machines.**

Always follow these safety guidelines and other safety rules set by your teacher. Using tools and machines can be dangerous. Think and act safety first.

---

and the calculator are tools that extend our abilities.

Generally, tools can be described as hand tools, power hand tools, or machines. A **hand tool** is the simplest form. The user holds it in the hand and moves it to perform work. It is powered only by the user. Hand saws, screwdrivers, and hand planes are examples of hand tools. **Power hand tools** are improved hand tools. The user holds one in the hand and moves it to perform work, but the power comes from an external source, such as an electric motor. Power circular saws, electric screwdrivers, and power planes are examples of power hand tools. **Machines** stay still during processing. They use a tool powered by an external source. The powered tool is fastened to the machine during the actual processing. Table saws, drill presses, and planers are examples of machines.

Another category of production tools is equipment. Equipment covers devices that cannot be defined as machines, power hand tools, or hand tools. Equipment stays still on a structure during processing and uses human or thermal (heat) power to process materials. Examples include the human-powered squaring shears used to shear metal, as well as ovens, kilns, and furnaces used to melt materials.

All tools, machines, and equipment are used to extend human abilities by increasing the power, speed, efficiency, accuracy, and productivity of processing materials or information.

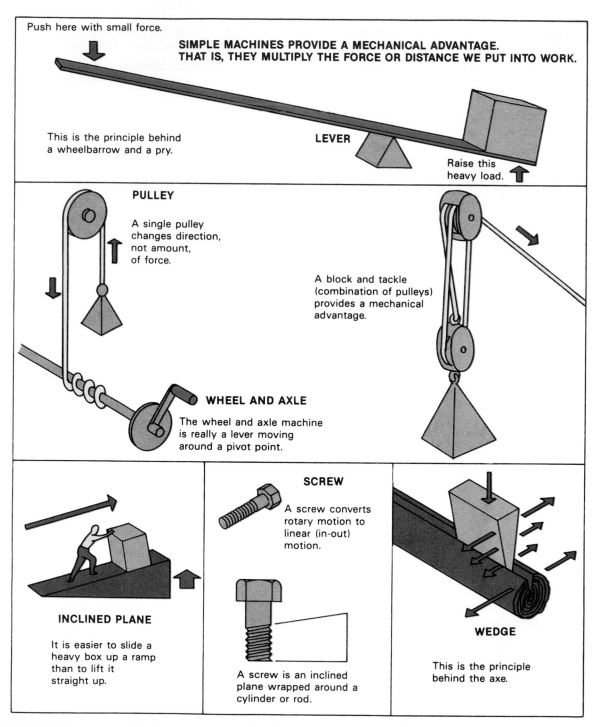

FIGURE 8–1 Workers use the six basic machines to gain a mechanical advantage of force in processing materials.
(Reprinted from LIVING WITH TECHNOLOGY *by Hacker and Barden, copyright © 1988 by Delmar Publishers Inc.)*

The Six Basic Machines

We base the principles that describe how tools work on the six basic machines — wheels, levers, pulleys, inclined planes, wedges, and screws, Figure 8-1. The purpose of these basic machines is to gain a mechanical advantage in doing work. A **mechanical advantage** is an increase in a force. Mechanical advantage of force is abbreviated MAf.

A simple example of mechanical advantage is using a hammer to drive nails in wood. Without a hammer, you would not be able to drive the nails. By placing the hammer in your hand, you are creating a lever that gives you a mechanical advantage of force over the nail. The nail itself uses wedge action to cut into the wood.

For another example, look at the drill press, Figure 8-2, often found in production labs. The drill bit uses a wedge for its cutting action. Inclined planes hold the drill bit in the chuck. Screw threads hold the drill press together. Pulleys transfer power from the motor to the drill bit. The handle on the drill press acts as a lever attached to a wheel and axle. Every time one of the six basic machines is used in a tool or machine, mechanical advantage is realized. Identify the six basic machines in other tools and machines in your production lab.

Classifying Tools and Machines

The tools and machines used in production can be classified in a number of different ways. Usually, they are put into groups based on the processes they are used to perform. In general, the four most important classes of production tools and machines are measuring and layout, forming, separating, and combining. Measuring and layout tools are used to process information on sizes and locations. Rulers, micrometers, and tape measures are examples. **Forming tools** change the outside shape or inside structure of materials. Bending, stretching, pressing, and casting are examples of forming. **Separating tools** cut down the size of materials by removing excess material. Cutting tools, sawing tools, and drilling tools are examples. **Combining tools** add one part to another. Nailing, gluing, welding, and painting are examples.

Separating tools and processes will be explained in detail in Chapter 9; forming and combining tools and processes will be explained in Chapter 10. For now, let's look at measuring and layout tools. In addition, we will examine some of the other tools and machines used in production.

Measuring Tools

Production workers use a variety of measuring tools including rulers, gages, and micrometers. Rulers or rules are used to measure short lengths, Figure 8-3. They are usually made of steel and measure 12, 24, or 36 inches long. **Tape measures** have flexible steel blades that come in a

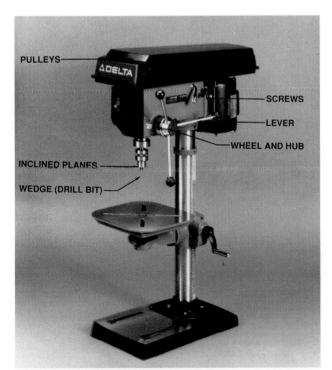

FIGURE 8-2 The six basic machines can be found in any machine. *(Courtesy of Delta International Machinery)*

FIGURE 8-3 A steel rule. *(Courtesy of L.S. Starrett Company)*

variety of sizes, Figure 8–4. Common tape measures may range from 8 to 20 feet in length. Others are available in sizes up to 200 feet and longer. The folding rule, Figure 8–5, is used where the flexibility of the tape measure is a disadvantage.

Gages are used to check sizes and dimensions on the parts. One of the more common gages is the **go-no-go** gage used during quality control inspections in manufacturing. When a product part is checked with this type of gage, it is either kept (a go) or rejected (a no-go).

Micrometers are used for accurate measurements to a thousandth of an inch on precision parts, Figure 8–6.

Layout Tools

Layout tools are used for laying out lines and measurements and checking angles. Squares, dividers, chalk lines, levels, and the plumb bob are common layout tools.

Squares

Squares are usually used to measure 90-degree (square) angles. One of the most commonly used squares in construction is the rafter or framing square, Figure 8–7. It includes several rulers, tables, and scales. The rafter square is used to lay out rafters and stairs and to check the squareness of corners. The try square is a small square with a 6- to 12-inch blade, Figure 8–8. The try square is useful for laying out and checking right angles on small surfaces. The combination square is similar to the try square, but has a

FIGURE 8–4 Steel tape. *(Courtesy of L.S. Starrett Company)*

FIGURE 8–5 Wooden folding rule. *(Courtesy of Sears, Roebuck & Co.)*

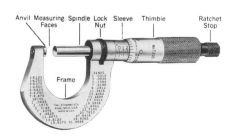

FIGURE 8–6 Micrometer. *(Courtesy of L.S. Starrett Company)*

FIGURE 8–7 Using the rafter square to lay out angles.

FIGURE 8–8 Try square. *(Courtesy of Sears, Roebuck & Co.)*

FIGURE 8–9 Combination square.

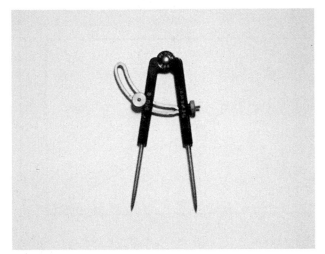

FIGURE 8–10 Dividers.

the surface and released. When the line hits the surface, it leaves a line of chalk. This marks a straight line, Figure 8–12.

Levels

The **spirit level** makes use of the fact that an air bubble rises to the top of a container filled with liquid. A small, slightly curved glass tube (vial) is nearly filled with alcohol (spirit). The vial is mounted in an aluminum or wood straightedge that measures from 2 to 6 feet in length. At least one vial is mounted in such a way that it

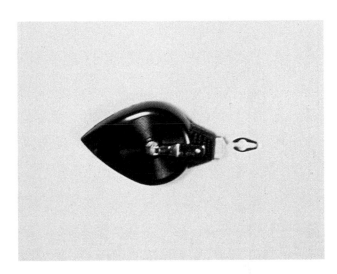

FIGURE 8–11 Chalk line reel.

movable head, Figure 8–9. The head of the combination square has a right-angle surface, a 45-degree angle surface, and a spirit level. The combination square can replace a try square in most cases.

Dividers

Circles and arcs are laid out with **dividers**, Figure 8–10. For a given radius, the dividers are set with a separate measuring device. Then one leg is held on the center of the circle or arc and the other scribes a curved line.

Chalk Line

A **chalk line** reel is used to mark long straight lines, Figure 8–11. The reel case contains powdered chalk and a string or line. The chalk-covered line is stretched tight between two points on a surface. The line is then pulled away from

FIGURE 8–12 Snapping a chalk line.

Math Interface: 6-8-10 Method

Any triangle with sides of 6, 8, and 10 units of length includes a right angle. This principle is used in construction by measuring 6 feet along one side of a corner, and 8 feet along the other side. When the distance between the 6-foot and 8-foot marks equals 10 feet, the corner angle will measure 90-degrees, Figure 1.

The squareness of a rectangle or square can also be checked by measuring the diagonals. When the diagonals are of equal length, all of the corners are 90 degrees, Figure 2.

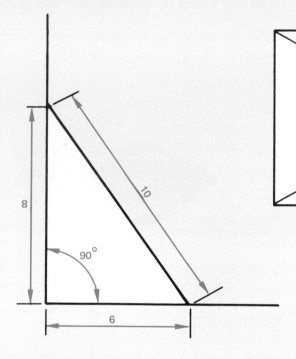

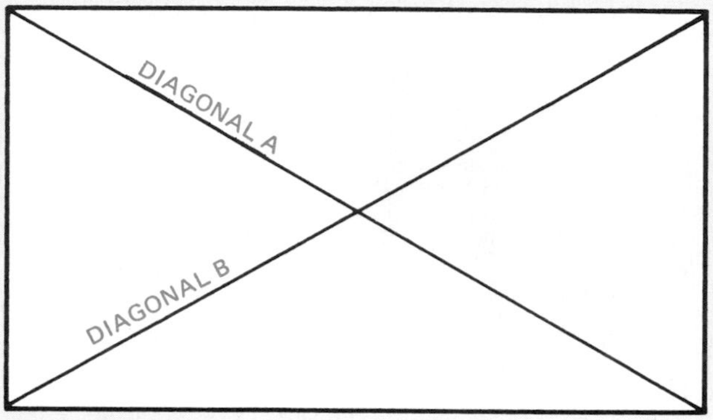

FIGURE 2. Measuring diagonals to check square corners on a rectangle.

FIGURE 1. triangle with sides of 6, 8, and 10 units of length includes a right angle.

can be used to measure plumb, or perfectly vertical, Figure 8–13.

The line level, Figure 8–14, is a small spirit level that can be hung on a stretched line. The line level is used to check level between two points that are too far apart to be checked with the spirit level.

Plumb Bob

The **plumb bob** is a pointed weight attached to a string, Figure 8–15. When suspended on its string, the force of gravity causes the string to

hang perfectly plumb (vertical). The plumb bob is used to check plumbness the way a level is used to check levelness.

Builder's Level

The **builder's level**, like a spirit level, is used to check levelness between two points. It can also be used for measuring angles on a horizontal plane. The basic parts of the builder's level are shown in Figure 8–16.

The builder's level must be mounted on a level tripod (3-legged stand) for use. A target rod is a

FIGURE 8–13 This carpenter uses a spirit level to check plumb. *(Courtesy of Morgan Products Ltd.)*

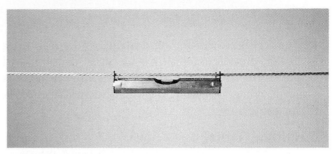

FIGURE 8–14 Line level hung on a stretched line.

separate device used with the builder's level. Construction workers view the target rod through the telescope of the builder's level to check the levelness of several points, the depth of holes dug in the ground, or the levelness of parts of a building.

FIGURE 8–15 Plumb bob.

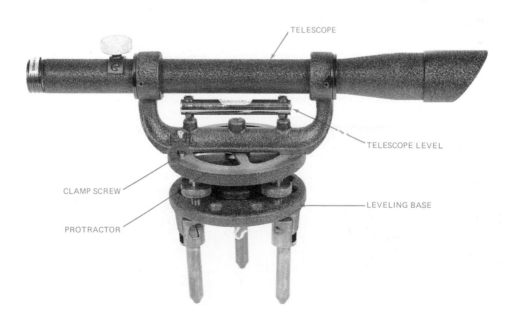

TELESCOPE

TELESCOPE LEVEL

CLAMP SCREW

LEVELING BASE

PROTRACTOR

FIGURE 8–16 Parts of a builder's level. *(Courtesy of L.S. Starrett Company)*

Other Important Production Tools

Measuring, forming, separating, and combining are the four most important types of tools and machines used in production. Some other tools are also used for holding, positioning, and moving materials.

Holding Tools

People are not strong enough to hold parts securely during certain processes. The two most common holding tools used in production are **pliers** and **clamps**. Pliers are hand tools used to hold small parts. The most common plier styles are slip joint, groove joint, locking, and needle-nose, Figure 8–17. Clamps hold pieces of material to a machine during processes such as drilling. They also hold pieces of material together dur-

Slip joint

Needle or long nose

Locking

FIGURE 8–17 Three common types of pliers. *(Courtesy of Stanley-Proto Industrial Tools, Covington, GA.)*

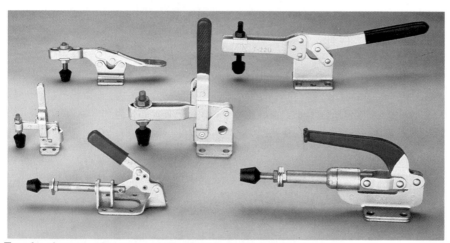

Toggle clamps. *(Courtesy of Woodworker's Supply Inc.)*

C-clamps. *(Courtesy of Stanley-Proto Industrial Tools, Covington, GA)*

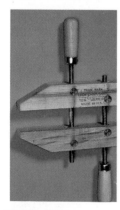

Hand screw. *(Courtesy of Woodworker's Supply Inc.)*

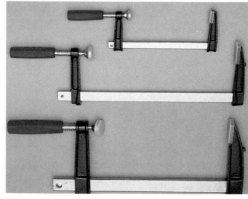

Bar clamps. *(Courtesy of Woodworker's Supply Inc.)*

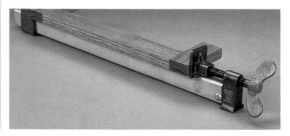

Bar clamp. *(Courtesy of Woodworker's Supply Inc.)*

FIGURE 8–18 Here are a few of the different types of clamps.

ing bonding or gluing. Some of the more common hand-held clamps are the bar clamp, C clamp, hand screw, and toggle clamp, Figure 8–18. For increased holding power, pneumatic (compressed air) or hydraulic (compressed liquid) cylinders can power clamps.

Positioning and Layout Tools

Workers must accurately hold the materials being processed. Two common positioning tools are jigs and fixtures. **Jigs** guide the path of a tool on a part being processed. They are attached to the part being processed, or the part attaches to the jig. **Fixtures** attach to a machine and position parts being processed in a specific location on that machine. Both jigs and fixtures improve accuracy in processing materials by eliminating the need for the worker to measure and lay out exactly where a hole is to be drilled or a material is to be cut, Figure 8–19.

To reduce errors in measurement and layout, patterns can be used. A pattern is a sample or outline of a finished product. It can be used to measure and lay out the size and shape of the product. For example, suppose you wanted to cut out a piece of wood that was the same size and shape as your hand. You could measure your hand and then lay out those measurements on the board. Or you could use your hand as a pattern. You could place your hand on the board and trace around your hand and fingers with a pencil. Tracing around your hand is much quicker and more accurate than measuring all the parts of your hand and laying out the sizes.

FIGURE 8–19 This centering jig is used to guide a drill bit. *(Courtesy of Woodworker's Supply Inc.)*

Transporting Machines

An important part of mass production manufacturing is transporting or moving materials or product parts from one machine to another. Manufacturers use a number of different transporting machines. These include conveyor belts, overhead rail conveyors, high lifts, and automatic guided vehicles (AGV), Figure 8–20.

FIGURE 8–20 Conveyors like this are a form of transporting machine. *(Courtesy of American Honda Motor Co., Inc.)*

Summary

Tools are used to extend human abilities in performing work. A variety of tools are used in production. These include hand tools, power hand tools, machines, and equipment. Before any tools are used, production workers and students must learn and follow safety rules.

Most tools and machines use the six basic machines for their operation. The six basic machines include wheels, levels, pulleys, inclined planes, wedges, and screws. The six basic machines help tools provide mechanical advantage.

The most commonly used production tools include measuring and layout tools, forming tools, separating tools, and combining tools. Rules, tapes, squares, dividers, chalk lines, and levels are the measuring and layout tools described in this chapter.

DISCUSSION QUESTIONS

1. How would you define tools?

2. What is the difference between hand tools, power hand tools, and machines?

3. Why is it important to know and follow safety rules when using tools and machines?

4. What are the six basic machines? Pick a machine in your production lab and identify how the six basic machines are used.

5. What are the four main classes of tools described in this chapter?

6. What measuring and layout tools were described in this chapter?

CHAPTER ACTIVITIES

 Learning Safety Rules for Tools and Machines

OBJECTIVE

Using production tools and machines safely is very important. First you must learn certain safety guidelines and rules.

Many tools and machines may be used in production. The materials that will be used in the finished product or structure will determine what tools and machines you will need. Your school's production lab has tools and machines that you and your classmates will use to make products and build structures. Your teacher has made safety guidelines and rules that every student must know before using tools and machines. In this activity, you will learn these guidelines and rules for the tools and machines in your school's lab.

EQUIPMENT AND SUPPLIES

1. Safety Guidelines and Rules (from teacher)

PROCEDURE

1. Make a list of the tools and machines in your lab.

2. Obtain a list of the safety guidelines and rules for the tools and machines in your lab from your teacher.

3. Learn the following information about tools and machines in your production lab:

   - Names of the tools and machines
   - Uses for the tools and machines
   - Names of the important parts on tools and machines
   - Specific safety hazards when using certain tools and machines
   - Specific guidelines and rules established by your teacher

4. Take a safety quiz on tools and machines that you will be using in the production class.

MATH AND SCIENCE CONNECTIONS

The six basic machines use basic physical science principles for their operation. One important principle is equilibrium. Equilibrium is also called balance. A common example would be two people playing on a see-saw. There are math formulas that can be used to calculate the equilibrium, or balance, between the two people. The same formulas can be used to explain why basic machines, like the wheel or lever, work.

RELATED QUESTIONS

1. Why is it important for production workers and students to know the safety rules for tools and machines?
2. Why do you think your teacher requires students to pass a safety quiz before allowing them to work with tools and machines?
3. What tools and machines do you think are the most dangerous in your school's production lab? Why?
4. Are there any safety guidelines or rules that apply to all, or mostly all, the tools and machines in your lab?

 Measurement and Layout

OBJECTIVE

Accurate measurement and layout are important factors in production. In this activity, you will practice measurement and layout with some of the measuring tools discussed in this chapter.

EQUIPMENT AND SUPPLIES

1. 1″ x 12″ x 24″ piece of softwood
2. Tape measure or folding rule
3. Combination square or rafter square
4. Compass or wing dividers

PROCEDURE

1. Two duplicate pieces of wood stock will be laid out for the ends of a tote carrier that will be produced in another chapter.
2. Using the square, check to see that one end of the lumber is square with one long edge. If it is not square, mark a square line near the end.
3. With the tape measure or folding rule, and using the blade of the square as a straightedge, mark a line 9½″ from the square edge.
4. Working from the square edge, mark a square line across the piece 10″ from the square end. Mark the center of this line (point A), Figure 1.
5. Using point A as the center, scribe a circle with a 1″ radius.
6. Mark points B and C 6″ from the square end.
7. Using the blade of the square as a straightedge, draw lines from points B and C to the edge of the circle.
8. Repeat steps 2 through 7 to lay out the second part on the opposite end of the lumber.
9. Keep the marked stock for the next chapter when you will cut out the pieces for the tote carrier.

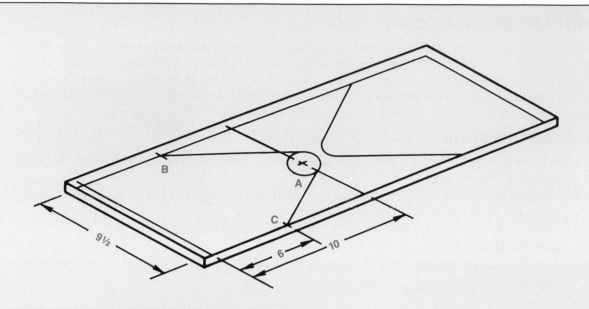

FIGURE 1. What your layout lines should look like.

MATH AND SCIENCE CONNECTIONS

When producing wood products, measurements and layouts should be accurate to less than $\frac{1}{16}$ inch. With metal products, it is not uncommon to require an accuracy of .001 inch ($\frac{1}{1000}$ inch). Metal has a denser molecular structure and can be easily worked to this accuracy. Wood has a porous structure. Some of the gaps in the wood structure can be larger than .001 inch.

RELATED QUESTIONS

1. What does it mean to check a part for "square"?
2. Why do you think it is important for production workers to measure and lay out sizes accurately?
3. Is there any way that the accuracy of your fellow classmates could be improved in measuring and laying out the sizes for the tote carrier? Are there any special tools that could be used other than a ruler and square?
4. Compare your lumber with that of your classmates. How well do your measurements and layouts compare?

CHAPTER 9

Separating Processes and Tools

OBJECTIVES

After completing this chapter, you will be able to:

■ Define separating processes and tools.

■ Name the three categories of separating processes.

■ Identify and describe various chip removing tools.

■ Identify and describe various shearing tools.

■ Identify and describe various special separating tools.

■ Name various safety precautions that should be taken when using separating tools.

KEY TERMS

Abrasive papers	Files	Saber saw
Chisels	Grinding wheels	Scroll saw
Crosscut	Hacksaw	Shearing
Coping saw	Milling	Shears
Cutters	Oxyacetylene torch	Turning
Dielectric	Punches	
Dies	Ripsaw	

Separating processes and tools cut down the size of materials by removing excess materials. They remove part of the original size and shape of a material, leaving the required size and shape. Cutting a piece of paper to size with scissors or drilling a hole are examples of separating processes. Separating tools must be made from materials that are harder than the material being separated. They are made from hard tool steel, carbide, ceramic, or industrial diamonds.

There are three categories of separating processes and tools: chip removing tools, shearing tools, and other special separating tools.

Chip Removing

Chip removing processes are used to cut excess materials away. Chip removing results in a loss or waste of material in the form of chips (small bits of the material).

There are six basic ways to separate materials by chip removing. These are sawing, hole machining, turning, milling, cutting, and abrading.

Sawing

Sawing is an example of chip removing. It removes sawdust chips from the material during separating. Saw blades separate materials along straight or curved lines. They can be used in hand tools, power hand tools, and machines. Blade shapes are straight, circular, or continuous-band. Straight-blade sawing involves moving the blade back and forth against the material. Circular blades are rotated in a power tool or machine to do the sawing. Continuous-band blades are revolved around two or more wheels. Steel is used for most saw blades, but some are carbide-tipped. There is a wide variety of saw blades available, based on the material being separated. There are special blades for separating wood, plywood, metal, plastic, and masonry. Generally, the following types of saw blades are used:

- **Crosscut** saw blades are used to separate wood across the grain.
- **Ripsaw** blades separate wood along the grain, Figure 9–1.
- **Hacksaw** blades separate metals.
- **Scroll, coping,** and **saber saw** blades separate materials along curved lines.

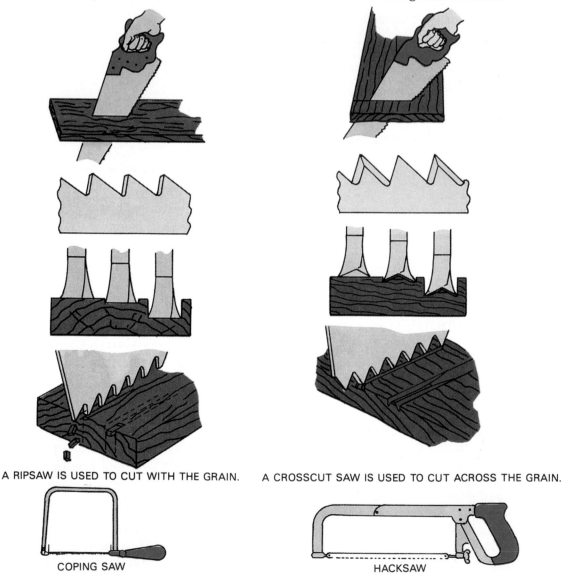

A RIPSAW IS USED TO CUT WITH THE GRAIN.   A CROSSCUT SAW IS USED TO CUT ACROSS THE GRAIN.

COPING SAW   HACKSAW

FIGURE 9–1 The two basic types of saws for cutting wood are the crosscut saw and the ripsaw.

CAUTION: A properly sharpened saw should be handled with care. Keep hands away from the saw teeth. Never allow the teeth to contact metal or any other hard materials.

FIGURE 9–2A Scroll saw. *(Courtesy of Ryobi America)*

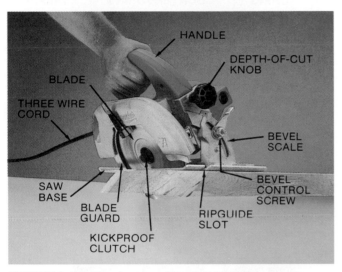

FIGURE 9–2B Circular saw. *(Courtesy of Milwaukee Electric Tool Corporation)*

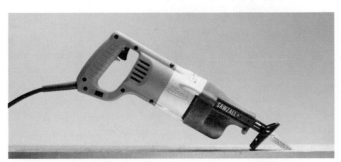

FIGURE 9–2C Reciprocating saw. *(Courtesy of Milwaukee Electric Tool Corporation)*

FIGURE 9–2 Power hand tools that use saws.

Three common power hand saws include the scroll saw, reciprocating saw, and portable circular saw, Figure 9–2. Some of the more common sawing machines used in production are table saws, radial arm saws, band saws, and scroll saws, Figure 9–3.

CAUTION:
1. Do not overtighten the blade screw on a circular saw. Most saws have a special washer that allows the blade to slip if it binds. This prevents the saw from kicking back toward the operator.
2. The stock to be cut off should not be held or supported. This may cause the blade to bind.

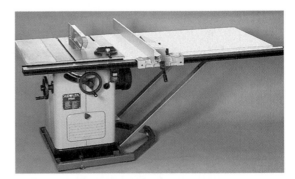

FIGURE 9–3A Table saw. *(Courtesy of Delta International Machinery)*

FIGURE 9–3B Band saw. *(Courtesy of Delta International Machinery)*

FIGURE 9–3 Machines that use saw blades.

Hole Machining

There are six common hole machining processes, Figure 9–4. These are drilling, boring, reaming, counterboring, countersinking, and tapping. Drilling uses a drill bit to make holes in materials. Boring is used to make the size of a drilled hole larger and more accurate. Reaming also makes the size of a drilled hole larger and more accurate, and makes a very smooth surface on the inside of the hole. Counterboring makes a hole larger only partway along its depth. This allows bolt heads or nuts to be hidden below the surface of the material. Countersinking produces a tapered surface in one end of a hole. This allows flathead screws to lie flush with the surface of the material. Tapping produces threads inside a hole.

There is a wide variety of drills for different materials and types of holes, Figure 9–5.

- Twist drills are the most common drills used for wood, metal, and plastic. They are made of solid hardened steel and have two cutting lips. The most common twist drill sizes range from $\frac{1}{16}$ inch to $\frac{1}{2}$ inch in diameter.
- Spur bits have a center point used to guide the bit; they are used to make clean, smooth holes. Spur bits come in the same sizes as twist drills.
- Spade bits make larger holes in materials. They come in sizes over $\frac{1}{2}$ inch.

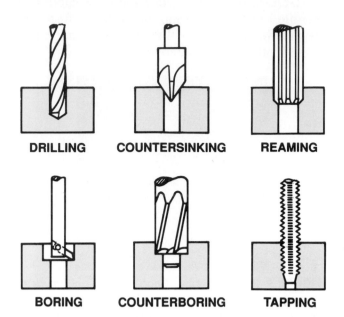

DRILLING   **COUNTERSINKING**   **REAMING**

BORING   **COUNTERBORING**   **TAPPING**

FIGURE 9–4 The six common types of hole machining processes are drilling, reaming, boring, counterboring, countersinking, and tapping.

- Forstner bits make larger flat-bottomed holes.
- Auger bits cut large holes in wood.
- Taps cut threads on the inside of a hole that has already been drilled.
- Reamers smooth out a drilled hole to specifications.

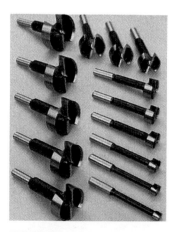

FIGURE 9–5A Twist drill. *(Courtesy of DoAll Company, Des Plaines, IL)*

FIGURE 9–5B Spur bits. *(Courtesy of Woodworker's Supply Inc.)*

FIGURE 9–5C Forstner bits. *(Courtesy of Woodworker's Supply Inc.)*

FIGURE 9–5D Spade bits. *(Courtesy of Irwin Co.)*

FIGURE 9–5 Here are several styles of drills and bits.

Drills and bits can be used in hand tools, power hand tools, or machines. The hand drill is used for smaller bits, Figure 9–6A. Auger bits are held in a bit brace, Figure 9–6B. Power hand drills are used for more power and larger bits, Figure 9–6C. The drill press is one of the most important production machines, Figure 9–6D.

CAUTION:
Power Tool Safety
1. Tools that are not double insulated should be connected to an electrical ground.
2. Wear eye protection when operating power tools.
3. Keep all guards and protective devices in place.

FIGURE 9–6A Hand drill.

FIGURE 9–6C Power electric drill. *(Courtesy of Milwaukee Electric Tool Corporation)*

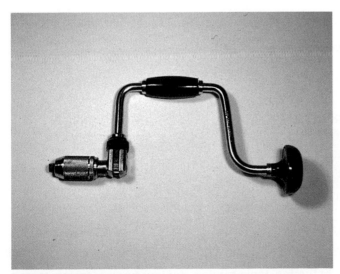

FIGURE 9–6B Bit brace.

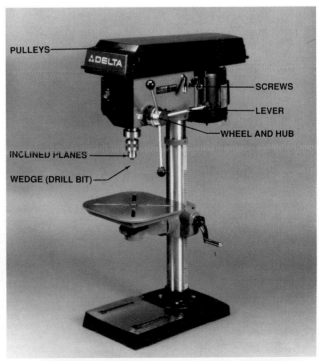

FIGURE 9–6D Drill press. *(Courtesy of Delta International Machinery)*

4. Do not use defective tools.

5. Unplug the tool when changing bits, blades, and attachments or when making adjustments.

6. Unplug the tool when left unattended.

7. Check to see that the cutting edge of the tool will have a clear path as it penetrates the workpiece.

8. Use power tools only after you have received instruction on their use from your teacher.

Turning and Milling

Turning processes are used to make cylindrical or cone-shaped parts with a machine called a lathe, Figure 9–7. There are seven ways to perform turning processes. These are straight turning, taper turning, contour turning, facing, thread turning, necking, and parting, Figure 9–8.

Lathe tools are usually held in the hand when cutting wood, but secured to a machine (a lathe)

FIGURE 9–7 Wood lathes are used to make cylindrical objects, such as legs for tables and chairs. *(Courtesy of Delta International Machinery)*

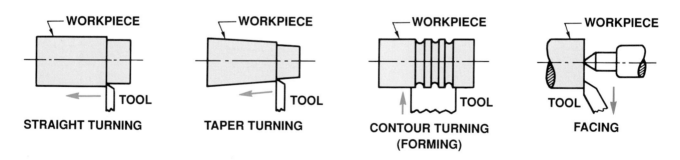

STRAIGHT TURNING TAPER TURNING CONTOUR TURNING (FORMING) FACING

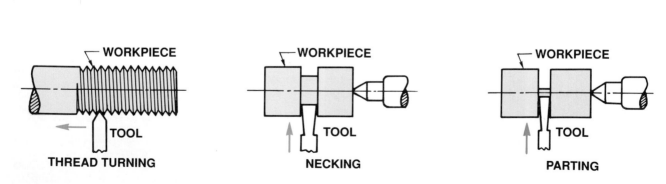

THREAD TURNING NECKING PARTING

FIGURE 9–8 Seven common turning processes are straight turning, taper turning, contour turning, facing, thread turning, necking, and parting.

when cutting metal. The lathe is one of the most important machines in manufacturing. It is used to add threads to or cut cylinder-shaped objects to size, Figure 9–9.

Milling processes are used to square or contour materials. Most milling processes use one or more revolving cutting tools. Metal milling, shaping, and planing processes are widely used to make very accurate flat surfaces, shoulders, slots, angles, and other features. Some common vertical and horizontal milling processes are shown in Figure 9–10. Common wood milling processes include jointing, planing, shaping, and routing, Figure 9–11. Jointing is used to true (remove warps from) an edge or surface. Planing is used to reduce the thickness of materials. Shaping and routing are used to produce contoured edges on materials.

FIGURE 9–9 The lathe is one of the most important machines in manufacturing. This lathe is computer controlled. *(Courtesy of Brodhead Garrett / Denford Machine Tools Ltd.)*

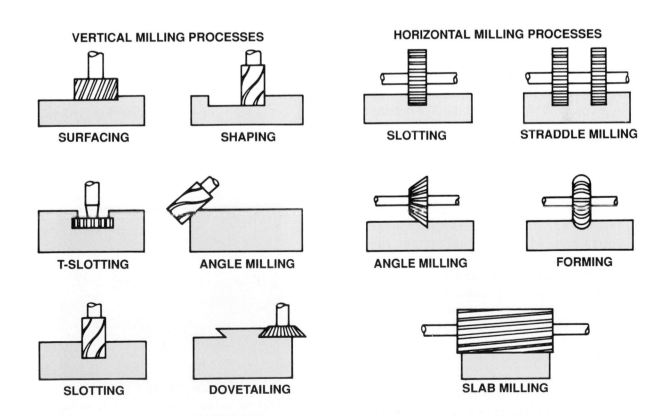

FIGURE 9–10 Here are some common vertical and horizontal metal milling processes.

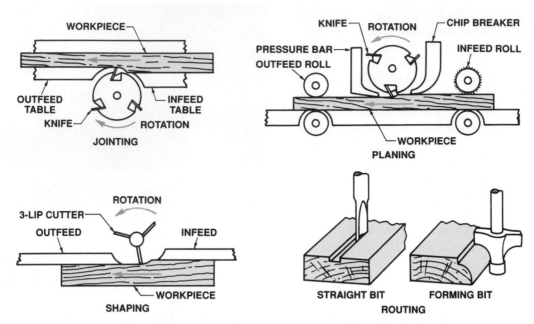

FIGURE 9–11 These are some typical processes for jointing, planing, shaping, and routing wood.

Cutting

Cutting tools are used to process materials to a certain thickness, cut grooves, or to make special edges. **Cutters** have two, three, or more cutting edges. Most cutters are made from steel; some are carbide-tipped. A few of the different types of cutters used in production follow.

- A router uses router cutters. Router cutters have two cutting points that make decorative edges or grooves in wood. They are sometimes called router bits.
- A shaper uses shaper cutters. Shapers perform processes similar to the router, Figure 9–12.
- A milling machine uses milling cutters, Figure 9–13. Milling cutters have many cutting points. Milling machines cut gears, grooves, and flat surfaces in metal.
- Planers and jointers are machines with three or four cutting joints called knives. Planers and jointers are mostly used to cut flat surfaces on wood.

Abrading

Abrading processes use small bits of abrasive material to provide cutting action. Sanding and grinding are two major types of abrasive machining processes, Figure 9–14. Sanding uses coated **abrasive papers** to smooth or finish the surface

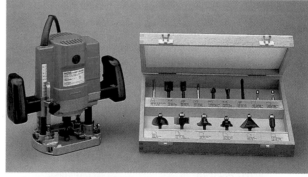

FIGURE 9–12A Router and cutter set. *(Courtesy of Woodworker's Supply Inc.)*

FIGURE 9–12B Shaper cutters. *(Courtesy of Delta International Machinery)*

FIGURE 9–12 Router and shaper cutters make decorative edges and grooves in wood.

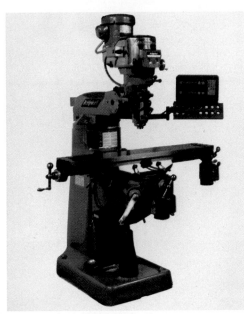

FIGURE 9–13 This milling machine has a multiple-point cutter that cuts gears, grooves, and flat surfaces in metal. *(Courtesy of Bridgeport Machines)*

FIGURE 9–14 This worker is working on a combination sanding and grinding machine. Notice the worker's safety glasses. *(Courtesy of Delta International Machinery)*

of materials. Abrasive material is glued to a paper or cloth backing. Some of the more common abrasive materials used are aluminum oxide, garnet, and flint. Abrasive papers come in the form of sheets, disks, belts, and sleeves. Sanders are power hand tools and machines used with abrasive papers. Sander styles include belt, disk, orbital, and drum sanders, Figure 9–15.

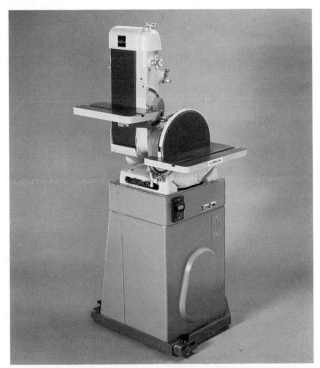

FIGURE 9–15A Disk/belt sanding machine. *(Courtesy of Delta International Machinery)*

FIGURE 9–15B Belt sander. *(Courtesy of Woodworker's Supply Inc.)*

FIGURE 9–15C Orbital sander. *(Courtesy of Woodworker's Supply Inc.)*

FIGURE 9–15 Here are three different types of sanders; each uses abrasive papers to sand and finish materials.

Grinding processes use abrasive bits glued together in the shape of a wheel, Figure 9–16. Grinding wheels are used on power hand tools or machines called grinders, Figure 9–17. The wheels are made from hard, abrasive (scraping) materials, such as aluminum oxide, silicon carbide, or diamond. Workers use grinders mostly on metals. Different grinding wheels are made to cut soft metals like brass or aluminum, tough metals like steel, and very hard metals like cast iron.

Files are hand tools with rows of cutting edges. They are used to remove small amounts of metal, wood, or other materials. Filing processes are similar to abrading. Single-cut files have chisel-shaped cutting edges, while double-cut files have diamond-shaped cutting edges, Figure 9-18. Files usually have flat, half-round, and round cross-sectional shapes.

Shearing

Shearing processes are used to cut excess materials with no loss of material. Let's say a twelve-inch-long material was separated into two six-inch-long pieces using shearing. If the two six-inch pieces were placed back together, they would measure the original twelve inches, Figure 9–19. Shearing is similar to cutting paper with scissors. The excess material cut away with shearing can be recycled or used again. That makes shearing processes very efficient because little, if any, material is wasted. The most com-

GRINDING

FIGURE 9–16 A grinding wheel.

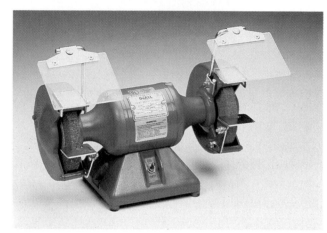

FIGURE 9–17 Benchtop grinder. *(Courtesy of DoAll Company, Des Plaines, IL)*

mon shearing tools used in production include shears, chisels, planes, utility knives, punches and dies, and rotary cutters.

Kinds of teeth
Single-cut

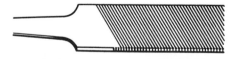

Single set of parallel, diagonal rows of teeth. Single-cut files are often used with light pressure to produce a smooth surface finish or to put a keen edge on knives, shears, or saws.

Double-cut

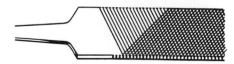

Two sets of diagonal rows of teeth. The second set of teeth is cut in the opposite diagonal direction and on top of the first set. The first set of teeth is known as the overcut while the second is called the upcut. The upcut is finer than the overcut. The double-cut file is used with heavier pressure than the single-cut and removes material faster from the workpiece.

FIGURE 9–18 Close-up views of single-cut and double-cut files. *(Courtesy of Cooper Tools)*

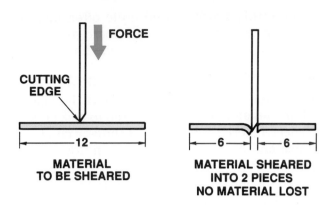

FIGURE 9–19 Shearing processes separate material without loss of any material.

FIGURE 9–21A Wood chisel.

Shears are large, heavy-duty scissors used to cut materials. Tinsnips and aviation snips are two hand tools that shear thin sheet metal, Figure 9–20. For thicker pieces of metal, squaring shears are used. Squaring shears are powered by a person or some other source.

Chisels have a single straight cutting edge. The most common chisels are those used as hand tools in woodworking, Figure 9–21A. They are fitted with wood or plastic handles. The thin, sharp cutting edge is ground with a bevel on one side. Cold chisels are used for cutting metal and masonry materials. They are all steel and blunter than wood chisels. A brick set is a wide cold chisel used to cut bricks and blocks, Figure 9–21B.

FIGURE 9–21B Brick set chisel.

Aviation snips

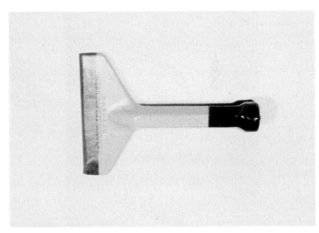

Straight cutting snips

Circular cutting snips

Combination snips

FIGURE 9–20 Here are four different styles of hand-held shears, also called snips. *(Courtesy of Stanley-Proto Industrial Tools)*

CAUTION:

1. Wood chisels in proper condition are very sharp. Never place either hand in front of the chisel. Always cut away from your body.

2. If the head of a cold chisel becomes mushroomed, the sharp edges should be ground off before the chisel is used. The sharp edges could cause injury.

Planes are used to smooth surfaces on lumber. There are a number of different types and styles of planes. Some of the common names include jointer, fore, jack, smooth, and block planes. Jointer planes are the largest and block planes the smallest. Figure 9–22 shows the proper steps for using a plane.

FIGURE 9–22 (Top) At the beginning of the stroke pressure is applied to the front. **(Middle)** At midstroke, pressure is applied to both the handle and front. **(Bottom)** At the end of the stroke, pressure is applied to the handle.

The **utility knife** is used quite often in construction to cut gypsum wall board and other materials. It includes a razor blade held secure in a metal handle.

CAUTION: Utility knives are very sharp. Never place your hand or body in front of the blade.

Punches and **dies** are often used to make holes in sheet materials. Figure 9–23 shows a common punch-and-die setup. The punch forces the material down into the die. The size and shape of the hole left behind are controlled by the size and shape of the punch and die. A paper hole puncher is a good example of how a punch and die works.

Rotary cutters produce round or irregular shapes. The cutters are like wheels with sharp edges. They are used to cut rolls of metal, plastic, paper, or other thin materials. Figure 9–24 shows the basic setup of a rotary cutter.

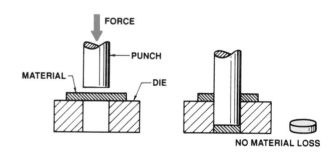

FIGURE 9–23 This is a common punch-and-die shearing setup.

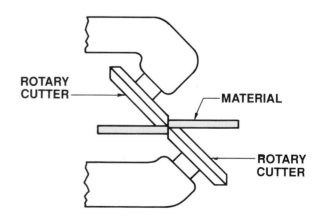

FIGURE 9–24 Here is a common rotary shearing setup for producing round or irregular shapes.

Special Separating Tools

This category includes other separating tools that do not fit into the chip removing and shearing categories. There are eight different ways to separate materials with special separating pro-cesses. They are thermal (gas torch), electrical (electrical discharge, electron beam, ion beam), electrothermal (plasma arc), chemical (acid etching), electrochemical, light wave (laser), liquid (water jet), and induced fracture separating, Figure 9–25.

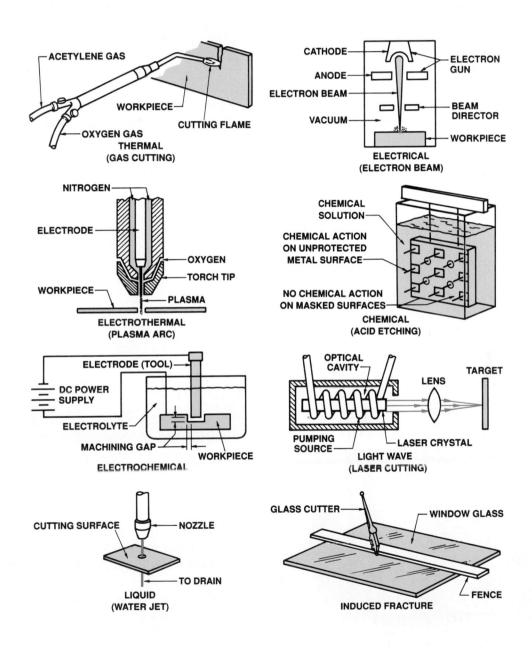

FIGURE 9–25 The eight major special separating processes are thermal, electrical, electrothermal, chemical, electrochemical, light wave, liquid, and induced fracture.

Flame cutting uses the heat from burning gases to separate materials. The most common flame cutting equipment uses an **oxyacetylene torch** to burn oxygen and acetylene gas. The oxyacetylene torch can produce flames with temperatures over 6000 degrees Fahrenheit, Figure 9–26.

Electrical discharge machining (EDM) uses an electric spark for accurate separation of metal materials. An electrode and the material being cut are placed in an oil bath called a **dielectric.** When an electrical current feeds through the electrode, a spark is generated that cuts away the unwanted metal.

Induced fracture separating is used to cut glass. A glass cutter is used to score a line on the glass, Figure 9–27A. Then the glass is fractured (broken) along the scored line, Figure 9–27B.

Lasers use high intensity light to cut a variety of materials.

FIGURE 9–27A The glass cutter scores the surface of the glass. (NOTE: Gloves are removed here to show hand positions, but they should be worn when cutting glass.)

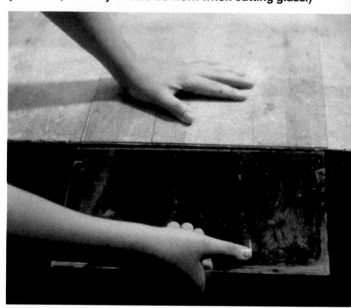

FIGURE 9–27B The glass is broken along the scored line. (NOTE: Gloves are removed to show hand positions.)

Summary

Separating tools cut down the size of materials by removing excess materials. The three types of separating tools are chip removing, shearing, and special.

Chip removing tools create small bits of waste material called chips. Chip removing processes discussed in this chapter include sawing, hole machining, turning, milling, and abrasive machining. Chip removing tools include drills and bits, saw blades, cutters, files, grinders, abrasive papers, shapers, and lathe tools.

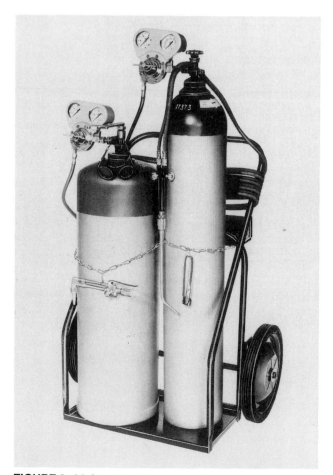

FIGURE 9–26 An oxyacetylene torch burns a mixture of oxygen and acetylene gas for flame cutting of metals. *(Courtesy of MECO — Modern Engineering Company, Inc.)*

Shearing tools do not create chips. Common shearing tools include shears, chisels, planes, utility knives, punches and dies, and rotary cutters.

Special separating tools do not fit the categories of chip removing or shearing. These special processes and tools include thermal (gas torch), electrical (electrical discharge, electron beam, ion beam), electrothermal (plasma arc), chemical (acid etching), electrochemical, light wave (laser), liquid (water jet), and induced fracture separating.

DISCUSSION QUESTIONS

1. What do separating tools and processes do?
2. What is the difference between chip removing and shearing tools?
3. Can you name and describe three different types of drills?
4. What is the difference between a crosscut saw, and a ripsaw?
5. Can you name two types of shears, chisels, and planes?
6. What are three special separating tools?
7. Identify and describe at least one shearing, one chip removing, and one special separating process that was used to manufacture a product found in your home and school.

CHAPTER ACTIVITIES

 ## Using Sawing and Drilling Tools

OBJECTIVE

In this chapter you read about separating tools and machines used in production. In this activity, you will develop skills in using several separating tools, including saws, drills, and abrasives. Using the layout done in the previous chapter activities, you will cut out the parts for the tote carrier and drill holes for the handle. You will combine the parts of the tote carrier after reading the next chapter.

EQUIPMENT AND SUPPLIES

1. Pieces of wood laid out in the previous chapter
2. Wood, ¾″ x 2½″ x 24″ 1 piece (finished size)
3. Wood, ¾″ x 9½″ x 26″ 3 pieces (finished size)
4. Try square
5. Crosscut saw or table saw
6. Ripsaw or table saw
7. Coping saw, bandsaw, or jigsaw
8. Block plane or jointer
9. Disk sander or belt sander
10. Tape measure

11. Files
12. Hand drill, bit brace, or drill press
13. 1-inch spade bit, ⅜″ twist drill bit or #6 auger bit
14. Vibrating sander and coated abrasive sheets
15. Safety glasses

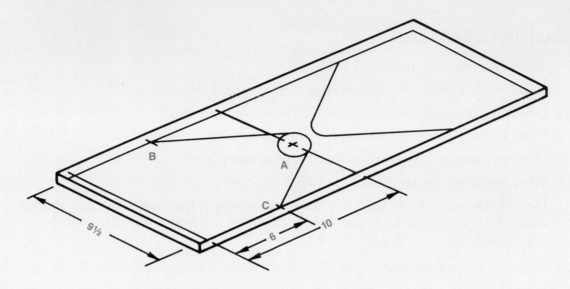

FIGURE 1. Layout of the ends of the tote carrier.

PROCEDURE

CAUTION: Wear your safety glasses, follow all safety rules, get your teacher's permission before using any tools or machines, and **Always Be Careful.**

In the previous chapter, you laid out the stock in Figure 1. Now, you will cut out the pieces of the tote carrier. First you will cut out the ends, Figure 2.

Cutting out the Ends

1. Use the crosscut saw or table saw to cut the ends across the grain to make the stock square.
2. Using the ripsaw or table saw, rip along the grain to within ¼″ of the marked width.
3. Plane the ripped edge to the line with the block plane or jointer. Check often with a try square to be sure the edge is square, and make small cuts.
4. Use the block plane, disk sander, or belt sander to smooth the end grain where it was sawed with the crosscut saw.
5. Lay out the two center points across the centerline of the ends as shown in Figure 2. The first is ½″ from the top, the second is ½″ from the first.
6. Drill a ³⁄₁₆″ hole at each of these points using the hand drill, bit brace, or drill press. Be sure to drill the holes square to the stock, Figure 3.

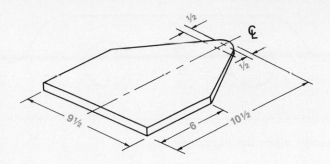

FIGURE 2. End of the tote carrier after being cut out.

Cutting out the Handle

1. Lay out the pattern shown in Figure 4 on the ¾″ x 2½″ x 24″ piece of wood. This will be the handle.
2. Saw the outside shape of the handle using the bandsaw, jigsaw, or coping saw.
3. Drill the 1″ holes in the handle using the bit brace or drill press.

FIGURE 3. A try square or combination square can serve as a guide for square holes.

4. Connect the outside edges of these holes with straight pencil lines as shown in Figure 5.
5. Saw out the remainder of the handle with the coping saw or jigsaw.
6. Smooth all sawed edges of the handle with a file or disk sander.

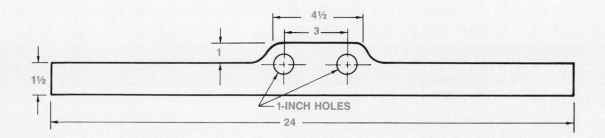

FIGURE 4. Tote carrier handle.

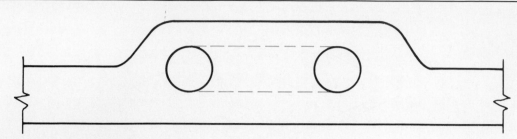

FIGURE 5. Saw out the handle along the dotted lines.

Cutting out the Sides and Bottom

1. Cut one end of each of the three pieces of ¾″ x 9½″ x 26″ wood square on the table saw.
2. Mark and cut a square line 24″ from the squared end on two pieces of the stock.
3. Rip the two pieces 6″ wide. These will be the sides.
4. Cut the third piece 25½″ long. Leave it 9½″ wide. This will be the bottom.
5. As with the ends and handle, plane, file, and sand smooth all cut parts.
6. You should now have the following pieces:
   - 2 end pieces
   - 1 handle
   - 2 side pieces
   - 1 bottom piece
7. Sand smooth all surfaces on each of the pieces. You will combine these pieces to make the tote carrier in the next chapter.

MATH AND SCIENCE CONNECTIONS

The size of auger bits is given as a number. The numbers range from 3 to 16. The number indicates the size in units of $\frac{1}{16}$ inch. The number 6 bit used in this activity cuts a $\frac{6}{16}$-inch ($\frac{3}{8}$-inch) diameter hole. The small number 3 bit cuts $\frac{3}{16}$-inch diameter holes. The number 16 bit cuts $\frac{16}{16}$-inch (1-inch) diameter holes.

RELATED QUESTIONS

1. What is the difference between a crosscut saw and a ripsaw?
2. What is the difference between a twist drill bit and an auger bit?
3. What tools did you use to smooth the ripped edges of the wood? What are the advantages or disadvantages of using that tool?
4. Which tool or process did you find most difficult in cutting out the parts for the tote carrier? Why did you find it difficult?

 Pyramid Puzzle

OBJECTIVE

The Pyramid Puzzle is an interesting product that you can produce using wood, metal, or plastic materials, Figure 1. Your teacher will decide what materials will be used to make the puzzle. In this activity, you will use various separating tools and machines to make the parts for the pyramid puzzle.

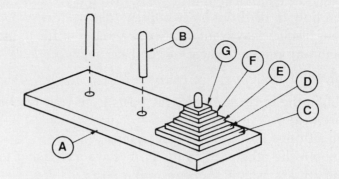

```
PYRAMID PUZZLE

PARTS LIST

Part   Qty.   Part Name        Description

A      1      Base             1/2T x 4W x 11L Wood, Plastic or Metal
B      3      Posts            1/4 Dia. x 1 1/2 L Wood, Plastic or Metal Rod
C      1      Piece #1         1/4T x 3 1/2W x 3 1/2L Wood, Plastic or Metal
D      1      Piece #2         1/4T x 3 1/8W x 3 1/8L Wood, Plastic or Metal
E      1      Piece #3         1/4T x 2 3/4W x 2 3/4L Wood, Plastic or Metal
F      1      Piece #4         1/4T x 2 3/8W x 2 3/8L Wood, Plastic or Metal
G      1      Piece #5         1/4T x 2W x 2L Wood, Plastic or Metal
```

FIGURE 1. Pyramid Puzzle and parts list.

CAUTION: Wear your safety glasses, follow all safety rules, get your teacher's permission before using any tools or machines, and **Always Be Careful.**

EQUIPMENT AND SUPPLIES

1. Wood, metal, or plastic parts (See the parts list for sizes. Your teacher will decide which materials will be used.)
2. Miter saw, crosscut saw, or table saw
3. Hand drill or drill press
4. ¼″ twist drill
5. Try square
6. Disk sander, belt sander, and/or vibrating sander
7. Abrasive sheets
8. Wood glue, plastic adhesive, or epoxy adhesive
9. Wood finish or paint
10. Safety glasses

PROCEDURE

Preparing the Base

1. Obtain the material for the base from your teacher. Cut the base to size using the saw specified by your teacher.
2. Check the squareness of the piece using a try square.

3. Lay out the location of the three holes in the base, Figure 2.
4. Using the hand drill or drill press, drill the ¼ ″ holes ⅜ ″ deep. Your teacher will show you how to drill the holes only ⅜ ″ deep.
5. Your teacher may want you to use a router or file to put a fancy edge around the base.
6. Sand any rough surfaces on the base smooth.

Preparing the Posts

1. Obtain the rod material for the posts from your teacher. Cut them to length with the correct saw.
2. Smooth one end of each post square.
3. Smooth the other end so the posts have a rounded top.
4. Smooth any rough surfaces on the posts.

Preparing the Puzzle Pieces

1. Obtain the material for the pieces from your teacher.
2. Lay out the sizes of each of the five pieces on the materials.
3. Cut the pieces to size using the correct saw.
4. Check the pieces for square.
5. Smooth the pieces.

Assembling the Puzzle

1. Glue the rods into the base using the correct glue or adhesive.
2. Check the rods to make sure they are square to the base using the try square.
3. Let the glue or adhesive dry.

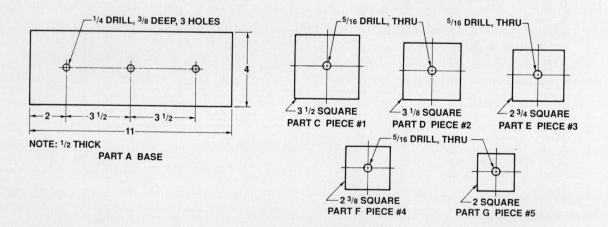

FIGURE 2. Dimensions for the base and puzzle pieces.

4. Finish the pieces of the pyramid puzzle with paint or other finish as directed by your teacher.

Playing the Pyramid Puzzle

Stack all the puzzle pieces on one post; largest to smallest — bottom to top. Your task is to move all the disks one by one to another post. They must be stacked on the new post in the original order — from largest to smallest. The puzzle problem is that no disk should ever rest on another disk smaller than itself.

MATH AND SCIENCE CONNECTIONS

The minimum number of moves needed to solve the Pyramid Puzzle is 31. A math formula can be used to find this number, no matter how many puzzle pieces are used. For x number of pieces, the number of moves needed to solve the puzzle can be found with the formula $2^x - 1$. For five pieces, the calculations look like this:

Number of moves = $2^5 - 1$
Number of moves = $32 - 1$
Number of moves = 31
For three pieces:
Number of moves = $2^3 - 1$
Number of moves = $8 - 1$
Number of moves = 7

How many moves would be needed if you used eight puzzle pieces?

RELATED QUESTIONS

1. What specific tools did you use to make the pyramid puzzle?
2. What safety rules did you follow for the tools you used?
3. Were you able to use the same tools when you cut wood, metal, and plastic?
4. What grit abrasive papers did you use?

CHAPTER 10

Forming and Combining Processes and Tools

OBJECTIVES

After completing this chapter, you will be able to:

■ Define forming and combining tools and processes.

■ Name the three categories of forming processes.

■ Identify and describe tools used in the three categories of forming processes.

■ Name four categories of combining processes.

■ Identify and describe tools used in the four categories of combining processes.

■ Name various safety precautions that should be taken when using forming and combining tools.

■ Use various forming and combining tools for production processes.

KEY TERMS

Adhesive	Common nails	Mechanical fastening
Annealing	Compressing	Mold
Arc welder	Conditioning	Penny size
Bonding	Countersink	Pneumatic
Box nails	Die	Pop rivet gun
Casting	Drying	Rolls
Catalyst	Finishing nails	Spot welder
Claw hammer	Firing	Soldering gun
Coating	Float	Stretching
Cohesive	Forming	Trowel
Combining	Hardening	Work hardening

In the previous chapter, you read about separating processes, tools, and machines. This chapter focuses on the other two types of production processes and tools — forming and combining.

Forming

Forming processes change the outside shape or inside structure of materials. The outside shape of materials can be changed through casting, molding, compressing, and stretching processes. The inside structure of materials can be changed with conditioning processes.

Casting and Molding

Casting and molding are processes similar to making ice cubes in a freezer. The ice cube tray is a tool called a **mold.** A mold is a cavity used for forming liquid or semiliquid materials, Figure 10–1. **Casting** is the act of pouring the liquid material into the mold. In molding, the liquid material is forced into the mold. Manufacturers use casting to form metals and plastics, Figure 10–2. Construction workers use casting to form concrete. The five basic steps in casting or molding any material are:

1. Preparing a pattern
2. Preparing the mold
3. Preparing the material
4. Putting the material in the mold
5. Removing the solid piece

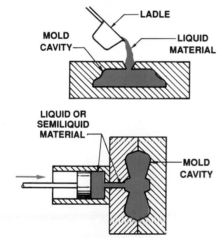

FIGURE 10–2 The molds in this photo form molten metal into brick-shaped blocks. *(Courtesy of FMC Gold Company)*

FIGURE 10–1 Casting occurs when liquid materials are poured into a hollow cavity (top). Molding takes place when liquid or semiliquid materials are forced into a hollow cavity (bottom).

A pattern of the final product is used to make the mold. Metals, plastics, and ceramic materials are often used with casting and molding processes. Metals and plastics must be melted into a semiliquid state before casting or molding. Ceramic materials must be mixed with water before casting or molding. Once the solid piece is removed from the mold, Figure 10–3, it must be finished to make a final product.

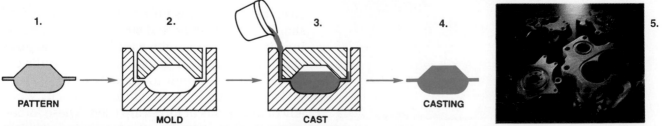

FIGURE 10–3 Here are the five basic steps involved in casting or molding any material. *(Photo courtesy of Cross and Trecker Corporation)*

Sand, metal, or wood usually form molds for casting. Sand molds are called one-shot because they are broken after use. Metal molds are called permanent molds. They can be used over and over again. Molds made from wood and plywood can also be used again. Both wooden and metal forms are used to cast concrete.

Cast concrete must be formed, shaped, and smoothed with special tools before it dries. Two of the forming tools used in construction include floats and trowels. A **float** smooths out some of the unevenness in the surface of freshly placed concrete, Figure 10-4. As soon as possible after the concrete is placed in forms, it is floated. This leaves a textured but uniform surface. Floats can be made of wood, aluminum, or magnesium. When the concrete is more firm, a **trowel** is used for a smoother surface, Figure 10-5. Trowels are made of steel. They produce a very smooth surface on the concrete. Power trowels are used on large concrete jobs, Figure 10-6.

When masons lay brick or block, they also use trowels and other forming tools. The mason's trowel, which is pointed, is used to apply mortar to bricks or blocks and to remove excess mortar, Figure 10-7. Another forming tool is the masonry jointer. It smooths the joints between bricks or blocks.

Compressing and Stretching

The second type of forming is compressing and stretching. **Compressing** processes squeeze and

FIGURE 10-4 Floating concrete. *(Courtesy of Portland Cement Association)*

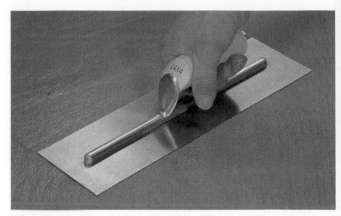

FIGURE 10-5 Cement trowel. *(Courtesy of Goldblatt Tool Company)*

FIGURE 10-6 Concrete masons using power trowels to finish a large concrete floor. *(Courtesy of Marshalltown Trowel Company)*

FIGURE 10-7 Mason's trowel. *(Courtesy of Goldblatt Tool Company)*

stretching processes pull materials into desired shapes. Compressing and stretching are like rolling pizza dough flat with a rolling pin. The pizza maker pushes down (compresses) and pulls out (stretches) the dough to make it fit the pan.

Compressing and stretching can occur when any of four kinds of forces are used. These forces, shown in Figure 10-8, are compression, tension, shear, and torsion.

There are seven ways to compress (squeeze) or stretch (pull) materials into desired shapes. These are squeeze forming, rolling, extruding, drawing, stretch forming, spinning, and bend-ing, Figure 10–9. Although specific process names may vary depending on the type of material be-ing formed, the seven basic processes are the same.

FIGURE 10–8 The four types of forces that can be applied to materials during compressing or stretching processes are tension, compression, shear, and torsion.

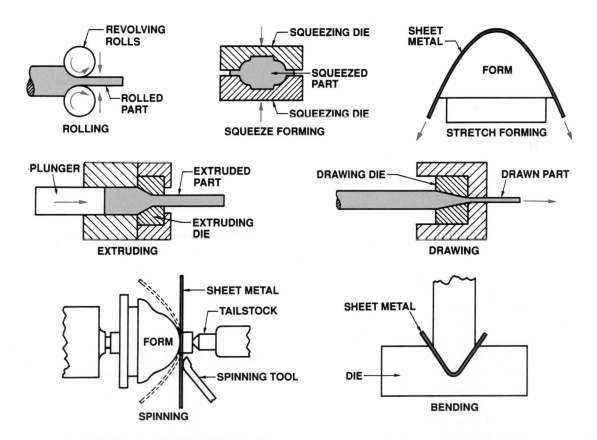

FIGURE 10–9 The seven basic ways to compress or stretch materials are rolling, squeeze forming, extruding, stretch forming, drawing, spinning, and bending.

In the rolling process, metal is forced under cylindrical tools called **rolls.** The rolls have a shape cut into their surface that forms the metal into shapes. Railroad rails are made by forcing metal through two matched rolls.

A second group of compressing and stretching tools are called dies. A **die** has a metal surface with a shape matching the surface shape wanted on the finished piece of material. Here is an easy-to-use example of die forming materials. Take two identical cereal bowls that fit inside each other. Place a piece of paper between them and press down on the top bowl. This is die forming, also called stamping. The two bowls are matched dies. Die forming makes car parts like fenders, hoods, and trunk lids. In actual manufacturing, large hydraulic presses power dies to supply the needed pressures.

Conditioning

The third type of forming is conditioning. **Conditioning** processes change the inside structure of materials. These inside changes are not visible. Materials are conditioned to make them stronger, harder, softer, more elastic, or easier to use.

There are three major ways to condition materials — thermal conditioning, mechanical conditioning, and chemical conditioning.

Thermal Conditioning. Thermal conditioning processes use heat to change the inside structure of materials. Metal, wood, and ceramic materials are often conditioned with heat. Two common thermal conditioning processes for metals are hardening and annealing.

Hardening involves heating a material to a certain temperature, and then cooling it rapidly. This process makes the metal harder. The manufacture of drill bits is a good example of hardening. Drill bits are made from steel. They are often used to drill holes in other steel parts. If the bit must drill holes in steel, it must be made harder than the steel. The drill bits are placed in an oven or forge at high temperatures. After the bits reach a certain temperature, they are cooled quickly in a bath of water or oil.

Annealing is just the opposite of hardening; it means to soften. To anneal a metal, it is heated to a certain temperature and then allowed to cool slowly.

A common thermal conditioning process for wood is **drying.** Drying removes moisture from materials. When wood is cut from a tree, it has a moisture content of about eighty-five percent. Wood is dried in a kiln (furnace) to remove this excess moisture. **Firing** is another thermal conditioning process similar to drying. Firing involves heating clay-based ceramic materials to make them hard.

Mechanical Conditioning. Mechanical conditioning processes use force to change the internal structure of materials. Mechanical conditioning processes may also be called **work hardening.** When certain materials are compressed or stretched, their inside structure can be changed, Figure 10–10. The grains inside the material can become elongated (made longer). This makes the material harder.

Chemical Conditioning. Chemical conditioning uses chemical reactions to make changes inside materials. Plastics are often conditioned by chemical reactions. A material that causes the chemical reaction to take place is called a **catalyst.** Epoxy adhesive is an example of plastic material conditioned by chemical reaction. The epoxy hardens when a catalyst is added to it. Other plastic materials are conditioned in this manner.

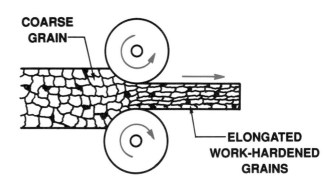

FIGURE 10–10 Work hardening causes the grains inside the material to be elongated. The rolls in this illustration are compressing the material.

Combining

Combining processes are used to add one part to another. Materials can be combined by mixing, bonding, coating, and mechanical fastening.

Mixing

Mixing is often used in food, drug, and chemical manufacturing. Solid, liquid, and gaseous materials can be mixed. Solid materials like food ingredients or concrete ingredients are mixed together by power-driven machines. Food mixing machines are large versions of home kitchen mixers. Liquid materials like paint can be mixed by pouring, shaking, or blending. Two different colors of paint can be blended together by pouring them into one container. The container is then put on a machine that shakes and mixes a new paint color. Gases are mixed by a process called metering. Metering means to measure by volume. When gases are to be mixed, each gas is metered. The correct amount of each gas is then fed into a closed container. Gases are easily mixed.

Solids, liquids, and gases can also be combined together by mixing processes. For example, solid materials like salt or sugar can be mixed with water or other liquids by a process called dissolving. Also, gases and liquids are commonly mixed. Many soft drinks have carbon dioxide gas dissolved into the liquid drink. This creates the gas bubbles in certain soft drinks.

Mortar and concrete are two materials in construction that must be mixed. For small jobs, they can be mixed by hand with a shovel or hoe. For larger jobs, power mixers can be used, Figure 10–11.

Paint and other finishes must also be mixed before use. These materials can be mixed by hand

FIGURE 10–11 Power mortar mixer. *(Courtesy of Stone Construction Equipment Inc.)*

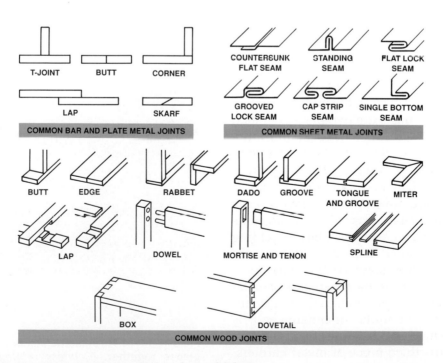

FIGURE 10–12 Here are some common metal and wood joints.

with a stir stick. Power stir sticks can be used with an electric drill.

Bonding

Bonding processes combine solid materials together. The area of the material where the bond takes place is called a joint, Figure 10–12. The type of joint selected depends on the strength and appearance wanted. The type of material being used must also be considered.

Bonding is a permanent combining process. Bonded joints are rigid so that parts cannot move or be separated. Materials can be bonded by using **adhesive** or **cohesive** processes.

Adhesive bonding processes use glues, cements, or some other adhesive to bond parts together. The adhesive is applied in a thin layer at the joint area of the mating parts. Adhesive bonding can be used to bond metal, plastic, wood, and ceramic materials. Special adhesives or glues must be used for the particular type of material being bonded.

Cohesive bonding processes use heat, pressure, or chemicals to melt and bond materials at the joint. The melted materials cohere, or mix together.

Two common cohesive bonding processes are welding, and soldering or brazing. Welding bonds two pieces of metal by melting them together. Two common pieces of welding equipment are **spot welders** and **arc welders,** Figure 10–13. The tongs or points of a spot welder are attached to two pieces of metal on one spot. Then an electrical current passes through the tongs. The electrical flow heats the two pieces of metal until they are bonded (melted) together on the one spot. Arc welders also use electrical flow to bond metals. In arc welding, a welding rod is positioned a short distance away from two pieces of metal and an arc jumps the gap between the electrode and the metal. This electrical arc melts the two pieces of metal together. The arc welding electrode moves along in a continuous path, while the spot welder stays in one spot.

In soldering or brazing, the two pieces of metal being bonded are not melted together. Instead, a third metal, called solder or brazing rod, is melted into a small gap between two pieces of metal. When this third piece of metal hardens, it bonds (solders or brazes) the two metal pieces.

FIGURE 10–13 Arc welders can bond two pieces of metal together by welding. *(Courtesy of Miller Electric Manufacturing Company, Appleton, WI)*

Soldering is usually done with an electric **soldering gun,** Figure 10–14. Brazing uses an oxyacetylene torch to melt the brazing rod into the gap between the two pieces of metal being bonded.

Plastics can be welded together by melting two parts at a joint. Special chemicals, such as ethylene dichloride, can be used for cohesive bonding of certain types of plastics.

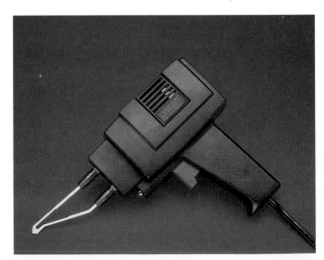

FIGURE 10–14 This electric soldering gun bonds metal pieces together. *(Courtesy of Ungar Division, Eldon Industries)*

FIGURE 10–15 Paint brush (left) and spray gun (right). *(Courtesy of Binks Manufacturing Company)*

Coating

Coating processes place one or more layers of one material on top of another material. Materials are coated to protect them from wearing out or to decorate them. The most important type of coating processes are done by applying a thin coating of one material over another material. Examples of applying a thin coating include painting with a brush or spraying a lacquer coat. The coating material sticks to the surface of the part being coated. The thin coat of paint or other finish material can protect and decorate the piece being coated. Dipping parts in a liquid coating material is another example of this type of coating.

Paint brushes, rollers, and spray guns are three tools that add finishes to materials, Figure 10–15. There are many different brushes and rollers available. The type of finishing material being used and the surface being covered are two factors to be considered when selecting a brush or roller. Molding plastic over metal tool handles or other metal surfaces is another important coating technique. The tool handles are heated in ovens and then dipped in liquid plastic. When the handle cools, a small amount of plastic coats the handle.

Mechanical Fastening

Mechanical fastening is one of the oldest processes used to combine materials. Early cave dwellers used vines to tie sticks and stones together to make tools and weapons. The two major kinds of mechanical fasteners are threaded fasteners and nonthreaded fasteners, Figure 10–16.

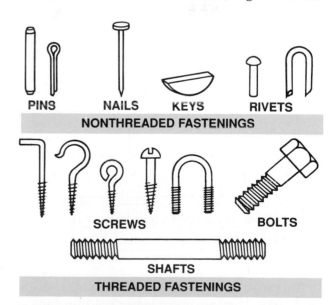

FIGURE 10–16 Here are common examples of threaded and nonthreaded mechanical fasteners.

Screws

Wood screws are listed according to the following:

- Material. Screws can be made of brass or steel.
- Gauge size. Gauge size indicates the diameter of the unthreaded part of the screw. The higher the number, the larger the diameter.
- Length. Wood screws are available in lengths from ¼ " to 4".
- Head type. Screws can have flat, round, oval, and pan heads, Figure 1.

Before screws can be installed, three holes must be drilled — the pilot hole, shank hole, and **countersink,** Figure 2. The shank hole is drilled first. It is just large enough to fit the unthreaded part of the screw. The top surface of the shank hole is drilled with a countersink for flathead screws. The pilot hole is slightly smaller than the threaded portion of the screw, and is drilled in the second piece of wood. A combination drill, Figure 3, makes the pilot hole, shank hole, and countersink in one operation.

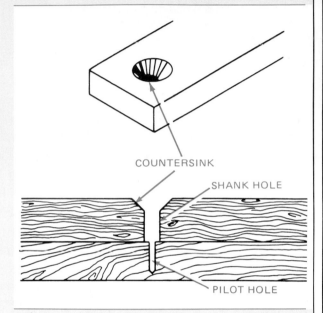

FIGURE 2. Countersink, shank hole, and pilot hole.

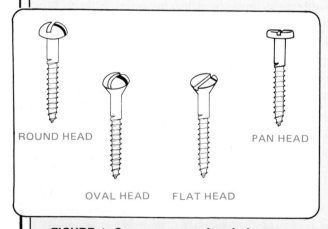

FIGURE 1. Common screw head shapes.

ROUND HEAD

PAN HEAD

OVAL HEAD   FLAT HEAD

FIGURE 3. Combination drill.

Threaded fasteners include screws, nuts, shafts, and bolts. They are much better holding devices than nonthreaded fasteners because they have a larger surface holding area. This larger surface allows greater friction between the two parts being combined. Threaded fasteners can also be removed easily.

The two most common tools used with threaded fasteners are screwdrivers and wrenches. A wide range of sizes and styles of screwdrivers is available. The most common styles are the straight slotted screwdriver and the Phillips head screwdriver, Figure 10–17. Wrenches are used to hold bolts and nuts so they can be threaded together to combine two materials. Once again, many sizes and styles of wrenches are available. The most common styles are the fixed-size open or box-end wrench and the adjustable-size wrench, Figure 10–18. Screwdrivers and wrenches are hand tools that have been improved by adding an outside power source. Electric and pneumatic (compressed-air)-powered screwdrivers and wrenches make the job of mechanical fastening with screws and bolts easier and faster, Figure 10–19.

Nonthreaded fasteners include nails, staples, rivets, pins, and keys. In all cases, workers combine two or more materials by forcing straight metal mechanical fasteners (nails, staples, etc.) through the materials being combined. The number one tool for forcing nails through materials

Adjustable size

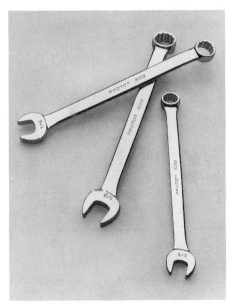

Fixed size

FIGURE 10–18 Here are the two basic wrench styles — the adjustable size and the fixed size. *(Courtesy of Stanley-Proto Industrial Tools, Covington, GA)*

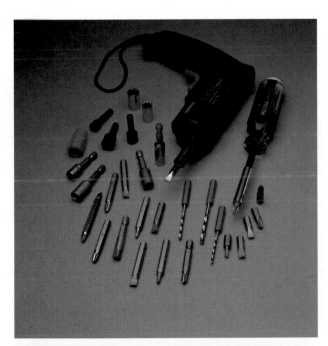

FIGURE 10–17 Here are an electric screwdriver and a hand tool screwdriver. They both adapt to a number of uses. Notice the different sizes of straight slotted and Phillips head screwdriver bits. *(Courtesy of Vermont American Corporation)*

FIGURE 10–19 Battery-operated drill. *(Courtesy of Porter-Cable Corporation)*

is the **claw hammer,** Figure 10–20. Claw hammers come with handles of wood, steel, or fiberglass, and in 13-ounce, 16-ounce, and 20-ounce weights. Workers use staple guns to staple. Both of these are hand tools that use human power to force the mechanical fasteners through the materials. In recent years, hammers and staple guns have been improved by adding an external power source, such as electricity or pneumatic power. Powered nail guns and staple guns greatly improve the speed of nailing and stapling.

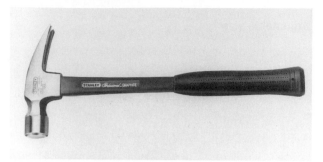

Claw hammer

Ball peen hammer

FIGURE 10–20 Claw hammers are used to drive nails. Ball peen hammers are used in metalworking. *(Courtesy of Stanley-Proto Industrial Tools, Covington, GA)*

Air-Powered Tools

Air-powered or **pneumatic** tools are often great timesavers. Many of the nailing jobs in construction can be done much more quickly with pneumatic nailers, Figure 1. Pneumatic nailers can hold 300 nails. *CAUTION:* Air may seem harmless, but it can be dangerous. Under the pressure normally used for pneumatic tools, air from an air hose can penetrate the skin or do permanent damage to the eyes. Never point a high-pressure air hose at a person or use it to blow dust from your clothes.

FIGURE 1. This carpenter is using a pneumatic nailer to build a wall frame. *(Courtesy of Duo-Fast Corporation)*

Nails

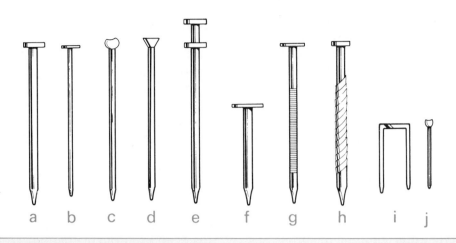

FIGURE 1. Some of the most common kinds of nails: (a) common nail; (b) box nail; (c) finishing nail; (d) casing nail; (e) duplex-head nail; (f) roofing nail; (g) ring-shank drywall nail; (h) screw-shank nail; (i) staple; (j) wire brad.

FIGURE 2. Nail set.

There are many different types of nails, Figure 1. Most nails are designed for a specific purpose, but a few are used for many different purposes. Some of the more popular nails are common nails, box nails, and finish nails. **Common nails** are used most often. They have a flat head and smooth shank and are used for framing work in construction. **Box nails** are similar to common nails, but have thinner shanks. The thin shanks make box nails less likely to split wood. **Finishing nails** have very small heads and thin shanks. The head of a finishing nail can be driven below the surface of wood with a nail set, Figure 2. The nail head and hole can then be covered with putty. Finishing nails are used where appearance is important.

Nail lengths are specified by **penny size.** The term penny is abbreviated with a letter "d." Figure 3 shows the length, in inches, for different penny sizes.

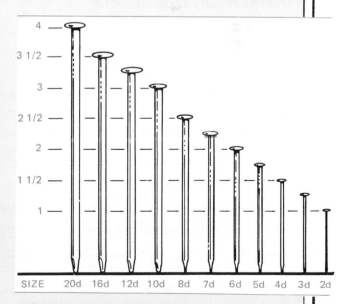

FIGURE 3. Penny sizes of nails.

Nails and threaded fasteners can often be removed from the materials that were combined. Combining materials with rivets, on the other hand, is permanent. A rivet is a metal pin with a head on one end and a straight shank on the other. It looks like a bolt without threads. Rivets are used to combine two or more pieces of metal. The rivet is placed in a hole drilled through the pieces. Then the straight shank end is flattened or spread so the rivet will not come out of the hole. The metal skin on the outside of an airplane is fastened to the frame of the plane with rivets.

Three common riveting tools are hammers, rivet sets, and pop rivet guns. Hammers and rivet sets are used together. The rivet set protects the rivet head and holds the rivet in place. The hammer is used to spread the straight shank end of the rivet. **Pop rivet guns** combine the processes so that workers do not need hammers and rivet sets. Pop rivets are often used on lighter pieces of sheet metal.

The speed and efficiency of riveting have also been improved by using outside power sources such as electricity and pneumatic power.

Summary

Forming processes include casting and molding, compressing and stretching, and conditioning. Casting, molding, compressing, and stretching processes change the outside shape of materials. Conditioning processes change the inside structure of materials.

Combining processes add one part to another. The added material will protect, decorate, or improve the usefulness of the first material. The four categories of combining are coating, bonding, mechanical fastening, and mixing. Painting is an example of coating. Welding and soldering are examples of bonding. Screws and nails are examples of mechanical fasteners. Paint, mortar, and concrete are mixed.

DISCUSSION QUESTIONS

1. How would you define forming tools? Combining tools?

2. What are the three categories of forming processes? What are the four categories of combining processes?

3. Can you name and describe at least one process example for casting and molding, compressing and stretching, and conditioning? What products are made by each of these types of forming processes?

4. Identify and describe at least one mixing, one bonding, one coating, and one mechanical fastening process that was used to produce a product or structure found in your home or school.

5. Describe specific safety rules and precautions for each of the forming and combining tools you named in questions 3 and 4.

CHAPTER ACTIVITIES

 ## Combining Parts for the Tote Carrier

OBJECTIVE

In the previous chapter you prepared the parts for a tote carrier. In this activity, you will combine the parts using nails, screws, and glue to make the finished tote carrier, Figure 1.

EQUIPMENT AND SUPPLIES

1. Parts for the tote carrier:
   - 1 handle
   - 2 ends
   - 2 sides
   - 1 bottom
2. Bar clamps
3. Wood glue
4. Try square
5. 4d common nails
6. Hammer
7. #8 x 1½ " round head wood screws
8. Screwdriver
9. Drill and drill bits
10. Safety glasses

PROCEDURE

CAUTION: Wear your safety glasses, follow all safety rules, get your teacher's permission before using any tools or machines, and **Always Be Careful.**

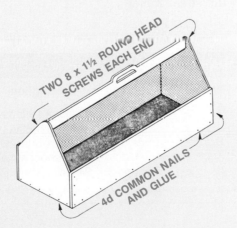

FIGURE 1. Tote carrier.

1. Assemble the pieces of the tote carrier dry (without glue) to check their fit. If needed, adjust pieces so the parts fit well.
2. Hold the end, bottom, and side pieces together with bar clamps. Remove the bar clamps and add glue to the joints.
3. Reposition the clamps. Check the squareness of the inside of the carrier. Hammer the nails through the sides into the ends and bottom.
4. Recheck the carrier for square. Nail the ends into the bottom.
5. Position the handle between the ends and mark the location of holes for the screws.
6. Drill the shank and pilot holes. Drive the screws through the ends into the handle.
7. Remove excess glue with a damp cloth or towel.
8. Have your teacher check your work. Your teacher may want you to add a finish to the tote carrier.

MATH AND SCIENCE CONNECTIONS

You used number 8 wood screws in this activity to fasten the handle to your tote carrier. The number 8 is an indication of the diameter of the shank of the screw. This is the wire gauge size.

RELATED QUESTIONS

1. What combining tools and processes did you use in this activity?
2. What was the most difficult part of this activity?
3. What other technique could have been used to combine the parts of the tote carrier other than screws, nails, and glue?
4. If you could make the tote carrier again, what improvements would you make?

 # Separating — Forming — Combining

OBJECTIVE

Making products from sheet metal requires separating, forming, and combining processes. In this activity, you will make a sheet metal planter box, Figure 1. You will separate (shear), form (bend), and combine (rivet) the sheet metal.

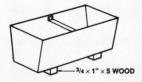

PARTS LIST			
Part	Qty.	Part Name	Description
A	1	Body	24 Ga. × 19 L sheet metal
B	2	Ends	24 Ga. × 6 1/2 × 6 1/2 sheet metal
C	1	Support	24 Ga. × 1 1/4 × 7 1/4 sheet metal

FIGURE 1. Planter box.

CAUTION: Wear your safety glasses, follow all safety rules, get your teacher's permission before using any tools or machines, and **Always Be Careful.**

EQUIPMENT AND SUPPLIES

1. 24 gauge sheet metal, get sizes from parts list.
2. Wood — ¾" x 1" x 11"
3. Sheet metal shears
4. Box pan and brake
5. Drill and drill bits
6. Pop rivets and gun
7. #8 x ½" round head wood screws
8. Crosscut saw
9. Try square
10. Safety glasses
11. Spray primer and paint

PROCEDURE

1. Lay out the dimensions for sheet metal pieces A, B, and C, Figure 2. Your teacher will give you the length dimension for XX on part A.

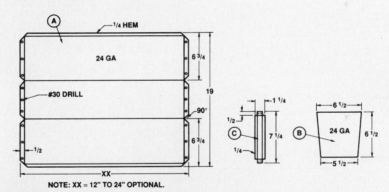

FIGURE 2. Planter box plans.

2. Cut the pieces to size with the appropriate shears.
3. Crosscut the wood piece into two 5-inch pieces.
4. Drill the holes for the pop rivets on parts A and C.
5. Bend the ¼" hems on parts A and C on the box pan and brake.
6. Assemble parts A and B and check square. Mark the location of the pop rivet holes through A onto B.
7. Drill the pop rivet holes in part B.
8. Reassemble A and B. Check for square. Combine the parts with rivets.
9. Position the wooden feet under the planter box. Mark and drill two screw holes for each foot. Fasten the feet to the box with wood screws.
10. Bend the ½" tabs on the end of part C. Insert inside the planter box and mark the location for rivet holes.
11. Drill the rivet holes in part A for part C. Fasten part C with rivets.

12. Prime and paint the planter box the desired color.
13. Have your teacher check your work.

MATH AND SCIENCE CONNECTIONS

Sheet metal thickness is given in a gauge number. The 24-gauge sheet metal used in this activity is approximately .0236 inches thick.

RELATED QUESTIONS

1. What separating tools and processes did you use in this activity?
2. What forming tools and processes did you use?
3. What combining tools and processes did you use?
4. What is one disadvantage of using rivets to make the planter box?

CHAPTER 11

Automating Tools: Computers, Robots, and Lasers

OBJECTIVES

After completing this chapter, you will be able to:

■ Describe why production is automating tools, machines, and processes.

■ Identify the many applications of computers in production.

■ Differentiate among CAD, CADD, CAM, CIM, JIT, and flexible manufacturing.

■ Describe the basic parts, classifications, and uses of robots in production.

■ Describe various applications of lasers in manufacturing and construction.

KEY TERMS

Anthropomorphic	CAM	Laser
Automation	CIM	Robot
CAD	Degrees of freedom	Software
CADD	Flexible manufacturing	Work envelope
CAD/CAM	JIT	

Automation means to perform work without the aid of people. Many tools, machines, and processes in production can be improved with automation. Usually the goal of automating tools, machines, and processes is to improve efficiency. Doing work in less time, using less materials, or with less money are examples of improved efficiency.

In this chapter, you will read about three different technologies that are used to improve efficiency and automate processes — computers, robots, and lasers.

Computers: Multipurpose Machines for Production

A computer is defined as a programmable calculating machine that can store, retrieve, and process data. Computers are not magical devices. They are machines or tools just like the other tools and machines described in this book. Computers have been applied to many jobs in production. They improve the efficiency, accuracy, and productivity of many production processes. Just like the other tools and machines described in this book, computers extend human abilities and make some jobs easier.

In manufacturing, every department has found a use for computers. In the management department, supervisors and managers gather information about the progress of work in all the other departments. In the engineering department, computers are used for computer-aided drafting (CAD) and computer-aided drawing and design (CADD). These systems increase the speed of designing and drawing processes. In marketing, researchers, advertisers, and salespeople use computers to get data on consumers, to quickly study market research, and to create advertisements. In the finance department, accountants and bookkeepers used computers to do the payroll, create cost estimate budgets, and keep track of other financial information. In the production department, computer-aided manufacturing (CAM) is used to control material processing machines. In all departments, computers can improve production efficiency by reducing mistakes and reduce production costs by replacing workers.

The real tool when using a computer is the software. **Software** is a set of coded instructions written to control the operations of the computer, Figure 11–1. Without software, a computer would

FIGURE 11–2 A CAD system lets drafters produce complex drawings faster and with more precision. *(Courtesy of Hewlett-Packard Company)*

be a useless machine. The list of written instructions in software is called a program. Many software programs are available for every job in production. Companies often use commercially available computer software for CAD, CAM, financial records, and other applications.

Computers in Engineering

Two uses of computers in engineering are computer-aided drafting (CAD) and computer-aided drawing and design (CADD).

CAD

Drafters and architects used **CAD** systems to make technical drawings of products or structures to be produced, Figure 11–2. In the past, drafters and architects worked with T-squares, scales, triangles, and pencils to create these drawings by hand on paper. Today, more and more drafters and architects use CAD systems to perform drafting operations. Doing drawings by hand requires a great deal of time. Each line, letter, and shape on a drawing must be created by the drafter. The drafter or architect must have great skill to draw these features exactly the same every time.

In CAD, most of these jobs are done by the computer. CAD software programs are written

FIGURE 11–1 This screen is showing the computer user a "menu" of options available in the software. *(Courtesy of Hewlett-Packard Company)*

so that a drafter needs only to identify the type of line, letter, or shape required. The computer then draws the feature perfectly on a computer screen. Most CAD systems also include a drawing library of commonly used symbols and shapes to allow faster drawing. As an example, the design for an electronic product might have hundreds of symbols to identify different standard electronic components. Before CAD systems, the drafter would draw each of these symbols by hand. With a CAD system, the drafter needs only to pull a copy of the needed symbol from the library and place it on the drawing. The same is true when an architect designs a new skyscraper. Standard symbols for windows, doors, plumbing and electrical fixtures, and heating/cooling systems can be selected quickly from a library and placed on the drawing. This feature alone can save hundreds of hours of work on large jobs.

Another advantage of CAD is ease in revising drawings. In the past, when a product was changed or improved, the drawings (on paper) had to be completely redrawn or traced. Today, CAD systems allow the drafter to redraw only those parts and features that have been revised.

As drafting tools, CAD systems reduce the amount of time it takes to make and revise working drawings. They can also reduce the number of mistakes made in drawing special symbols and features in a product or structure.

CADD

CADD systems are used by engineers who design and test product and structure plans. The difference between CAD and CADD systems is the simulation ability in CADD. CAD is used to make two- and three-dimensional drawings of objects. CADD is used to make those drawings, as well as to simulate testing procedures. In the past, when a product or structure was designed by an engineer, scale models and full-sized prototypes were often made. These models and prototypes were then tested for durability, strength, and performance. This process involved making dozens of different models and prototypes until the engineers found the combination of materials, parts, and design they wanted. CADD systems have simulation programs built in. This allows engineers to design the product or structure on

FIGURE 11–3 CADD drawings often look like wire models of the finished product. *(Courtesy of Gillette Company)*

the computer, test the performance of the product or structure, and change parts that don't work properly. Designing and testing computer models can be done in a small portion of the time needed to make and test real models. Engineers can change materials and parts on a product. They can simulate the stresses, strains, and wear on structures like a bridge. The drawings used in CADD often look like wire models, Figure 11–3.

Designing and testing products and structures this way takes less time. It also saves the company money that used to be spent on models and prototypes. Using CADD systems improves the quality of products and structures.

Computers in Manufacturing

The use of computers in processing materials is called computer-aided manufacturing or **CAM**. CAM involves controlling tools and machines with computers, Figure 11–4. In the past, all production machines were operated and controlled directly by human operators. Then, numerical control (NC) machining was introduced. With NC, operators code the machine operations onto a punched tape. The punch tape is then used to control the machines. Recently, CAM has been replacing NC. Instead of coding operations onto

FIGURE 11–4 The machines in this CAM system are controlled by computer. *(Courtesy of Cincinnati Milacron, Inc.)*

FIGURE 11–5 A CAD/CAM system lets designers store plans in the computer, and then send them to a computer-controlled machine where the product is made. *(Courtesy of Grumman Corporation)*

a punch tape, CAM operators write software programs for computers that are used to control machines.

CAD/CAM

The next step in computerizing production operations was to tie the engineering and production processes using CAD/CAM systems. With **CAD/CAM**, a drafter in engineering draws the plans for a product on a CAD system. The design information is then sent directly to machines in production where the product is made, Figure 11–5.

Computer-Integrated Manufacturing

After the engineering and production processes were connected with CAD/CAM, computers were used to integrate (tie together) all processes and departments. In **CIM**, the management, engineering, production, finance, and marketing workers are all linked by computer, Figure 11–6. Information about the progress in any part of the company can be seen at once by any worker on the CIM system. Company-wide planning can be done with CIM. Workers can be prepared for specific jobs; finance workers can buy materials when needed; marketing can plan advertising to match the completion of the product; engineering and production can communicate with CAD/CAM; and management can direct the whole company by following the progress of each department. With CIM, all the employees can work toward the single goal of producing and selling a quality product.

FIGURE 11–6 Workers using a CIM system can get a broad picture of the entire operation by looking at a computer screen. *(Courtesy of International Business Machines Corporation)*

Just In Time (JIT)

Once a company has started working with CIM, it is not far from JIT manufacturing. With **JIT**, a company can save money by carefully timing the different parts of the manufacturing process so that they all fit together "just in time."

For example, manufacturing companies have always ordered materials and supplies in large quantities before they were needed. These materials and supplies were then stored in nearby warehouses and delivered to the plant when needed. Renting a warehouse, storing supplies, and moving supplies from the warehouse can be expensive. Since CIM helps a company know just when it needs certain supplies or materials, the company can have supplies delivered just in time at the plant. This saves the company money that would have been spent to rent the warehouse. Also, the company will finish the products and deliver them to the market immediately. This saves the costs of storing finished items in the warehouse.

Flexible Manufacturing

Another important new use of computers is flexible manufacturing. **Flexible manufacturing** lets a company make many versions of the same product to meet specific consumer needs and wants. The General Electric Company uses flexible manufacturing to make 2,000 versions of a basic electric meter.

In flexible manufacturing, complex combinations of machines, called flexible manufacturing centers, make product parts, Figure 11–7. A typical center has several machines. Included are drills, mills, lathes, robots, and conveyors. All these machines are controlled by one computer. The center computer is programmed to make several different versions of the same basic product. As an example, an automobile manufacturer can use one flexible manufacturing center to make engines with four, six, and eight cylinders. As an engine moves toward the center on a conveyor, a laser or other tool reads a code on the engine. This code tells the computer whether

FIGURE 11–7 The flexible manufacturing centers in this system let the company make several versions of the same product. *(Courtesy of Cincinnati Milacron, Inc.)*

the engine should have four, six, or eight cylinders. The computer then sends the right instructions to each machine in the center.

Robots: Computer-Controlled Machines

One computer-controlled machine we hear so much about is the robot. The Robot Institute of America defines a **robot** as a "reprogrammable multifunctional manipulator designed to move materials, parts, tools, or specialized devices through variable programmed motions for the performance of a variety of tasks." Notice the words "reprogrammable" and "multifunctional." Reprogrammable means different software can be used with a robot. Different software makes the robot multifunctional (able to do several jobs). Robots are very accurate and reliable machines that can often do the job of a production worker faster and better, Figure 11–8A.

Basic Parts of a Robot

The basic parts of a robot include sensors, a control mechanism, actuators, and drives. Robots replace human workers and can be likened to humans. Robot sensors are like the human senses of seeing, hearing, and touching. Robots have pressure sensors, touch sensors, limit switches, light switches, and even video camera-based vision sensors. The robot control mechanism is like the human brain. A computer and software are used to control robots. Robot actuators are like a human hand. Robots have a variety of actuators, called end affectors. These include grippers, suction devices, and special attachments for welding, spray painting, and drilling. Robot drives are like human muscles. Robots have electrical, pneumatic, and hydraulic motors which drive the robot and its actuators.

Robot Classifications

The most common way to classify robots is by their degree of freedom. The most complex robot,

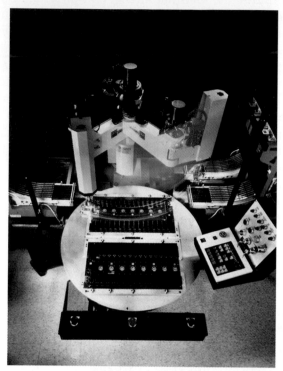

FIGURE 11–8A Although most robots look nothing like human beings, they can perform some human tasks with speed and accuracy. *(Courtesy of AMP Incorporated)*

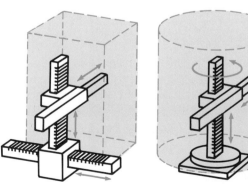

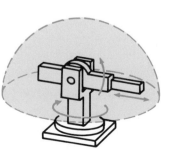

FIGURE 11–8B The space in which a robot can reach to do work is called a work envelope. The number of degrees of freedom will determine the shape of a robot's work envelope.

an **anthropomorphic** robot, duplicates complete human arm motions. Anthropomorphic means having human characteristics. This robot has six **degrees of freedom**:

1. Arm sweep; rotation about the base axis
2. Shoulder swivel; rotation about the shoulder axis
3. Elbow extension; rotation about the elbow axis
4. Wrist pitch; up and down at wrist movements
5. Wrist yaw; side to side wrist movements
6. Wrist roll; rotation about the wrist joint

Human arms have six degrees of freedom. You can sweep your arm, swivel your shoulder, extend your elbow, and pitch, yaw, and roll your wrist. Anthropomorphic robots are designed to duplicate the actions of human workers.

The more degrees of arm freedom a robot has, the more complex processes it can perform. Degrees of arm freedom is important to robot work envelope. **Work envelope** is the area in which the robot can reach and work. An anthropomorphic robot would have a work envelope shaped like a sphere. Less complex robots with fewer degrees of arm freedom have work envelopes which are shaped like rectangles, cylinders, or other shapes, Figure 11–8B.

Using Robots in Manufacturing

Robots can replace production workers in dangerous or unhealthy work conditions. They can also replace workers where highly accurate, consistent work is needed. Robots have a high degree of repeatability. That means they can do one job over and over again with consistent results.

The primary uses for robots in manufacturing are welding (spot and continuous-path arc), spray painting, drilling, and pick-and-place operations such as feeding parts to and from other machines. In the future, robots will continue to replace production workers in manufacturing, Figure 11–9.

Robots have not been applied in construction as often as in manufacturing. The skills and actions required by construction workers are not as easily reproduced with robots.

FIGURE 11–9 Robots have been used in manufacturing to do spot welding. This robot is spot welding the frame of an automobile. *(Courtesy of Ford Motor Company)*

Lasers in Manufacturing

Lasers are important new tools in manufacturing. **Laser** stands for "**l**ight **a**mplification by **s**timulated **e**mission of **r**adiation." Lasers provide a perfectly straight and monochromatic light that can be focused in very tiny diameters. Have you ever concentrated the sun's rays with a magnifying glass and burned paper? This is the basic principle behind using lasers in manufacturing. High-powered laser light is concentrated down to a tiny diameter of light. This tiny light beam can reach a temperature of 4800 degrees Fahrenheit — hot enough to melt steel. Even with this power, lasers can cut cloth, paper, and plastic without burning, Figure 11–10.

In manufacturing, lasers can cut, drill, and measure. A cutting laser melts and vaporizes the material being cut. When drilling, the holes are actually burned in the material. In measuring, lasers can accurately measure to one ten-millionth (.0000001) of an inch. (If the Empire State Building were equal to twelve inches, one ten-millionth of an inch would be equal to the thickness of a piece of paper!) Automobile manufacturers use lasers to measure car assemblies for straightness.

Several of the advantages and disadvantages of using lasers in manufacturing are listed in the following table, Figure 11–11.

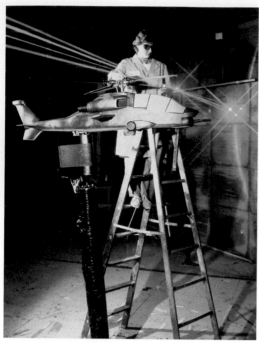

FIGURE 11–10 Lasers are used in many manufacturing processes, including precision measurement. *(Courtesy of NASA)*

ADVANTAGES OF LASERS
■ Cut materials are vaporized, no scrap is left over
■ Tiny light concentration during cutting, little wasted material
■ Fast cutting, which means lower production costs
■ Can cut or drill virtually any material
■ Highly accurate cutting, drilling, and measuring
■ Do not dull or wear out like drill bits or saw blades

DISADVANTAGES OF LASERS
■ Lasers are very expensive pieces of equipment
■ Trained, skilled workers are needed to set up and maintain lasers
■ Drilled holes are not straight and smooth like conventional drilled holes
■ High heat laser beam can be dangerous to workers and other equipment

FIGURE 11–11 Before lasers can be used in manufacturing, workers must think about their pros and cons.

Lasers in Construction

Lasers are also used in construction. They are used mostly for measuring and not for cutting and drilling processes. One common measuring application is to provide a laser reference beam. The reference beam can be used to guide machines that dig tunnels or excavate foundations, Figure 11–12. Another use is during the installation of dropped ceilings. A laser level can be used to project a horizontal reference beam around the entire room, Figure 11–13. Construction workers use the line to locate the support system for the

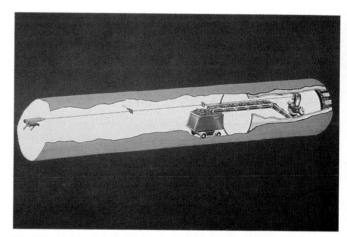

FIGURE 11–12 A laser reference beam being used to control the machine that digs a tunnel. *(Courtesy of Spectra-Physics)*

FIGURE 11–13 The laser level projects a reference beam around the room for a drop ceiling. *(Courtesy of Spectra-Physics)*

suspended ceiling. Whenever lasers are used for measuring, they improve accuracy and reduce time.

Summary

Production tools, machines, and processes are often automated to improve efficiency. Computers, robots, and lasers are examples of automating tools.

Computers are used in all aspects of production, from design and engineering to management and finances. Some of the applications of computers include CAD (computer-aided drafting), CADD (computer-aided drawing and design), CAM (computer-aided manufacturing), and CIM (computer-integrated manufacturing).

Robots are defined as "reprogrammable multifunctional manipulators designed to move materials, parts, tools, or specialized devices through variable programmed motions for the performance of a variety of tasks." Robots are made of sensors, control mechanisms, actuators, and drives. They are classified by the number of degrees of freedom they have. More complex robots have more degrees of freedom. Robots have been applied in a number of different tasks including welding, spray painting, and pick-and-place operations.

Laser stands for light amplification by stimulated emission of radiation. The concentrated light beams emitted by lasers have been used to cut and drill materials as hard as steel. They are also used for several measuring tasks.

DISCUSSION QUESTIONS

1. Why are production tools, machines, and processes automated?
2. Name four applications of computers in production.
3. What is the difference between CAD, CADD, CAM, CIM, JIT, and flexible manufacturing?
4. What are the basic parts of a robot? How are robots classified? Name three uses of robots in production.
5. How are lasers used in manufacturing and construction?

CHAPTER ACTIVITIES

 ## Homemade Robotic Arm

OBJECTIVE

In this activity, you will work as part of a team to design and produce a model robotic arm. When complete, the robot should be able to pick up and put down small objects, move up and down, and to the left and right.

   You will also learn basic hydraulic concepts in this activity. Hydraulic systems can be used to make robots move and work, Figure 1.

EQUIPMENT AND SUPPLIES

1. Six 60cc syringes and 6′ of aquarium tubing
2. 12″ x 12″ x ¾″ plywood for base
3. Scrapwood, dowels, and assorted fasteners
4. Assorted production tools and machines
5. Sink or pan to fill syringes
6. Safety glasses

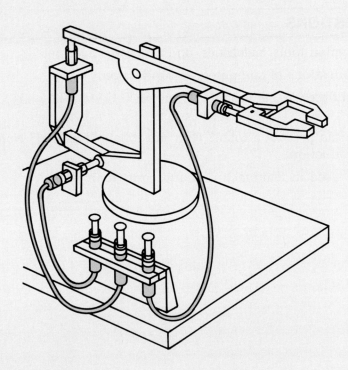

FIGURE 1. A hydraulic robotic arm, with three pairs of syringes. *(Figure continued on next page)*

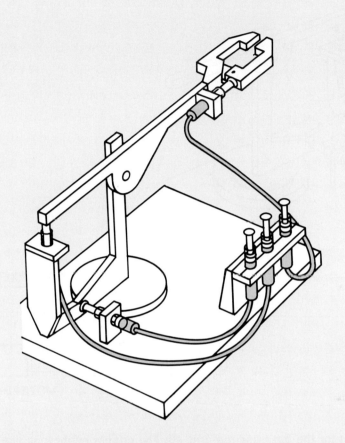

Figure 1. (continued)

PROCEDURE

Learning Basic Hydraulic Concepts

1. Form groups of three students.
2. Connect a pair of syringes with twenty-four inches of tubing. Remove the plungers.
3. Fill the syringes by immersing them in water. Replace the plungers and remove them from the water. You have assembled a simple hydraulic system.
4. Test the system by holding one syringe in each hand. Push down on one plunger while resisting the motion of the other one. Try pulling out on the plunger. What happens?

Designing the Robotic Arm

1. Plan to use three pairs of syringes to control the robotic arm. Use one pair for the gripper. The gripper will pick up and put down objects. The second pair will move the arm up and down. Use the third pair to move the base right and left.

2. One syringe of each pair will be the controller. The other will move one part of the robot (the arm, gripper, or base). The drawings below give ideas to help you get started, Figure 2.

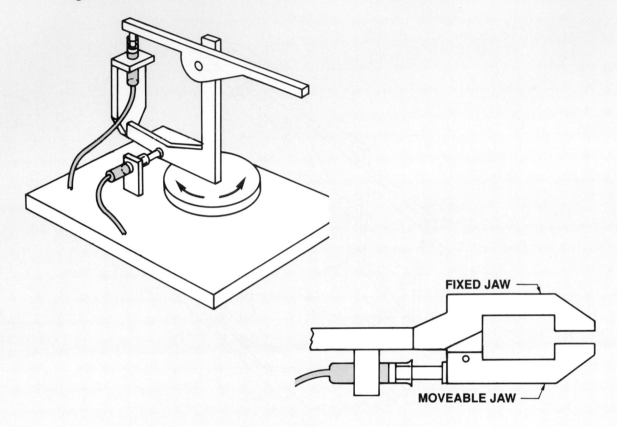

FIXED JAW

MOVEABLE JAW

FIGURE 2. (A) Ideas for the base and the arm. (B) The gripper might look like this.

3. One member of the team should design the gripper. Another team member should design the arm. A third person should design the movable base. Each team member should prepare several idea sketches.
4. Meet as a group to exchange ideas. Combine the best ideas and submit a final sketch to your teacher. Make necesary changes.
5. After approval, use the supplies and equipment made available by your teacher to produce the robotic arm. NOTE: Use only the equipment specified by your teacher. Follow safe practices at all times.
6. Work as a team to solve problems as they occur.
7. When the robotic arm is complete, demonstrate it to the class.

MATH AND SCIENCE CONNECTIONS

Water can be used in the hydraulic system of the robot because it cannot be compressed. When you push on one end of the cylinder, the water exerts the same amount of force at the other end of the connecting hose. If you tried to use air in the cylinder and connecting hose it would compress and not exert enough force to push the opposite cylinder.

RELATED QUESTIONS

1. Name three uses of robots in industry.
2. How could a robot help a handicapped person?
3. Describe how teamwork helped in the design of your robot.
4. Describe one change that could be made to improve the way your team's robotic arm works.

 **CADD**

OBJECTIVE

CADD stands for computer-aided drawing and design. It is used by engineers who design and test products and structures. CADD systems can be used to simulate testing procedures. In this activity, you will use a CADD system to design and test a product on a computer. The software you will use is called Car Builder ©. The product you will make is an automobile, Figure 1.

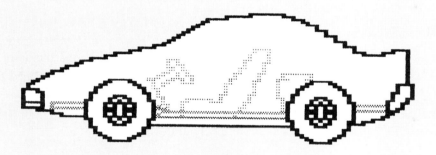

EQUIPMENT AND SUPPLIES

1. Computer and disk drive
2. Dot matrix printer and paper
3. Car Builder © software program

PROCEDURE

Designing Your Car

1. You can design any type of car you want with Car Builder ©. Your teacher may want you to design a specific type of car, such as the fastest car, the car with the most miles per gallon, or the car with the least wind drag.
2. Load the Car Builder © software program. Your teacher will show you the process used to design and save the parts for your automobile.
3. Design the mechanical system for your car first. Select all the components for your chassis and drive train, suspension and steering, and tires and brakes.

4. Design the body for your car. Select a front end, rear end, and passenger compartment.
5. Modify the body parts, if desired. The part of the Car Builder © program used to modify the body is like a CAD system.
6. Complete your car by installing windows and a name decal.

Conducting Simulated Tests

1. You can simulate two tests on your car — wind tunnel test and road test.
2. Conduct the wind tunnel test. Watch for the amount of drag your car has.
3. Conduct the road test. Read the results as they appear on the screen.

Printing Out the Results

1. After the road test, save your car on the disk.
2. Print out a copy of your car design. Your printout will include the results from the wind tunnel and road tests.
3. Your teacher may hold a class discussion to see who designed the ''best'' cars.

MATH AND SCIENCE CONNECTIONS

The efficiency of your car as measured in the wind tunnel is given as the drag coefficient. The lower the drag coefficient, the more efficient a car will be. The drag coefficient is a measure of the ability of a car to cut through the air.

RELATED QUESTIONS

1. How does the Car Builder © CADD system compare with those used in industry?
2. If CADD systems were not available, how would the product design and testing procedure be different?
3. Do you think CADD systems save time? Why or why not?
4. What workers would use CADD systems in manufacturing?

SECTION FOUR

MANUFACTURING SYSTEMS

(Courtesy of Ford Motor Company)

CHAPTER 12

Manufacturing Enterprise

OBJECTIVES

After completing this chapter, you will be able to:

- Define and describe the concept of manufacturing enterprise.
- Describe why many activities may be running at the same time in a manufacturing enterprise.
- Explain the importance of good planning when organizing a manufacturing enterprise.
- Identify the various steps taken to set up and run a manufacturing enterprise.
- Describe the importance of working safely with tools, materials, and processes in a manufacturing enterprise.

KEY TERMS

Advertising campaign
Continuous manufacturing
Custom manufacturing
Design engineering
Division of labor

Enterprise
Intermittent manufacturing
Investors
Marketing
Mass production

Production engineering
Subcontract
Tooling

In the earlier sections of this book, you were introduced to the basics of production technology. You read about the history of production, careers in production, production materials (metals, polymers, and ceramics), and the various processes and tools (separating, forming, and combining) used in production. You also learned about the use of computers and other automated tools for production. In this section of the book, you will learn how to run a manufacturing **enterprise**. An enterprise is a business or company.

Most of the information presented in this section of the book is about the processes used in a manufacturing enterprise. You can apply this information in a typical technology laboratory. Whether your laboratory area is designed for working wood, metal, or plastic, or has a general shop format with a little of everything, your class can create and run a manufacturing enterprise, Figure 12–1.

If a manufacturing enterprise followed exactly the order presented in this book, it might never get off the ground. To start and run a manufacturing enterprise, several processes should be running at the same time. Every situation is unique. This chapter describes one way to organ-

170

FIGURE 12-1 You and your classmates can set up a manufacturing enterprise in your school.

The Importance of Good Planning

Planning a manufacturing enterprise begins with making decisions about what to make; how to form, organize, and manage the company; how to obtain needed money; how many products to make; and other factors. Good planning is important to a smooth, efficient enterprise. Many times, several groups of students will be working on different parts of the enterprise, Figure 12-2. One group may be planning what product to make while another group is trying to obtain the needed money. When several groups are doing different jobs like this, it takes good planning to get the work done. Each student in the class is important. If you or one of your fellow classmates does not do your job, the entire enterprise might suffer. Each and every student-worker is important to the success of the enterprise.

Ownership and Management

One process begun early in the planning stage is making decisions about ownership and management of the company. The first step may be to form the company, complete with elections of president, vice presidents, treasurer, secretary,

ize and run a manufacturing enterprise in your school laboratory. It is certainly not the only method. Every class is different in size, course length, meeting schedule, laboratory set up, and other factors. That makes each classroom enterprise different. With that in mind, here are some suggestions on running a classroom manufacturing enterprise.

and so on. If the corporation model of organization is selected, the jobs of the elected officials should be stressed. There are other forms of ownership than the corporation. However, most large industrial enterprises use this approach for ownership and management. Review Chapter 4 for more detailed information on ownership and management.

Hiring Employees

After the officers are elected, employees (workers) must be hired. People are the most important input for any manufacturing enterprise. Some of the people who help in the early stages of planning include supervisors and engineers. It is their job to get the rest of the workers organized, design the product, and plan the manufacturing line.

Each student in your class will have different skills and experiences. This is good because manufacturing requires workers with a variety of skills, Figure 12–3. The best person for each job in the enterprise should be hired. Chapter 3 describes some of the various types of jobs and careers in production and manufacturing technology.

FIGURE 12–3 Workers with a variety of skills are needed in manufacturing enterprises. People who like to work with their hands are just as important as people who like to work with machines like computers.
(Courtesy of Ford Motor Company)

At some point during the classroom enterprise, it is possible that nonmanagement workers could unionize. If they feel that working conditions are not good, they may decide to strike. This is a good learning experience if the problem is solved through negotiation in a day or two.

The workers who actually make the products and the people who manage their work are the backbone of any manufacturing enterprise. Be sure to choose people who you can count on to get the job done.

Engineering: Designing Products and Planning Production

Deciding what product to make, what materials to use, what tools to use, and how to make the product are probably the most important decisions made as part of an enterprise. A product that is easy to make, attracts customers, and can be sold at a profit is needed if the enterprise is to be successful. If time permits, you might spend several weeks researching and developing product ideas. The processes associated with creating product ideas are called **design engineering**, Figure 12–4. Design teams of two or three students may come up with product ideas to present as possible products to be made. Ideas for products can come from your home, books, catalogs, magazines, or local stores. Each team could pick their best product idea and share it with the class. The class could then vote on which product to make. Another option is to have the teacher select the product that will be made. After the product idea is chosen, the design teams should make working drawings of products and provide a detailed bill of materials. A bill of materials lists all the parts on a product, their sizes, and their costs. To save time, some classes may **subcontract** with a CAD or drafting class in the school to do some of the working drawings. A model of the final product could be made to make sure the necessary tools and machines are available. More detailed information on design engineering is provided in Chapter 14.

After the product has been selected, production engineering processes can begin. **Production engineering** is responsible for planning which manufacturing system will be used to

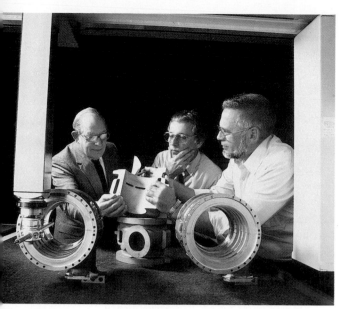

FIGURE 12–4 These design engineers are working on plans for a new product idea. *(Courtesy of Barnes Group, Inc.)*

ing of a product. In Chapters 9 and 10, you learned about the three categories of materials processes — forming, separating, and combining — as well as the tools and machines used for each of these processes. Forming, separating, and combining processes can be used to make any product. Many of the details related to the processing of materials in your enterprise will be determined by which product you decide to make. The list of materials, as well as the detailed list of steps needed to make the product, will tell you which materials processes you need. Tools, machines, and processes must be set up in the most efficient manner, Figure 12–5. It is important that the setup be in an order that will move the product down the line as smoothly as possible. In the laboratory setting, it might be hard to rearrange machines that are fixed in one spot. However, you will need to plan carefully to get the greatest efficiency from your manufacturing line.

make the product. A procedure, or list of steps that includes the necessary tools, machines, and processes, will be needed. A good way to present the steps needed to make a product is with operation process and flow process charts. Also needed will be special tools called **tooling**, materials handling systems like conveyors, and quality control inspection devices. Designing and making these devices is also the job of production engineering. More detailed information on production engineering is provided in Chapter 15.

A final engineering decision is which type of manufacturing system to use to make the products. Products can be made one at a time by individual workers with **custom manufacturing**. **Mass production** techniques can also be used. **Intermittent manufacturing** and **continuous manufacturing** are two examples of mass production. In order to choose the best manufacturing system for your enterprise, read Chapter 13.

Processing Materials and Making Products

Processing materials changes them into a more usable form. It is the actual manufacture or mak-

FIGURE 12–5 Tools, machines, processes, and people must be set up in the most efficient manner. These two student-workers are assembling a product. Their use of division of labor helps make production more efficient.

The tools and machines available in your laboratory will also determine which materials processes can be used, Figure 12–6. Your teacher will help you decide which forming, separating,

FIGURE 12–6 The tools and equipment available in your school laboratory will determine what types of processes you will perform and what types of products you will make. *(Courtesy of Miller Electric Manufacturing Company of Appleton, WI)*

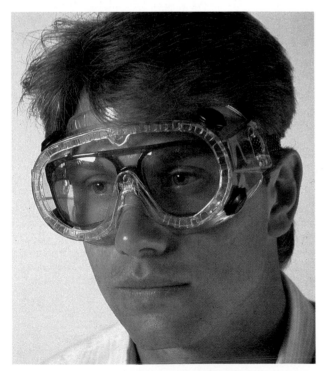

FIGURE 12–7 Wearing safety glasses and goggles is one important safety rule that every student-worker should follow. *(Courtesy of 1989 Lab Safety Supply Catalog)*

and combining processes to use. There are a number of different types of tools, machines, and processes available. You may not be able to learn all the materials processes that are available in your laboratory. To get the product made, you may have to use a **division of labor**, where each worker performs a different material process.

Safety and Processing Materials

Your teacher will also stress the importance of safety when performing materials processes. It is very important that you know and follow all the safety rules and regulations set up by your teacher. Working in manufacturing can be dangerous if people do not follow the safety rules, Figure 12–7. Tools, machines, and equipment can cause minor cuts, bruises, and scrapes, but they can also cause severe injury if not used properly and safely. Safety is covered in the beginning of the book and again in Chapter 8. Be sure to read these safety sections and learn all the safety rules set by your teacher.

Marketing

Marketing is selling the product. Students who will market the product must begin their work early in the enterprise. As soon as the product ideas are chosen, you can begin doing market research to determine how many people might buy the product. Market surveys could be used as a tool to select what product to manufacture. In addition, after the product has been chosen, marketing workers design an **advertising campaign**. This might include ads in the school newspaper, flyers passed out in study hall or classes, radio commercials over the school intercom system, or a television video ad shown during lunch in the cafeteria. Designing an eye-catching name and package for the product might also be the job of students involved in marketing. Chapter 16 discusses marketing processes.

Money and Manufacturing

Having enough money is key to any enterprise, Figure 12–8. Students in charge of the money should be organized and familiar with numbers. The first process will be to determine how much money will be needed. The bill of materials for

FIGURE 12–8 Setting up a manufacturing plant like this one can cost millions of dollars. Your school production laboratory will cost less to set up because the tools and machines have already been purchased. *(Courtesy of Johnson Controls, Inc.)*

the chosen product will give you an idea of how much money is needed to make one product; simply multiplying that cost by the number of products your class decides to make will give a total cost for the products. After you know how much money it will take to run the enterprise, the next step is to get the needed money. Enterprises can obtain money through three sources: loans from banks, selling bonds, and selling

stock. People who provide money to enterprises are called **investors**. An investor provides money to an enterprise with the hope of making more money back. Each of these sources of money is explained in Chapter 17.

Students in charge of the money will also need to keep records of all the money spent. Money will be spent to buy materials, special tools, machines, product parts, and supplies. In addition, your teacher may decide to pay the student-workers in the class. A certain amount of pay may be set by the teacher. Money earned in class may be used to buy extra points on tests or one of the products made by the class. Orders and payments received from customers must also be recorded. Records should be kept of the investors who buy stock or bonds or give loans to the enterprise. After the classroom enterprise is finished making and selling products, reports on how much money was made and where the money was spent must be created.

Impacts of Manufacturing

As in industry, the impacts of any classroom enterprise include the manufactured product and all the scrap, waste, and pollution made. The kinds of scrap, waste, and pollution you create are important to consider. Leftover materials should be recycled or reused rather than thrown out with the garbage, Figure 12-9. If not, they may cause pollution that will impact the air, water, and land environments.

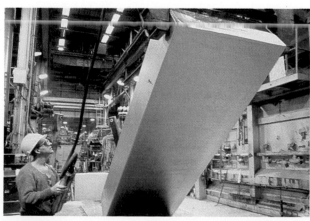

FIGURE A

FIGURE B

FIGURE 12–9 The large block of aluminum (Figure A) was made from recycled beverage cans. The aluminum will be used to make new cans (Figure B). *(Courtesy of Alcan Aluminum Company Ltd.)*

A large amount of scrap, waste, and pollution is costly for an enterprise. An enterprise that makes large amounts of these outputs may not be as running efficiently as possible. Finding ways to reuse scrap and dispose of waste materials can improve the efficiency of an enterprise. Throwing away excess materials and supplies costs money and can create pollution. Pollution is a very real and serious problem faced by many manufacturing enterprises. Its impact on people and our environment can be dramatic. You must weigh the effects of your pollution against the gains of the entire enterprise.

Your teacher may encourage you to use scraps to make other smaller products or projects. Or scrap and waste may be donated or sold — as kindling wood, for example. In any event, your class should plan carefully to dispose of its "garbage." Chapter 26 will discuss scrap, waste, and pollution, and their impact on our environment.

Summary

You can apply the information in this book to real-life manufacturing enterprises as well as your classroom experience. No matter what type of tools and equipment are in your lab, an enterprise can be organized and operated.

Following the exact order of the chapters presented in this book will probably not work for running an enterprise. In reality, several processes run at the same time or overlap. The order of the processes will also differ from classroom to classroom. Class size, meeting schedule, laboratory setup, available money, and other factors will affect the enterprise.

Action in an enterprise can begin with these processes:

- organizing an enterprise
- electing officers and hiring workers
- selecting a product idea
- creating working drawings and a bill of materials
- subcontracting work (if needed or desired)
- planning for the production of the products
- choosing custom, intermittent, or continuous manufacturing
- learning and following safety rules
- processing materials (forming, separating, and combining) to make products
- marketing (advertising and selling) the products
- raising money, spending money, keeping track of money
- studying the impacts of scrap, waste, and pollution.

This is just one order in which a manufacturing enterprise can be studied. Your teacher may want you to organize the enterprise in a different manner. All of the processes described in the chapters that follow may not need to be completed. As mentioned earlier, sometimes several of these processes may be happening at the same time. This requires detailed planning and all the students in the class to work together to get the job done. Good luck as you begin studying a manufacturing enterprise.

DISCUSSION QUESTIONS

1. What is a manufacturing enterprise?

2. Why might several activities or processes be running at the same time in a manufacturing enterprise?

3. Why is it important to have good planning when organizing a manufacturing enterprise?

4. What are the various steps involved in setting up and running a manufacturing enterprise?

5. How is safety important in a manufacturing enterprise?

CHAPTER 13

Types of Manufacturing Systems

OBJECTIVES

After completing this chapter, you will be able to:

- Define and describe custom manufacturing.

- Define and describe intermittent manufacturing.

- Define and describe continuous manufacturing.

- Define and describe just-in-time manufacturing.

- Compare custom, intermittent, continuous, and just-in-time manufacturing and the products that are made with each system.

- Describe three factors to be considered when selecting a type of manufacturing system.

KEY TERMS

Automation
Batch manufacturing
CAM
CNC
Continuous manufacturing

Custom manufacturing
Flexible manufacturing
Intermittent manufacturing
Job lot
Just-in-time manufacturing

Kanban
Life cycle
Lot
Mass production

Is All Manufacturing the Same?

Most people think of mass production and assembly lines when they think about manufacturing, Figure 13–1. But are there other types of manufacturing systems? Is the type of manufacturing system used to produce cars the same as that used to make designer clothes? Is a pencil made in the same manner as custom cabinets in your kitchen at home? Do other countries use the same systems we use in the United States? Because of the large number of different items that are produced, several types of manufacturing systems have evolved. No one system can be used to make all the different types of products.

FIGURE 13–1 Most people think of mass production when they think about manufacturing. *(Courtesy of Rockwell International Corporation)*

FIGURE 13–2 Special work vans are custom manufactured. *(Courtesy of Grumman Corporation)*

Types of Manufacturing Systems

There are three basic types of manufacturing systems:

1. custom manufacturing
2. intermittent manufacturing
3. continuous manufacturing

Let's look at each type more closely.

Custom Manufacturing

Custom manufacturing makes one-of-a-kind products or just a few products at a time. Each product is unique and different, Figure 13–2. One person or several people may work on a product from start to finish. Workers usually have many skills. Solving problems is a necessary skill because each product is different. In some cases a

FIGURE 13–3 Sailboats, like those raced in the America's Cup races, are custom designed and manufactured. *(Courtesy of Miller Electric Manufacturing Company of Appleton, WI)*

FIGURE 13–4 This paper company keeps a large supply of paper in storage. *(Courtesy of Hammermill Papers)*

single product may take several months to make. For example, large boats, airplanes, and custom tools may take a long time to manufacture, Figure 13–3. Other products that are custom manufactured include jewelry, pottery, custom furniture, and signs.

Custom manufacturing workers don't get bored with their jobs. Products that are custom made are often of the highest quality because of the time, care, and pride taken by the highly skilled workers. As a result, the price of custom goods is usually quite high. Because of their high quality, there is usually a buyer for a custom-made product even before work begins.

Intermittent Manufacturing

Intermittent manufacturing is also called **job lot**, **batch**, and **flexible** manufacturing. A certain number of products, called a **lot**, is made with this system. A lot is less than 500 products. A company that uses intermittent manufacturing needs a large space to store its supply of raw materials, Figure 13–4.

An intermittent production line may be used to make different products without changing tools, materials, or workers. If one product is not

FIGURE 13–5 Clothes are seasonal products made using the intermittent manufacturing system. *(Courtesy of Burlington Industries, Inc.)*

a good seller, the production line can be changed to make another product that might be selling better. This allows the company to adjust quickly to meet changes in consumer needs.

Products made with this system include those that are sold only at certain times of the year, such as clothes, paint, and holiday items, Figure 13–5. Products like musical instruments or special parts for industry, which are needed only by a small group of people, are also made with this system, Figure 13–6.

Because the product made is always changing, workers don't get bored that often. Skilled or semiskilled workers are needed.

FIGURE 13–6 This thrust reverser goes on a DC-9 jet. It is an example of a special part needed by industry that is made using the intermittent manufacturing system. *(Courtesy of Rohr Industries, Inc.)*

Continuous Manufacturing

Continuous manufacturing is also called **mass production.** A large number of the same product is made, usually twenty-four hours a day. The product moves on a continuous conveyor past the workers, almost like water flowing in a river past a person sitting on the bank, Figure 13–7. Companies using this system make only one or a few types of products.

Examples of products made with this system include televisions, radios, refrigerators, food, and cars, Figure 13–8. In other parts of the world,

FIGURE 13–8 Automobiles are produced using mass production (continuous manufacturing) techniques. *(Courtesy of Toyota Motor Manufacturing, USA, Inc.)*

Continuous Manufacturing and Computers

Many jobs on a continuous manufacturing line can be done easier and faster with computers. Running machines with computers is a form of **automation**. Automation means things are done without the help of people. Two types of computer automation are computer numerical control (**CNC**) and computer-aided manufacturing (**CAM**), as discussed in Chapter 11. With CNC, numerical information (numbers) is fed into a computer. The computer then uses the numbers to control a lathe or other machine. CAM involves controlling several processes at one time with computers. For example, conveyors, machines, and even CNC systems can be controlled by one computer.

This computer numerical control (CNC) machine is used to make wheels for large trucks. (Courtesy of Simmons Machine Tool Corporation)

the kinds of items made might be quite different. In China, chopsticks are made with continuous manufacturing.

In continuous manufacturing, there are many workers, but each does only one specific job. For example, if you worked on a continuous manufacturing line for cars, your specific job might be to put the bolts on a wheel. This type of work can cause workers to get bored with their jobs. Unskilled and semiskilled workers are needed for continuous manufacturing. While not always the case, continuous manufacturing is known for making medium-quality products at low prices.

Just-in-Time Manufacturing

Just-in-time (JIT) manufacturing is a system that can be used in custom, intermittent, or continuous manufacturing. JIT was developed in Japan. The goal of JIT is to make and deliver products *just in time* to be sold. Every step in the process of making a product is done just in time: Raw materials are bought just in time to be made into parts; work on parts is finished just in time to start final assembly; final assembly is completed just in time for the product to be finished; and finished products are ready just in time for delivery to the store.

Just-in-time manufacturing requires all the people, machines, and tools involved in making a product to work together very closely. Even the suppliers of raw materials and standard supplies like screws and nails must work closely with the company using JIT.

JIT is sometimes called **kanban**. Kanban is the name of a Japanese inventory system developed by Toyota, Figure 13–9. In kanban, a very small inventory is kept at the plant. Managers must plan well ahead to make sure suppliers deliver needed materials just in time. To keep track of inventory, special cards are placed on parts, supplies, and other materials used in the production line. When parts are removed from inventory, the information taken from these cards is given to managers who send orders for new materials to the suppliers. Keeping a smaller inventory reduces the cost of making products. It also requires workers on the production line to work very closely with management to keep them informed of shortages.

FIGURE 13–9 The box in this photo shows the cards used in the kanban inventory system. *(Courtesy of New United Motors)*

Another feature of JIT is the time it takes to set up the tools and machines needed to manufacture a product. In JIT manufacturing, this setup can be done very quickly. Part of the reason for this is the use of new technology like robots and other automated machines, Figure 13–10.

JIT has been shown to increase worker interest in the quality of the finished product. For this reason, JIT is being used more often by custom,

FIGURE 13–10 JIT manufacturing uses robots and other automated technology. *(Courtesy of New United Motors)*

intermittent, and continuous manufacturing companies. Workers in JIT are also given more responsibility for the quality of the finished product, Figure 13–11. With this increased responsibility, workers often feel more pride in the product and more important to the company. How workers feel about their worth to the company is referred to as the quality of work life, Figure 13–12.

Choosing a Manufacturing System

Choosing which manufacturing system to use can be a difficult task. Each system has certain features that make it best suited for a particular manufacturing need. When a company is trying to decide which system to use, several factors are considered, including the following:

1. the type of product to be made
2. the number of products to be made
3. product life cycle

Type of Product

The type of product to be made often helps the company decide which system to use. For example, a regular automobile may require a different type of manufacturing system than a racing dragster, Figure 13–13. The regular car may be mass-produced continuously because there is a large number of people who are willing to buy new cars. On the other hand, there are only a few people who own racing dragsters, so custom manufacturing is a better system for such a special product.

Number of Products

The number of products to be made is a second factor that must be considered. For example, you just read that an automobile could be made using custom or continuous manufacturing. Automobiles could also be made using intermittent or JIT manufacturing. However, if an order is placed for 50,000 automobiles, continuous manufactur-

FIGURE 13–12 The quality of work life affects the final quality of manufactured products. *(Courtesy of Scott Paper Company)*

FIGURE 13–13 This racing dragster was probably custom manufactured. *(Courtesy of Schiefer Sports Marketing)*

ing is probably the best choice. Making that many autos using a custom or intermittent system could take too much time and cost too much money, Figure 13–14. Continuous manufacturing is the best system for large quantities of one product.

Product Life Cycle

Product **life cycle** is the number of months or years the product is expected to be wanted by consumers. Some products, like clothing, have a short life cycle. Clothes are seasonal and they go in and out of fashion very quickly. Other products, such as light bulbs and food, will always be wanted by consumers, Figure 13–15. These products have a long life cycle. Intermittent manufacturing is best for short-life-cycle products, while continuous manufacturing is best for products with long life cycles.

FIGURE 13–15 These popular food items are examples of products in high demand. They have a long life cycle and are continuously manufactured. *(Courtesy of The Pillsbury Company)*

FIGURE 13–14 Products made in limited quantities, like this jet plane, are made with intermittent manufacturing. *(Courtesy of Grumman Corporation)*

Summary

There are three basic types of manufacturing systems; custom, intermittent, and continuous. Custom manufacturing produces one-of-a-kind, high-quality products using highly skilled workers. Intermittent manufacturing makes products in small amounts called lots. Skilled and semiskilled workers are needed for intermittent manufacturing. Continuous manufacturing makes a large number of one type of product using workers who specialize in one specific task. Computers are being used in continuous manufacturing to control machines and processes with CAM and CNC.

Just-in-time manufacturing is a system that can be used in custom, intermittent, or continuous manufacturing. The goal of JIT is to produce and deliver a product to the market just in time to be sold.

When a company is trying to decide which system to use, a number of factors are considered. Three important factors include: the type of product to be made, the number of products to be made, and the product life cycle.

DISCUSSION QUESTIONS

1. How would you describe custom manufacturing systems? What type of products are made? What type of workers are needed?

2. How would you describe intermittent manufacturing systems? What type of products are made? What type of workers are needed?

3. How would you describe continuous manufacturing systems? What type of products are made? What type of workers are needed?

4. How would you describe just-in-time manufacturing systems? What type of products are made? What type of workers are needed?

5. Why do you think the quality of work life is important to the quality of a finished product?

6. What three factors do companies consider when trying to decide which type of manufacturing system to use?

7. Identify local manufacturing companies and identify the type of manufacturing system each uses.

CHAPTER ACTIVITIES

 ## Comparing Custom and Continuous Manufacturing

OBJECTIVE

Three common types of manufacturing are custom, intermittent, and continuous. This activity focuses on custom and continuous manufacturing.

Custom manufacturing makes one-of-a-kind products with highly skilled workers. Continuous manufacturing makes a large number of one product with semiskilled or unskilled workers.

In this activity, you will make a product twice — first with custom manufacturing, and then with continuous manufacturing. You will compare the products made and the manufacturing systems used by looking at: (1) Product Quality, (2) Interchangeable Parts, (3) Production Times, and (4) Product Appeal. The product is a yo-yo, Figure 1.

FIGURE 1. Finished yo-yo.

EQUIPMENT AND SUPPLIES

For Custom Manufacturing

1. Two precut squares of hardwood per student, size .563″ thick by 2.50″ square
2. One dowel rod per student, size .25″ diameter by 1″ long
3. 36″ kite string per student
4. Wood glue
5. 100 and 150 grit sandpaper
6. Assorted hand tools: saws, files, rasps, drills, vises, rulers
7. Finishes and brushes
8. Safety glasses for each student
9. Stopwatch
10. Yo-yo product drawing (see Figure 2)

For Continuous Manufacturing

1. Precut hardwood boards, .563″ thick by 2.50″ wide by 2.50″ long (you will need two pieces for each student)
2. One dowel rod per student, .25″ diameter by 1″ long
3. 36″ kite string per student
4. Wood glue
5. 100 and 150 grit sandpaper
6. Power tools and machines: radial arm saw, band saw, router, drill press, finish sander

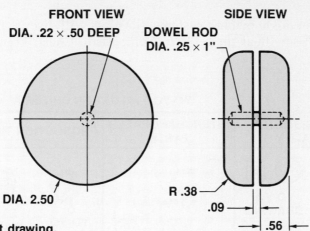

FIGURE 2. Yo-yo product drawing.

7. Spray finish
8. Safety glasses for each student
9. Stopwatch
10. Operation Sheets (from your teacher)
11. Jigs and Fixtures (from your teacher)
12. Yo-yo Comparison Chart

PROCEDURE

Custom Manufacturing

1. Obtain the yo-yo product drawing, necessary materials, and hand tools from the teacher.
2. Custom manufacture one yo-yo using hand tools and following the plans in the yo-yo product drawing. Each student will make a yo-yo. Perform all the steps needed to make the yo-yo, from cutting the wood into a round shape, to drilling the hole, to finishing and assembling the yo-yo.

NOTE: Be sure to follow all safety rules when using hand tools and to wear your safety glasses.

3. Each student should keep track of the time it takes to make the yo-yo using the custom manufacturing system.

Continuous Manufacturing

1. Help your teacher set up a continuous yo-yo manufacturing line. Special aids called jigs and fixtures will be attached to certain machines or used with power tools. These jigs and fixtures will help you manufacture the yo-yo using the continuous manufacturing system.
2. Your teacher will select one student to work on each of the processes on the continuous manufacturing line. Each student will be trained by the teacher to do a special job on the continuous manufacturing line.
3. After your teacher is sure all students know their jobs, continuous manufacturing of the yo-yo can begin.
4. Use the stopwatch to keep track of the time it takes to make one yo-yo for each student in the class using the continuous manufacturing system.

Discussion

When yo-yos have been made by both systems, the class will discuss the products. Compare the yo-yos and write your ideas on a Yo-Yo Comparison Chart, Figure 3.

YO-YO EVALUATION CRITERIA

SYSTEM	CONSISTENT QUALITY	INTERCHANGEABLE PARTS	PRODUCTION TIMES		PRODUCT APPEAL
	YO-YO YO-YO DRAWING RULER	YO-YO PARTS	TOTAL	1 YO-YO	YO-YO YES NO
CUSTOM					
CONTINUOUS					

FIGURE 3. Yo-yo chart.

1. **Product Quality.** Collect all the yo-yos made by custom and continuous manufacturing. Measure each yo-yo and compare its size with those on the yo-yo product drawing. Keep track of the number of products that match and don't match the drawing.
2. **Interchangeable Parts.** Select any three yo-yos from the custom and continuous manufacturing groups. Interchange parts among the three yo-yos in each group. Again, compare the yo-yos for size and quality after parts have been interchanged.
3. **Production Times.** For both custom and continuous manufacturing, compare the total production time for all yo-yos to the time it took to make each yo-yo.
4. **Product Appeal.** Finally, take a poll of the students in your class. Ask them to rate the product appeal of custom-manufactured and continuous-manufactured yo-yos. Which product would you be more willing to purchase if you saw it in a store?

MATH AND SCIENCE CONNECTIONS

Here is how you can calculate Total Production Time for Custom Manufacturing:

Total Time = Time to make each yo-yo x number of students

Here is how you can calculate One Yo-Yo Production Time for Continuous Manufacturing:

$$\text{One Yo-Yo Time} = \frac{\text{Total Time on Stopwatch}}{\text{Number of Products}}$$

RELATED QUESTIONS

1. How is custom manufacturing different from continuous manufacturing?
2. Which manufacturing system produced the products with the best quality? Why?
3. Which manufacturing system produced the products that could be most easily interchanged?
4. Which manufacturing system produced the products with the best production times? What makes a production time "best"?
5. Which manufacturing system produced the products your classmates found most appealing? What were their reasons?

CHAPTER 14
Design Engineering

KEY TERMS

Applied research
Assembly drawing
Bill of materials
Brainstorming
Computer simulation
Designer

Design standards
Detail drawings
Development
Mock-up
Prototype
Pure research

Rendering
Rough sketches
Thumbnail sketches
Working drawings

If you had an idea for a new product and wanted to get it manufactured, what would you do? Well, first you would have to convince a manufacturing company that your product was worth making. To do that, you would need a set of plans that could be used to communicate your product idea to the company. You would need to be able to provide the company with answers to the following questions:

- Can the product be manufactured with company tools?

- Will consumers buy the product?
- Can the product be sold at a profit?

To get these plans, you would turn to a **designer**. Designers create ideas for products. They also turn their ideas into plans that tell a company if the product can be made and sold at a profit, Figure 14–1. A designer could be an engineer, a technician, or an artist. Designers use their communication skills. The design process begins with research and development of new product ideas.

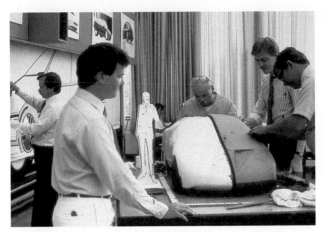

FIGURE 14–1 Designers create product ideas by making sketches, drawings, and models. *(Courtesy of General Motors)*

FIGURE 14–2 NASA's pure researchers study the space environment. *(Courtesy of NASA)*

Research and Development

Workers in research and development (called R&D) find new ideas for products, materials, and processes. Most of the products that are called "new and improved" are the result of R&D. R&D is made up of two different areas: research and development.

Research

Research is a step-by-step search for new ideas, facts, or information. Scientists, such as chemists, work in R&D doing much of this research. They use the scientific method to find and gather new ideas.

There are two types of research: pure and applied. **Pure research** (also called basic research) is the search for new ideas without thinking of any real uses for the ideas. **Applied research**, on the other hand, is the search for practical uses for the new ideas found through pure research.

NASA's (National Aeronautics and Space Administration) search for ways to make products in space can be used to demonstrate the differences between pure and applied research. Before any products can be made in space, scientists must learn more about the effects of space on people, materials, and processes. Their research in this field is an example of pure research. NASA's pure research is used to find out how things work in the space environment, Figure 14–2.

Once NASA knows how people and processes will work in space, research can begin to find new ways of using that information. The new information found through pure research is used to find new products that can be made in space. Finding new, useful products is an example of applied research.

On earth, pure and applied research is being conducted in manufacturing every day to learn more about materials, processes, products, machines, and even customers. Asking consumers about their needs and wants is another example of pure research.

Development

Development is a type of applied research that tries to make new products. Development gets ideas for new products from researchers. When plastics were first discovered from pure research, ideas for products that could be made from plastics were needed. Workers in development took the new plastic materials and used them to create new products. Designers work in the development process. This is also called the design process, Figure 14–3.

Steps in the Product Design Process

The product design process contains eight steps. The steps in this process can take days or even years to complete. This depends on how complex

FIGURE 14–3 The research scientist on the left is doing pure research. He is studying the chemistry of a new plastic material. Designers take information from this researcher to develop new products like the sailboard shown on the right, which is made from plastics. *(Courtesy of Amoco Corporation)*

the product is and the size of the company. When products are designed, the designers often go through the steps listed in Figure 14–4.

Step 1: Listing Design Standards

The goal of the designer is to make quality products that meet consumer needs at a fair price. Designers must think about many factors when they make a product idea. They must think about the materials that might be used, the tools and machines needed, and the skills of the workers in the company. Designers usually follow a set of **design standards** when they are working on a product idea, Figure 14–5. The design standards guide the designer. They are like rules that must be followed. The design standards are described as follows.

Function. Function means how the product works. The function of a product is one of the most important design standards. After all, if a product does not work (function), it is useless to the consumer. Designers must make products that work. Of course, there are many different product functions. The designer must think about how consumers will use the product.

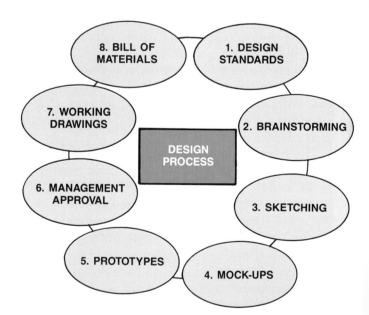

FIGURE 14–4 The eight steps in the design process.

Form. There is a saying in design that "form follows function." This means the form (size and shape) of a product must match its function. The size and shape of a table and chair are designed to match their function as, say, a place to work on a computer. A table that is too high or too low

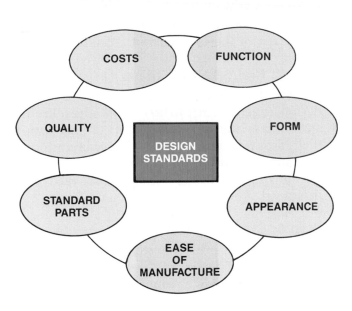

FIGURE 14–5 The design standards give designers a set of rules to follow.

for the worker does not have a form that follows its function. Designers try to find a product form that matches its function, Figure 14–6.

Appearance. Appearance is the beauty or looks of a product or other object. Most products are designed to attract the attention of consumers by having a pleasing appearance.

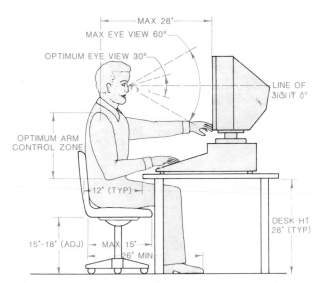

FIGURE 14–6 The form of a product must match its function. *(Reprinted from* DRAFTING IN A COMPUTER AGE *by Wallach and Chowenhill,* © *1989 by Delmar Publishers Inc.)*

Engineering places less value on appearance than on function and form. A beautiful product that does not work is useless. Many companies make good-looking products that do not work as they should. Cheap products that look like the original are always being made and sold. Sometimes these products do not work like the original.

Ease of Manufacture. Designers must think about the tools and machines needed to make a product. They must be sure workers have the skills needed to make the product. Of course, the company may decide to buy new tools if they think enough people will buy the product.

How difficult it will be to make the product is also considered. The fewer and the simpler the parts, the easier a product will be to make. Manufacturing a product that needs hundreds of parts is not easy.

Standard Parts. A standard part is one that can be bought easily. Examples of standard parts are the lumber, screws, nails, drill bits, and plywood that can be bought at the lumber yard or hardware store. Designers try to make products that can be made with standard parts. A product that is made from standard parts costs less to make, Figure 14–7.

Quality. A quality product is one that functions well and lasts a long time. Quality products can cost more to make than products that wear out quickly. Big, very expensive cars have quality. They can run for many years without wearing out. But they can cost ten times as much money as a smaller, less expensive car.

FIGURE 14–7 Even this car of the future will use standard parts for bolts, screws, wheels, and other parts. *(Courtesy of General Motors)*

Some products are designed to break after a shorter time. Parts are designed to wear out or break down after a few months or years. When products wear out, consumers will need to buy another. If products were made to last forever, manufacturers would never sell another product.

Cost. Cost is probably one of the most important design standards. The cost of making a product must allow for the company to sell it at a profit. Designers make sure product costs are kept low. Trade-offs are usually made between costs and the other standards. A trade-off is when we choose to give up one thing to get another. Designers usually choose to give up quality and other standards to get lower costs.

Also related to cost is the type of workers needed to make a product. Products made by skilled workers cost more to make. Products made by automated equipment cost less. Labor costs are a big part of the total cost of a product.

Step 2: Brainstorming Ideas

After the list of design standards is done, designers get together to brainstorm ideas. **Brainstorming** is a group of designers talking about their ideas for a new product. During brainstorming, any and all ideas are spoken out and written down. The list of ideas is then read and discussed. The best ideas are then chosen.

Step 3: Sketching Ideas

The best ideas from brainstorming are sketched out on paper. The first sketches made are called **thumbnail sketches** or **rough sketches**. Thumb-

Trade-Offs and the Design Standards

Designers may be given a set of design standards for a product from management. Or they may make their own set of standards. By looking at one product, you can think of trade-offs that were made. For example, think of the standards used to design automobiles. The standards and trade-off decisions are listed in Figure 1.

Designers must make trade-off decisions about the product design. The trade-off decisions made in designing cars result in a wide range of different automobiles. One of the most important steps in designing a quality product is to come up with a list of design standards.

FIGURE 1. Here are some design standards and trade-off decisions for automobile manufacturers. *(Courtesy of Pontiac Division, General Motors)*

STANDARDS	TRADE-OFF DECISIONS
Function	Speed, economy, passengers
Form	Compact, full-size, wagon
Appearance	Styling, colors
Ease of Manufacture	Accessories, options available
Standard Parts	Standard engines, wheels, parts
Quality	Reliable, breaks down
Costs	Economy or luxury

nail sketches are small (they may be bigger than a thumbnail, though). They can be of a complete product or of just a few parts of the product, such as handles, knobs, or corners. For every part of the product, designers sketch all their ideas, Figure 14–8.

Designers usually work as part of a design team. Each team member creates his or her own rough sketches. They then share their designs with the team. As a team, they select the best design ideas for the product.

Once the best ideas are picked, the rough sketches are improved. These improved sketches

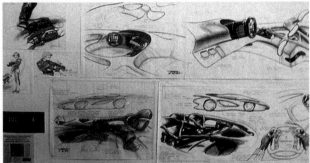

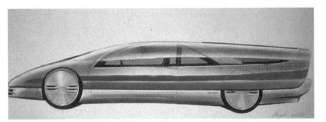

FIGURE 14–9 Renderings are sketches of the finished product. (Courtesy of General Motors)

show parts of the product in more detail. One type of improved sketch is called a **rendering**. A rendering shows the final product with all its details. Color is added and the product is shown as it will look when it is finished, Figure 14–9.

Management looks at the renderings and other sketches and asks these questions:

- Do we have the tools and people needed?
- Can we make the product in time?
- Will consumers buy the product?
- Can we sell the product at a profit?

Management then makes the final decision to continue or to scrap the product idea. If they decide to continue, the next step is to make mock-ups.

Step 4: Making Mock-ups

A **mock-up** is a scale model of the finished product. Mock-ups show the appearance of the product. They are made from easily worked materials such as cardboard, balsa wood, styrofoam, or clay. Mock-ups do not have any real working parts. For example, a mock-up of an automobile may have wheels, but they do not turn. They are just there for appearance, Figure 14–10.

FIGURE 14–10 Here are three different automobile mock-ups. Clay mock-ups for shown in the left and center photos. A plywood mock-up of the driver's seat is shown on the right. (Courtesy of General Motors)

Designers make several different mock-ups. The mock-ups are used to check the design standards. The ease of manufacture, the use of standard parts, and the final cost of the product are checked. One of the most important reasons for making mock-ups is to judge the appearance. The best-looking parts of the different mock-ups are chosen for the next step.

Step 5: Making a Prototype

The next step is making a prototype. A **prototype** is the first full-size working model of the product. Prototypes are made with the actual materials and working parts that will be used in production. Engineers test the prototype to make sure it functions properly. A number of design standards are checked. These include form, function, and quality, Figure 14–11.

FIGURE 14–11 A designer adds the final touches to a prototype automobile before road testing begins. *(Courtesy of General Motors)*

Reduce Design Time with Computers

Sketching, making, and testing mock-ups and prototypes can take weeks, months, or even years. To reduce the time needed, designers use computers. Computer-aided drawing and design (CADD) systems speed up the design process. CADD systems have software with all the information needed to make a model on the computer. The computer model is designed to function just like a mock-up or prototype. After a product design is fed into the CADD system, the designer can test the product on the computer. Designing and testing products this way is called **computer simulation**. Simulation means to create the appearance of something.

CADD systems are also used to figure the total costs of making the product. Prices for all the parts are fed into the computer as the design is made.

CADD systems reduce the time needed to get a final set of product plans. If a part does not function, it can be replaced on screen and

FIGURE 1. Designers can reduce the time it takes for the design processes by using CADD systems and computer simulations. *(Courtesy of Honeywell, Inc.)*

retested. With mock-ups and prototypes, replacing parts can take a long time. Computer simulations with CADD reduce the time involved in designing products.

Step 6: Getting Management Approval

After the prototype has been tested, the design team must get final approval from management to go to production. Management meets with the design team and people from production, marketing, and finance to discuss the product design. Each department gives its opinion of the product. If management approves, a set of working drawings and a bill of materials are made.

Step 7: Drafting Working Drawings

Working drawings are a set of drawings and plans that contains all the size, shape, and other information needed to make the product. A set of working drawings usually includes two different types of drawings: detail drawings and assembly drawings.

Detail Drawings. Detail drawings are usually multiview (more than one view) drawings of all the parts of a product. They contain all the information needed to make each part, Figure 14–12. Detail drawings are done by drafters using manual drafting tools or computer-aided drafting (CAD) systems.

Detail drawings can include one, two, or more views. Three views are common. The drawings show the shape and form of each part. Also included are the dimensions, with exact information on sizes, shapes, and locations of parts, such as holes, slots, or cuts. Notes may also be added to give other important information.

Assembly Drawings. An **assembly drawing** shows what the final product will look like when

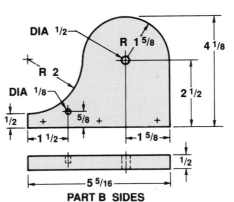

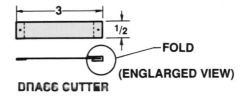

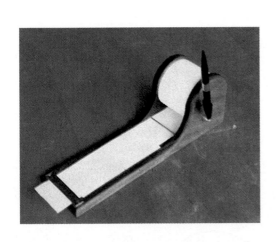

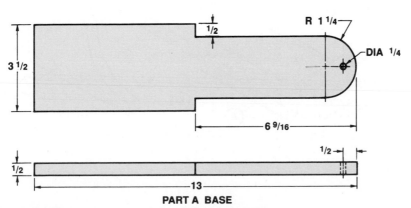

FIGURE 14–12 Detail drawings are made for each part of a product.

it is assembled (put together). Assembly drawings may be multiview or pictorial (picture) drawings. Pictorial drawings show the three dimensions (height, width, and depth) of a part or product with one view, Figure 14–13.

Another common assembly drawing is the exploded assembly, Figure 14–14. In an exploded assembly, all the parts of a product are drawn as if they had exploded from the finished product. Each part appears as a pictorial drawing. Lines are drawn between parts to show how the parts are assembled. This type of drawing is used in owner's manuals for products, such as bicycles and appliances, that are put together at home by the consumer.

Step 8: Creating a Bill of Materials

Creating a bill of materials is the last step in the design process. A **bill of materials** lists all the parts, components, and hardware needed to make the product. It is used to tell finance workers what materials must be ordered so the product can be made. The bill of materials may also be used to estimate the costs of a product, Figure 14–15.

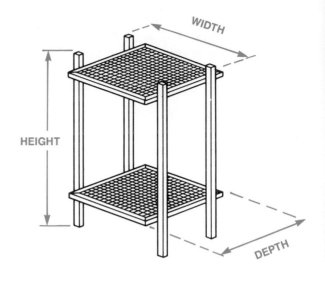

FIGURE 14–13 Pictorial assembly drawings show the height, width, and depth of a product in one view.

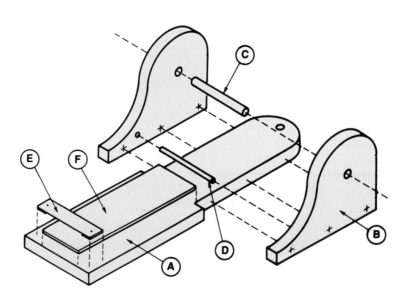

FIGURE 14–14 An exploded assembly drawing shows how all the parts of a product go together.

XYZ Manufacturing Co.

PRODUCT CODE	X-125-A	BILL OF MATERIALS		MODEL	08
QUANTITY	450			DRAWING	125

PART NO.	QTY.	PART NAME	DESCRIPTION	COST	
				UNIT	TOTAL
1	1	Base	walnut, $1^1/8 \times 4^1/4 \times 10^3/8$	1.64	1.64
2	2	Sides	walnut, $5/8 \times 3^3/8 \times 7^7/8$	.48	.96
3	1	Case Top	walnut, $1/2 \times 4 \times 10^3/8$	.62	.62
4	1	Middle Top	walnut, $1/2 \times 3^5/8 \times 9^3/8$	.50	.50
5	1	Cap	walnut, $1/2 \times 3 \times 8^1/2$	.35	.35
6	4	Frame	walnut, $1/2 \times 1/2 \times 8$	.06	.24
7	2	Retainer	pine, $1/2 \times 1/2 \times 7$	.02	.04
8	1	Handle	#H-750, brass finish	.75	.75
9	1	Clock Face	#547CF, painted, Clock Co. Intl.	1.25	1.25
10	1	Movement	#629MV, battery, Clock Co. Intl.	2.50	2.50
11	1	Glass	$1/8 \times 8^3/16 \times 7^7/8$, City Glass	.28	.28
APPROVED BY: _____			DATE: _____	TOTAL	$9.13

FIGURE 14–15 The bill of materials lists all the parts needed to make a product.

Summary

Designers create product ideas. The design processes usually begin with research and development. Research can be pure or applied. Pure research gathers new ideas without trying to find any practical uses. Applied research takes the new ideas gained through pure research, and tries to find practical uses like new product ideas.

The steps in the design process include the following: listing design standards, brainstorming ideas, sketching ideas, making mock-ups, making a prototype, getting management approval, drafting working drawings, and creating a bill of materials. These steps do not occur in a step-by-step order. Rather, they may occur in any order. All of the steps are important to the design of a quality product.

Once a company finishes the design processes, they are ready to begin the production engineering processes. Production engineering involves setting up a production line to manufacture a product.

Biotechnology: Human Factors Engineering

Biotechnology is a combination of biology and technology. Human factors engineering is a form of biotechnology. It designs products to help make life easier for people. Human factors engineering has developed some amazing technology! For example, people born with a hearing loss can be helped. By wearing very small devices within their ear canal they can regain the ability to hear.

Your ear actually hears noise by detecting vibrations in the air with small hairs inside the cochlea. The cochlea is the part of the ear canal where the eardrum is found. The nerve cells inside the eardrum are connected to these small hairs that change vibrations into signals. These signals are sent to your brain and processed as sound. People born with hearing loss usually have the nerve cells intact, but have not developed any of the tiny hairs that transmit the sound. The Cochlear Corporation has developed a "Mini System 22" that uses tiny electrodes implanted in the cochlea area to help transmit sound.

A small receiver is worn on the outside of the ear which is connected to a computer processor. The computer processor is about the size of a personal cassette player and is worn on the body. When a sound is detected it is sent to the computer processor. The processor compares the sound with known sounds. If a match is made, it stimulates the implanted electrodes to "reproduce" the sound in the ear. The nerve cells receive this information and send it to the brain, where it is "heard" by the individual.

The unique feature of this system is that it is programmable. If one of the twenty-two electrode implants causes discomfort, it can be shut down. Also, if an incoming sound is not recognized by the processor, then it doesn't send any message to the ear.

This reduces any unwanted confusion for the wearer.

Currently, Human factors engineering research is being completed for an implanted "artificial vision" system that would aid vision-impaired people. Another system is in the development stage for people who suffer from paralysis or an inability to move their limbs (legs, arms, hands, and feet). More research is needed to find workable solutions to these problems. Who knows, your generation may be the first to solve these problems!

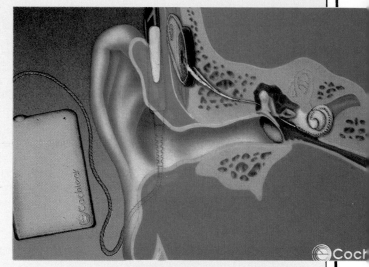

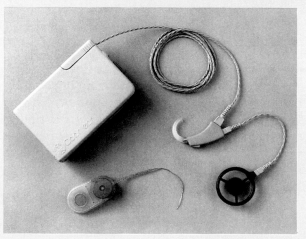

DISCUSSION QUESTIONS

1. Describe the function of a designer. Identify the basic processes performed by a designer.
2. What is the difference between research and development?
3. What are the eight steps in the design process? Describe each step.
4. What are the design standards described in this chapter? Which do you think are most important?
5. What is the difference between a mock-up and a prototype?
6. What are the two different types of drawings included in a set of working drawings?
7. How are computers being used to reduce the time needed to make new product designs?

CHAPTER ACTIVITIES

 ## Sketching Product Ideas

OBJECTIVE

An important step in the product design process is sketching product ideas. During this stage, designers often draw sketches to record their product design ideas.

In this activity, you will sketch your ideas for a product to be manufactured by your class.

EQUIPMENT AND SUPPLIES

1. Drawing and sketching tools
2. Grid paper or drawing paper
3. Design standards worksheet (from your teacher)
4. Water colors or magic markers

PROCEDURE

1. As a class, decide what type of products you will want to manufacture.
2. Study similar products in local stores or catalogs. If possible, cut out photos of products you would like to make. The more products you study, the more ideas you will have.
3. In small design groups or as a class, list the design standards for your product. Use the design standards worksheet supplied by your teacher. Refer to Figure 1 as you fill out your worksheet. Your teacher may set the standards.
4. Two of the most important standards may be ease of manufacture and costs. Pay special attention to these standards.
5. Once you have a list of standards, begin to create thumbnail sketches of your ideas. Make your sketches freehand. Use grid paper to make the sketching easier.
6. Meet with the members of your design group and review thumbnail sketches. Pick the best ideas.

DESIGN STANDARDS WORKSHEET	
DESIGN STANDARDS	**TRADEOFF DECISIONS**
Function	How will the product operate, be used?
Form	What size and shape will the product be? How will people use the product? What Consumer safety factors must we consider?
Appearance	Will the product have an appealing appearance?
Ease of Manufacture	How many parts are needed? What tools and materials are needed?
Standard Parts	Will standard materials and supplies be used?
Quality	How long will the product last?
Costs	What is the product's price range?

FIGURE 1. Refer to this example as you fill out your worksheet.

7. Meet with the class production planners or your teacher to make sure the product can be made with the tools and machines in your lab.
8. Once the design has been approved by your teacher, draw renderings of the final product design. Be sure to include all details, such as color and hardware.
9. Your renderings may be used in the next activity, "Making Mock-ups and Prototypes."
10. If you do not have enough time to finish your renderings in your production class, ask drafting or art students to make the sketches.

MATH AND SCIENCE CONNECTIONS

To sketch product ideas to scale, try using graph paper. Make each square on the graph paper equal to one inch or some other unit of measurement. This can be easier than sketching a large product full-sized.

RELATED QUESTIONS

1. Besides looking at other products, what can a designer do to find ideas?
2. Which design standard do you think is the most important?
3. Why are the first sketches made by designers called "thumbnail" sketches?
4. Do you think sketching can be improved with computers? Why?

 # Making Mock-ups and Prototypes

OBJECTIVE

After product sketches are made, designers create mock-ups and prototypes. A mock-up is a scale model that does not function. It is used to check the appearance of a product. A prototype is the first fully functional, full-size model of the product. It is made with the actual materials and parts that will be used in the final product. Engineers conduct tests with the prototype to make sure it functions properly.

In this activity, you will make mock-ups and a prototype of your product ideas. Based on these models, management will decide if the product should be manufactured.

FIGURE 1. Designing products is an important part of manufacturing.

EQUIPMENT AND SUPPLIES

1. Product design sketches and renderings
2. Mock-up modeling materials:

   - clay
   - cardboard
   - styrofoam
   - plywood

3. Design and modeling tools:

   - measuring scales
   - knives, razors, other cutting tools
   - hot glue guns, tape, other fastening tools
   - water colors, magic markers, paints

4. Prototype manufacturing tools (similar to the tools and machines used to manufacture products)
5. Safety glasses

PROCEDURE

1. Choose the best product design sketches and renderings developed in the "Sketching Product Ideas" activity.
2. Work in small groups or alone to make a mock-up of your design. Use easily worked materials, such as clay or cardboard.

3. Mock-ups are usually smaller than the actual product. If the product is small, the mock-up may be made full size. Remember, mock-ups do not have any real working parts.

4. Hold a meeting with other students who made mock-ups. Each student should present his or her mock-up to the class. As a class, pick the best product design.

5. Produce a prototype of the best product design. Remember, a prototype is fully functional and full-size. The prototype should be made from the actual product materials. All operating parts should work.

6. NOTE: Your teacher may choose a small group of students from the class to make the prototype. This frees other students to make working drawings, conduct market research, or begin production planning processes.

7. After the prototype is complete, test it to make sure all the parts work as planned.

8. Market research can be done with the prototype. Set up a table in the cafeteria during lunch. Ask students and teachers if they would buy the product and how much they would be willing to spend.

9. Finance workers in the class could try to decide the cost of making the product. They could also try to decide how much profit would be made.

10. Hold a meeting with management or your teacher. Decide if the product should be mass-produced.

11. If management gives their approval, create a set of working drawings and a bill of materials.

RELATED QUESTIONS

1. Why are mock-ups made smaller than the finished product? Why are clay and cardboard used?

2. What are the differences between a mock-up and a prototype?

3. How well did your prototype meet your design standards?

4. How can computers be used to reduce the time need to make mock-ups and prototypes?

 Product Design Brief

OBJECTIVE

Managers often ask designers to create a new product. They can communicate their idea for a new product to designers in a design brief. A design brief is a short description of the product. It includes design standards, sizes, materials, costs, and other factors.

In this activity, you will design a simple product following a design brief. You will role play a designer by sketching ideas, and making mock-ups and a prototype. Your prototype will be tested for its function.

EQUIPMENT AND SUPPLIES

1. Product Design Brief (see page 442)
2. Product Design Standards (see page 443)
3. Grid paper or drawing paper
4. Drawing tools
5. Materials (to be picked later)
6. Tools (to be picked later)

PROCEDURE

1. Work in small design groups of two or three students.
2. Read the Product Design Brief.
3. Follow the design ideas on the Product Design Brief.
4. Work with your group and brainstorm your ideas for the product.
5. All members of the group should make thumbnail sketches of their ideas.
6. Pick the best ideas from the sketches and make one improved design.
7. Make renderings of the improved design.
8. As a group, make a mock-up of your best idea.
9. As a class, pick the best mock-up ideas. Your teacher may hold a contest to decide which design is "best."
10. Your teacher will ask one student, one group, or the entire class to make a prototype of the best design.
11. Your class may decide to mass produce the product.

MATH AND SCIENCE CONNECTIONS

It is important in this activity to consider the effect of heat on the materials you use. The materials you choose for this product must stand temperatures of 500 degrees Fahrenheit. What effect will this temperature have on the materials? Pick a material that will not burn, catch fire, or melt at 500 degrees.

RELATED QUESTIONS

1. Did you have any problems working with the members of your design group?
2. What was the most important stage of the design process for this product?
3. Who do you think created the design standards for this product?
4. When your class picked the best design ideas, what standards were used most often?

CHAPTER 15

Production Engineering

OBJECTIVES

After completing this chapter, you will be able to:

■ Describe processes production engineers use to plan for the manufacture of products.

■ Identify and describe the various charts and sheets engineers use to plan for production.

■ Explain the uses for a plant layout, materials handling systems, and tooling.

■ Discuss the importance of planning for quality control.

■ Plan for the manufacture of a given product by completing various production engineering processes.

KEY TERMS

AGV	Materials handling	Productivity
Bottlenecks	Methods engineers	Product layout
Debugging	Operation process chart	Quality control engineers
Fixed-position layout	Operation sheet	Tooling
Flow process chart	Pilot run	Tooling up
Manufacturing engineers	Plant layout	Work station
Materials flowcharts	Process layout	

Planning for production is done by three different types of engineers — methods engineers, manufacturing engineers, and quality control engineers. Each of these engineering groups is important. It is their job to plan the most efficient production system possible. To do this, they must find the best combination of tools, machines, processes, and people. Two key factors in production engineering are productivity and costs. Engineers try to improve productivity and lower costs. **Productivity** is found by dividing the number of products made by time, usually worker hours, Figure 15–1.

Methods Engineering

Methods engineers plan the methods or processes needed to make the parts in a finished product. These planners create three charts to show how a product is made: the **operation process chart**, the **flow process chart**, and the **operation sheet**.

The operation process and flow process charts use symbols created by the American Society of Mechanical Engineers (ASME), Figure 15–2. These symbols are for the five processes needed to make any product. Operations include material processes such as drilling, cutting, and nail-

208

$$\text{PRODUCTIVITY} = \frac{\text{PRODUCTS MANUFACTURED}}{\text{TIME (WORKER HOURS)}}$$

FIGURE 15–1 Productivity is a measure of manufacturing efficiency. It is the number of products made per worker hour.

ing. Inspections are checks to make sure that product parts match the sizes on the working drawings. Transportations are the moving of parts from one place to the next. Delays are times when parts are waiting for the next process. Storages are when parts or products are held and protected. Methods engineers use these symbols during their planning.

Operation Process Chart

The operation process chart lists all the processes needed to make one product. The operation process chart in Figure 15–3 is for a yo-yo.

◯	OPERATION	Object is changed in any of its physical or chemical characteristics, is assembled, or is disassembled.
▢	INSPECTION	Object is examined for quality or quantity in any of its characteristics.
⇨	TRANSPORTATION	Object is moved from one place to another, except when such movements are part of an operation or inspection.
◗	DELAY	Object is held for the next operation, inspection, or transportation.
▽	STORAGE	Object is kept and protected against unauthorized removal.
◉	COMBINED ACTIVITY	When activities are performed concurrently or by the same person at the same work station, the symbols for those activities are combined.

FIGURE 15–2 The American Society of Mechanical Engineers designed the five symbols used on operation and flow process charts.

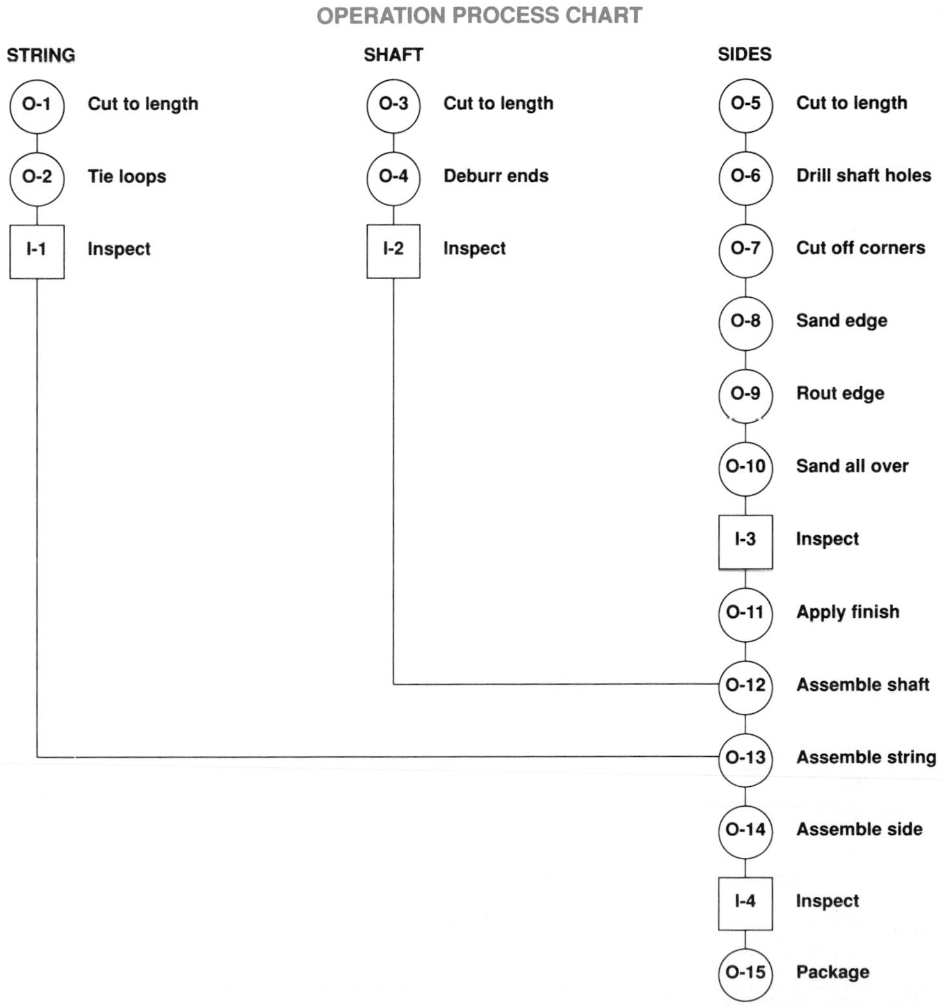

OPERATION PROCESS CHART

FIGURE 15–3 The operation process chart lists all the processes needed to make one product.

A yo-yo has three parts: the string, a shaft, and the sides. Notice the three columns on the chart. Each column lists all the operations and inspections done on each part. Transportations, delays, and storages are not put on an operation process chart. The bottoms of the columns are connected to show the assembly of parts into a finished product.

Operation process charts are used to identify the number of tools and people needed to make the product. Methods engineers can make several changes to this chart. The changes are made to improve productivity. To find the best productivity can take years of practice.

Flow Process Chart

The operation process chart provides only part of the information needed for production planning. The second chart created is the flow process chart. Several flow process charts are made. Each flow process chart lists all the processes needed to make one part of a product. The flow process chart in Figure 15–4 is for one side of a yo-yo. Every operation, transportation, inspec-

FLOW PROCESS CHART

SUMMARY	NO.	TIME		PRODUCT	Widget	DATE
◯ OPERATIONS	7			JOB/PART	Side #2	
⇨ TRANSPORTS	6			CHART BEGINS	O-1	
☐ INSPECTIONS	1			CHART ENDS	T-6	
▽ DELAYS	1	25		CHARTED BY		
▽ STORAGES	0			APPROVED BY		

NO.	SYMBOLS	TASK DESCRIPTION	MACHINE	TOOLING
O-1	●⇨☐▽	Cut to length	Table saw	Stop block
T-1	◯➡☐▽	Move to drill press		
O-2	●⇨☐▽	Drill shaft holes	Drill press	Drilling Fixture
T-2	◯➡☐▽	Move to bandsaw		
O-3	●⇨☐▽	Cut off corners	Bandsaw	Bandsaw Fixture
T-3	◯➡☐▽	Move to disc sander		
O-4	●⇨☐▽	Sand outside edges	Disc sander	Sanding Fixture
T-4	◯➡☐▽	Move to routing work station		
O-5	●⇨☐▽	Rout edges	Router	Routing Jig
O-6	●⇨☐▽	Sand all over	Finish sander	
I-1	◯⇨■▽	Inspect shape, size, routing		Gage I-1
T-5	◯➡☐▽	Move to finishing work station		
O-7	●⇨☐▽	Apply spray finish	Spray gun	Spray/Dry Rack
D-1	◯⇨☐▶▽	Delay for drying		Spray/Dry Rack
T-6	◯➡☐▽	Move to assembly work station		
	◯⇨☐▽			
	◯⇨☐▽			

FIGURE 15–4 The flow process chart identifies all the processes needed to make one part on a product.

tion, delay, and storage process is listed. This gives the methods engineer a more complete picture of the tools and people needed to make a product.

As with the operation process chart, engineers usually make several revisions to the flow process chart. With each revision, they try to cut out time spent in transportations, delays, and storages. These processes lower productivity.

Operation Sheets

An operation sheet details the work to be done at each **work station**. Work stations are the places in a production line where operations occur, such as machines and benches. Workers read the operation sheet to learn how to do their job. The operation sheet lists the steps needed to do the operation. Also listed are the machines, tools, and equipment needed. A drawing is given to show any special parts of the operation, Figure 15–5.

OPERATION SHEET

Operation No.:	O-5		Product:	T.D.V.-12

Operation Name: **Storage Hole Drilling**

Part Name:	**Block**
Part No.:	**1**
Quantity:	**50**
Stock/Material:	**Pine**

Tools/Equipment:	**Drill Press**
	3/4" Spade Bit
	Brush
Jigs/Fixtures:	**Jig O-5**
Supplies:	**Inspection Gage O-5**

Drawing of Major Details:
Center is 5/8" from end, on center

3/4" Dia. Spade Bit

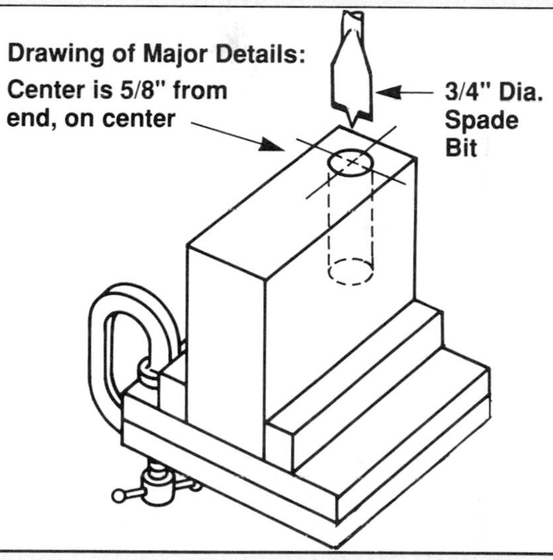

Operation Description:

Step No.	Description of Step	Materials	Machine	Tools and Equipment
1	Insert part in Jig O-5, turn on machine	Pine	Drill Press	Jig #2 3/4" Spade Bit
2	Drill hole 2.50 inches deep		Drill Press	Jig #2 3/4" Spade Bit
3	Remove part from jig, rotate 180° and drill other hole		Drill Press	Jig #2 3/4" Spade Bit Brush
4	Remove part from jig, inspect hole depth, diameter			Inspection Gage O-5
5	Place acceptable parts on conveyor			

FIGURE 15–5 The operation sheet details the work to be done at a work station.

Manufacturing Engineering Processes

Manufacturing engineers plan the layout of tools, machines, and people on a production line. Once the methods engineers have made the operation process chart and the flow process charts, the manufacturing engineer can begin work. Manufacturing engineers plan three things: plant layout, materials handling, and tooling.

Plant Layout

The **plant layout** is the way tools, machines, and other work stations are arranged. Manufacturing engineers draw the plant layout on a factory floor plan. Figure 15–6 shows a plant layout. Every machine and bench used is drawn to show its location on the factory floor. Plant layout drawings like this are called **materials flowcharts**. The lines and arrows drawn between machines and benches show the flow of materials

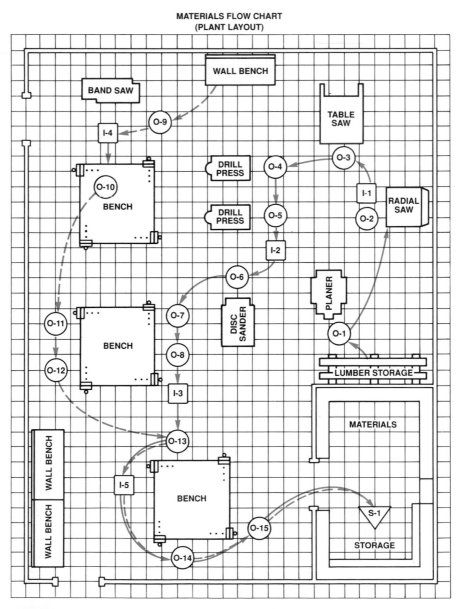

FIGURE 15–6 The flow of materials through a factory is drawn on a plant layout.

from one work station to the next. The manufacturing engineer tries to design a smooth flow of materials between work stations.

There are three types of plant layouts: fixed position layout, process layout, and product layout. Each system has certain advantages.

Fixed-Position Layout. The **fixed-position layout** is used to manufacture very large and complex products, such as airplanes and ships, Figure 15–7.

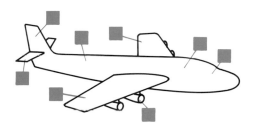

FIGURE 15–7 The fixed-position layout is used to make large, complex products like airplanes.

The main part of the product, such as an airplane body, remains in a fixed position on the factory floor. All the workers, their tools, and the airplane parts are brought to the main body. Before Henry Ford started mass production in the early 1900s, cars were made with fixed-position layout.

Fixed-position layout is used for custom or intermittent manufacturing systems. Highly skilled workers are needed. Portable tools and machines are used so workers can move them around the product.

Process Layout. In **process layout**, tools and machines used for similar processes are grouped together, Figure 15–8. All the saws are grouped in one area, all the lathes in another area, and so on. In a process layout, parts move through the plant in small batches or lots. A batch of parts is machined in one area. Then the batch is moved to the next area for more processing. The batch keeps moving from one area to the next until the product is complete. The three lines on Figure 15–8 show the material flow for three different parts.

The process layout is used with intermittent (batch) manufacturing. Skilled machinists are needed to run the machines in each area.

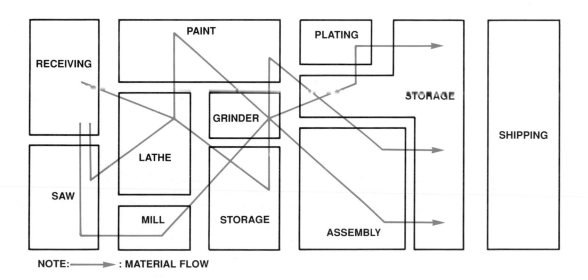

FIGURE 15–8 Machines that perform similar processes are grouped in a process layout.

Product Layout. The **product layout** has a smooth flow of materials from start to finish, Figure 15–9. Product layouts can be the most efficient type of layout.

Henry Ford used the product layout with his mass production system. Operations at each work station are broken down to their simplest parts. This creates a division of labor. Machines and work stations are arranged in the order of processes needed to make the product. Long conveyor belts are built between work stations to give a smooth flow of parts.

The product layout is used with continuous manufacturing. Unskilled or semiskilled workers are needed to perform the simple operations at each work station.

Engineering a Plant Layout. Manufacturing engineers can make many revisions to the materials flowchart. With each revision, they are trying to create a smoother flow of materials from start to finish. Because of the many revisions, they often use templates like those in Figure 15–10. The templates are cut out of heavy paper and placed on a drawing of the floor plan. The engineer can

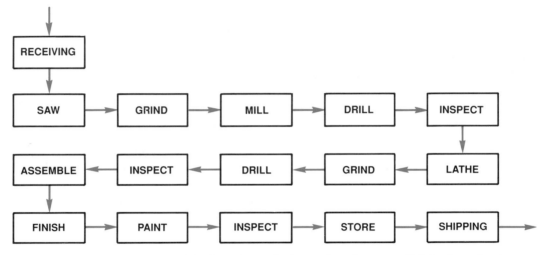

FIGURE 15–9 In a product layout, materials flow through the entire plant to a finished product.

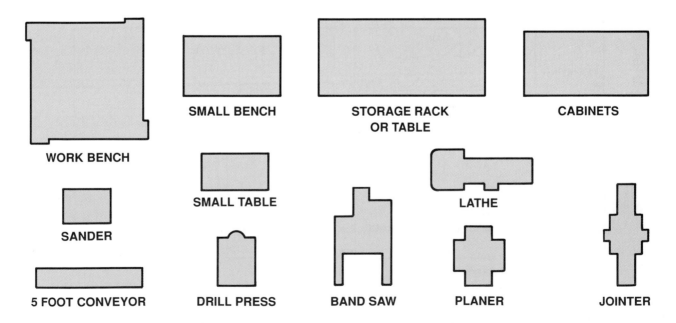

FIGURE 15–10 These templates can be moved around on the factory floor plan to engineer a plant layout.

move the templates around until a smooth flow is found.

For very complex or automated manufacturing systems, computer-controlled scale models are often used. The engineer in Figure 15–11 is checking the timing of events on a plant layout modeling system.

No matter what method is used to engineer a plant layout, the following factors must be considered:

1. **Safety.** The system should keep workers safe from accidents with machines.
2. **Utilities.** Work stations that need electricity, gas, and other utilities should be located close to a source.
3. **Ventilation.** Good ventilation must be provided for work stations where fumes or dust are produced. Painting, sanding, and welding areas are examples.
4. **Lighting.** Certain operations need extra lighting.
5. **Waste disposal.** Operations that make scrap materials must have a method of waste disposal. Good waste disposal is needed for good housekeeping. Good housekeeping is related to safety too.

FIGURE 15–11 Many manufacturing companies use computer-controlled models like this one to design and test a plant layout. *(Courtesy of Ford Motor Company)*

Materials Handling

Manufacturing engineers also plan the materials handling system. **Materials handling** is the transportation system used to move materials and parts through the plant. Conveyors and forklifts are examples.

The two types of materials handling systems are fixed-path and variable-path. Fixed-path systems, such as conveyors, rail cars, and elevators move materials on paths that cannot be changed. Variable-path systems, such as forklifts and trucks, move materials on paths that can be changed.

Fixed-path systems work better for continuous manufacturing or product layouts. Variable-path systems work better for intermittent manufacturing or process layouts.

Safety and Materials Handling. Safety is important for materials handling systems. Fixed-path systems should be located so they are not in the path of workers' movements. Moving parts, such as pulleys, belts, and motors, should be covered with guards. Safety shut-off buttons are also needed near work stations so workers can stop a system in case of an accident. Variable-path vehicles, such as forklifts, often have bright warning lights and loud bells.

Tooling

Tooling is special tools and devices used by production workers. Tooling helps the workers make accurate parts of high quality. Manufacturing engineers design the tooling.

Types of Tooling. The most common types of tooling include jigs, fixtures, patterns and templates. Jigs are used to guide the path of a tool on a part being processed. They are usually attached to the part being processed, or the part can be attached to the jig. Fixtures are attached to a machine and are used to position a part for processing. A fixture may be attached to a drill press table so holes are located accurately. A pattern looks like the finished product, but a little extra material is added. The patterns used to make molds for casting are an example. A template is a flat outline of a part. A template is used to guide the path of a tool during a layout or machining process. Drawing letters with a stencil template is an example.

AGV: Robotic Materials Handling Systems

One of the newest materials handling systems is the AGV. An **AGV**, which stands for automated guidance vehicle, is a robotic materials handling system. A computer controls the AGV as it moves through the plant. AGVs are programmed to follow fixed paths. The path can be identified with a wire in the floor or a line painted on the floor. Sensors on the AGV follow an electrical current in the wire. AGVs that follow painted lines have sensors that see the line. Some AGVs can be programmed to move on different paths by following lines of different colors.

FIGURE 1. This is a materials handling AGV (automated guidance vehicle).

Tooling Design Standards. Some of the standards manufacturing engineers follow when designing tooling include the following.

1. Ease of Use — The best tooling devices will be easy to use and easy to set up. Tooling with few moving parts is easy to use.

2. Accuracy — Accurate tooling will reduce human errors in measuring, laying out, and processing parts. Tooling must be made by skilled workers to be accurate, Figure 15–12. Being able to adjust tooling is also important. The stop in the drawing on the left in Figure 15–13 cannot be adjusted. The stop on the right can be adjusted by using a thumb screw. Another factor related to accuracy is chip relief. Tooling should be designed to eliminate chips when they occur during operations, Figure 15–14.

3. Safety — Engineers must always consider the safety of tooling. Tooling that clamps parts should not catch the fingers of an

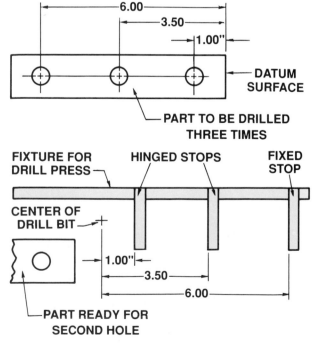

FIGURE 15–12 Locating drilled holes from one surface makes tooling more accurate.

FIGURE 15–13 Adding adjustment screws makes tooling more accurate.

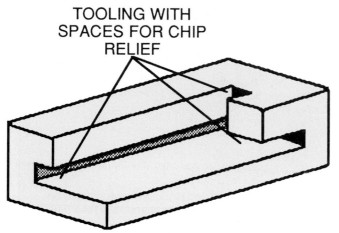

TOOLING WITH
SPACES FOR CHIP
RELIEF

FIGURE 15–14 Tooling devices should have openings to eliminate chips.

operator. Also, clamps should apply enough pressure to keep the part from being pushed out during cutting or drilling. Figure 15–15 shows several examples of clamping.

4. Speed—The time needed to set up and use tooling must be kept short. The best way to keep speed up is to keep tooling simple.

5. Costs—The cost of making tooling devices must not be too high. Simple tooling with few parts costs less to make and use.

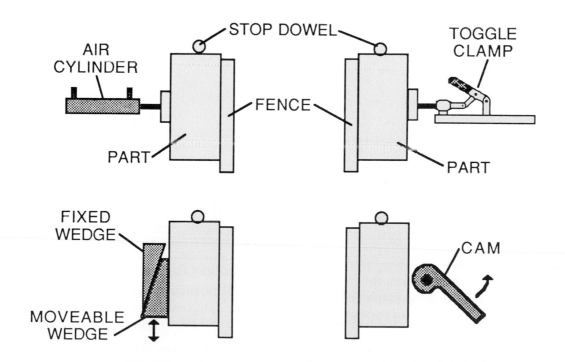

FIGURE 15–15 Here are four ideas for clamping devices.

The Importance of Tooling

For an example of the importance of tooling, consider the following situation. A company needs 100 of the part shown in Figure 1. Without tooling, a worker has to measure and lay out the location of each hole before it can be drilled. With tooling, the worker can position the part in a jig or fixture and drill the holes. The measuring and layout operations are built into the tooling.

What are the results? First, any human errors in measuring and laying out four different holes 100 times are reduced. Second, since human errors are reduced, product quality will probably increase. Third, the time spent in measuring and layout is reduced. This improves productivity and lowers costs. Fourth, without tooling the worker must be skilled in measurement and layout. With tooling, a less skilled worker who receives less pay can still do the job.

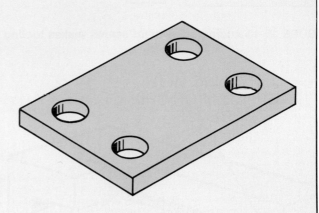

FIGURE 1. Locating and drilling the four holes on this part can be more accurate and take less time with the use of tooling.

Quality Control Engineering

The third type of production planning is quality control engineering. **Quality control engineers** make sure the finished product matches the working drawings. This is done by inspecting (checking) parts and products.

Quality Inspection

Quality must be built in during the manufacture of a product. It cannot be inspected into a product. Inspections are used to spot problems so they can be fixed.

Inspection involves three jobs: (1) measuring a part, (2) comparing the size of the part with the sizes given on working drawings, and (3) making a decision about the quality of the match. There are three possible decisions: accept, rework, or reject. To accept a part means it matches the sizes on the working drawings. Rework means the part does not match the sizes, but it can be fixed. A reject part does not match sizes, cannot be fixed, and must be junked.

Inspection Gages

Quality inspection is made easier by using gages. The two types are indicating gages and fixed gages. Both types are used to measure parts.

Indicating Gages. With an indicating gage, an inspector measures a part, reads a size off the gage, compares the size with the working drawings, and makes a decision to accept, rework, or reject the part. Common indicating gages include rulers, micrometers, and dial indicators.

Fixed Gages. When an inspector measures a part with a fixed gage, it either matches the right sizes or it does not. Fixed gages have the minimum and maximum tolerance built in. Go/no-go gages are examples, Figure 15–16.

Fixed gages reduce the chance of human error in reading measurements. They also reduce the time needed for each inspection. Fixed gages must be made accurately to match the sizes being measured. Adding a means of adjusting is often helpful. Screws can be added to allow the inspector to adjust the gage for accurate measurements.

Tooling Up: Preparing for Manufacturing

Tooling up is the final production planning process. All the workers, the plant layout, tooling, and materials handling systems are put together. Engineers and production workers lay out the plant, set up work stations, and install jigs and fixtures. Once the tools and equipment are in place, the company has a pilot run. A **pilot run** is a practice session when the production line is tested to make sure all systems are working. During the pilot run, a number of things are checked. These include the timing of events and the accuracy of tooling. Engineers want to find and remove any bottlenecks that might develop. A **bottleneck** is a point in a production line where parts back up and are delayed. Bottlenecks add time to the production process and lower productivity. The process of tuning up the production system is called **debugging**. Once the system has been debugged, the actual manufacturing of products begins.

Summary

Production planning includes three types of engineering: Methods engineers plan the operations needed to make a product. Manufacturing engineers plan the plant layout, materials handling system, and tooling. Quality control engineers plan for the inspection of parts and products. Each of the production engineering processes tries to improve productivity by reducing time and costs.

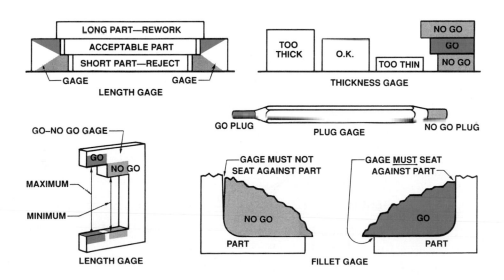

FIGURE 15–16 Here are fixed gages for quality inspection. Notice the red (no-go) and green (go) colors on the gages.

Quality Inspection by Laser

Lasers are being used in manufacturing for quality control inspections. The laser light can be used to detect flaws or mistakes in parts.

The simplest method is shown in Figure 1. The laser lights up the part. Lenses focus the light on a sensor. The sensor sends information about the light pattern to a computer. The computer compares the light pattern with programmed quality standards.

Figure 2 shows a laser being used to check the position of parts in an assembly. Laser light bounces off the part and goes through the lense onto a computer sensor. The location of the reflected light landing on the sensor tells the computer if the part is within tolerance (maximum and minimum sizes).

Laser inspection systems are very accurate in measurement. Many manufacturers are using laser inspection systems to improve product quality.

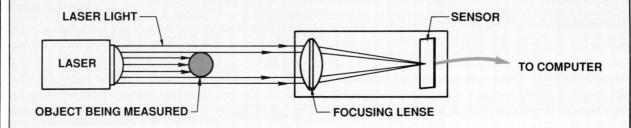

FIGURE 1. This is the simplest type of laser quality control inspection.

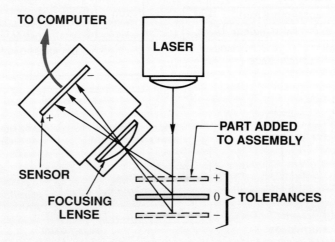

FIGURE 2. Quality control inspection of assembly can be done with a laser.

DISCUSSION QUESTIONS

1. What processes are involved in planning for production? What are the three different types of engineering processes discussed in this chapter?

2. What is the definition of productivity? How can productivity be improved?

3. What are the five symbols used on methods engineering charts? What do the symbols stand for? Who created the symbols?

4. Methods engineers create three charts. What are they? What are the differences among these charts? Why are the charts created?

5. What are the three types of plant layout? How are they the same or different?

6. What is quality control? What are the two types of inspection gages used in quality control?

CHAPTER ACTIVITIES

 ## Methods Engineering: Planning Processes

OBJECTIVE

Methods engineering involves planning all the steps needed to make a finished product. As you read in this chapter, methods engineers do their planning by creating charts. These charts give the engineer a graphic picture of the processes needed.

In this activity, you will plan the steps needed to make a product by creating an operation process chart, flow process charts, and operation sheets.

EQUIPMENT AND SUPPLIES

1. Working drawings for a product to be mass-produced
2. Graph paper
3. Blank flow process chart (from your teacher)
4. Blank operation sheet (from your teacher)
5. Drawing tools

PROCEDURE

1. Work in small methods engineering groups. This will help you divide the workload. Your class may be running some manufacturing engineering and quality control engineering processes at the same time. Each of these groups might need to work together to plan the production system.
2. You can make many charts and sheets. In general, the following are required:
    - ▪ One complete operation process chart
    - ▪ One flow process chart for each part in the product
    - ▪ One operation sheet for each important work station

3. Plan the steps needed to make each part on the product. Include all operations, inspections, transportations, delays, and storages.

4. Sketch a rough draft of the operation process chart. Include only the operations and inspections for each part.

5. Meet with the students in your work group. Review the rough draft operation process chart. Ask your teacher how the draft can be made better.

6. Draw the final operation process chart for the product.

7. One flow process chart is needed for each part. Your teacher may pick one student for each flow process chart to divide the workload. Use the blank flow process chart provided by your teacher.

8. Work with your teacher to decide which work stations need operation sheets. Work stations with simple processes may not need an operation sheet.

9. Remember three things when making the operation process chart, flow process chart, and operation sheets:
   - You may have to revise these sheets several times to plan the most efficient method of production.
   - Try to plan the most efficient system possible. Limit the number of delays, storages, and transportations.
   - Pick a simple product for your first planning efforts.

MATH AND SCIENCE CONNECTIONS

The major goal of methods engineering is to improve the efficiency of production. Production efficiency is called productivity. Productivity is measured with the following formula:

$$\text{PRODUCTIVITY} = \frac{\text{PRODUCTS PRODUCED}}{\text{TIME (WORKER HOURS)}}$$

RELATED QUESTIONS

1. Why do methods engineers create operation process charts, flow process charts, and operation sheets? Could a company manufacture a product without making these?

2. How are an operation process chart and a flow process chart alike and different?

3. What group of engineers made the symbols used on operation and flow process charts?

4. Why is it important for methods engineers to work closely with manufacturing engineering and quality control engineering?

Manufacturing Engineering: Plant Layout and Materials Handling Systems

OBJECTIVE

Two manufacturing engineering tasks are planning the plant layout and the materials handling systems needed. The plant layout should make for a smooth flow of parts through the production line. Materials handling systems can also help smooth the flow of parts. You can make simple materials handling systems for your lab.

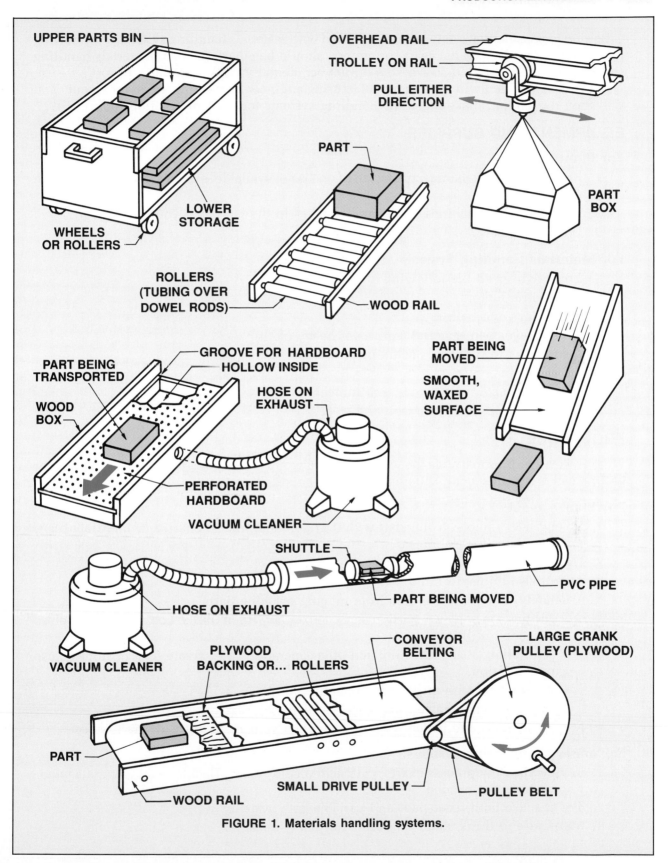

FIGURE 1. Materials handling systems.

UPPER PARTS BIN

WHEELS OR ROLLERS

LOWER STORAGE

PART

ROLLERS (TUBING OVER DOWEL RODS)

WOOD RAIL

OVERHEAD RAIL

TROLLEY ON RAIL

PULL EITHER DIRECTION

PART BOX

GROOVE FOR HARDBOARD

HOLLOW INSIDE

PART BEING TRANSPORTED

WOOD BOX

HOSE ON EXHAUST

PERFORATED HARDBOARD

VACUUM CLEANER

PART BEING MOVED

SMOOTH, WAXED SURFACE

SHUTTLE

PVC PIPE

HOSE ON EXHAUST

PART BEING MOVED

VACUUM CLEANER

PLYWOOD BACKING OR... ROLLERS

CONVEYOR BELTING

LARGE CRANK PULLEY (PLYWOOD)

PART

SMALL DRIVE PULLEY

PULLEY BELT

WOOD RAIL

Be sure to consider the safety hazards involved with materials handling systems. Any moving parts, such as pulleys, belts, gears, or motors, should be guarded. Also, materials handling systems should be located so workers do not accidentally walk into them.

In this activity, you will act as a manufacturing engineer by designing a plant layout. You will then design and make materials handling systems for your production line.

EQUIPMENT AND SUPPLIES

For Plant Layout:
1. Floor plan for the manufacturing lab (if one is not available, sketch a floor plan on graph paper)
2. Templates for machines, tools, and benches in the lab (from your teacher)
3. Drawing tools

For Materials Handling Systems
1. You might need the following:
   - electric motors
   - pulleys and belts
   - conveyor belt material (canvas or heavy cloth)
   - rollers (dowel rods, plastic pipe, conduit, coffee cans)
   - casters
   - various sizes of plywood and lumber
   - various sizes of angle iron and steel

PROCEDURE

Work in small groups to divide the workload. One group can engineer the plant layout. Other groups can work on several different materials handling systems.

For Plant Layout:
1. To engineer a plant layout, start with the operation process chart, a factory floor plan, and templates for machines.
2. Design a materials flowchart.
3. Keep these things in mind:
   - **Safety.** Keep workers safe from moving machine parts.
   - **Utilities.** Do not lay extension cords across the path of traffic. Locate machines close to needed utilities.
   - **Ventilation.** Place finishing and other operations that create dust or fumes close to ventilation.
   - **Lighting.** Be sure to provide enough lighting.
   - **Waste.** Provide waste cans where needed.
4. Move machines and benches into place when your materials flowchart is done.

For materials handling systems:
1. Review the flow process charts and materials flowchart. Decide what materials handling systems are needed.
2. Try to make both fixed-path and variable-path systems.
3. Make sure to think about the following safety hazards:

- A conveyor belt that moves too fast can shoot parts off the end and hit workers.
- Moving parts should be guarded.
- Safety shut-off switches should be provided.
- Variable-path vehicles should have warning lights or signals.

MATH AND SCIENCE CONNECTIONS

Conveyors are a common materials handling system in manufacturing. Manufacturing engineers must know the speed at which a conveyor travels. Conveyor travel speed is measured in the number of feet it moves in one second. This is called feet per second. If the conveyor travels too fast, the parts may shoot off the end of the conveyor or move past a work station too quickly. A fast-moving conveyor would be a safety hazard.

RELATED QUESTIONS

1. Which engineers are responsible for planning the plant layout and materials handling systems?
2. What factors should be kept in mind when making a materials flowchart?
3. What are the advantages and disadvantages of fixed-path and variable-path materials handling systems?
4. What are some safety factors related to materials handling systems?

CHAPTER 16

Marketing Processes

OBJECTIVES

After completing this chapter, you will be able to:

- Describe the role of marketing processes in manufacturing.
- List the five major marketing processes.
- Identify and describe the three different methods of market research.
- Describe the relationship between product planning and marketing.
- Identify various advertising methods used in marketing products.
- Describe the role of salespeople in manufacturing.
- Appreciate the importance of product service to a manufacturing company.

KEY TERMS

Advertising	Market research	Product service
Experimental studies	Mass media	Sales forecast
Historical studies	Product planning	Surveys

The hope that people will want to buy a product is too important to be left to chance alone. Too much is at stake for a company to focus only on making the product. A great deal of time must also be spent on trying to sell, or market, the product to consumers. Marketing processes involve many activities that occur during all stages of manufacturing. In a sense, all activities included in the process of marketing can be considered selling, Figure 16–1.

There are five major marketing processes: market research, product planning, advertising, selling, and product service, Figure 16–2.

Market Research

In order to sell a product, the manufacturing company needs to know several things about consumers. Finding out information about consumers is the job of **market research**. Some of the questions market research seeks to answer include:

- Who are the potential consumers?
- Where do the potential consumers live?
- What products and services do consumers want?

FIGURE 16-1 Coca-Cola, the world's best known trademark, is a symbol for the company in more than 155 countries. *(Courtesy of Helena Frost Associates, Inc.)*

- What products and services will consumers buy?
- How much money will consumers pay for the product?

The answers to these questions are very important for the manufacturing company. There must be a consumer who is willing to buy the product being made if the company is to stay in business.

Market researchers conduct studies to learn the answers to the questions about consumers. Finding out why a person chooses one product over another is a difficult task. To make their job easier, market researchers do a number of different kinds of studies. The three most common

types of market research studies are historical studies, experimental studies, and surveys. **Historical studies** look at how well similar products that were made before have sold. If a product was a good seller for the past few years, it may sell even better today. **Experimental studies** are done when a company is thinking about making and selling a new or different product. For example, a toy company may make several different prototypes of a new toy. Each toy prototype may be slightly different in color, appearance, or function. Children are observed playing with the toys, Figure 16-3. Researchers watch

FIGURE 16-3 Toy manufacturers often make several prototypes of a product and observe children playing. Experimental studies like this can tell the company which toys will sell. *(Courtesy of Fisher-Price)*

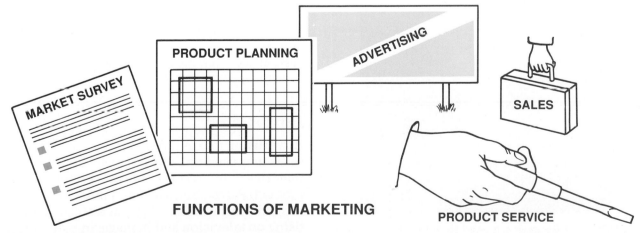

MARKET SURVEY

PRODUCT PLANNING

ADVERTISING

SALES

PRODUCT SERVICE

FUNCTIONS OF MARKETING

FIGURE 16-2 Marketing involves these five processes.

the children playing and note which toy is liked best. This type of experiment can give the company a very good idea if the new product will sell. **Surveys** are a very popular type of market research. A survey usually includes a set of questions written on a piece of paper, Figure 16–4. Potential consumers of a new product are asked to fill out the survey by answering the questions. The questions on the survey are often like the questions mentioned on pages 226–227. The company is trying to decide what products consumers want, which products they will buy, and how much they are willing to pay.

In addition to studying consumers, market research also studies other manufacturing companies. It is important for a company to know what products their competitors are planning to make. Beating the competition to the market with a new product is very important.

Product Planning

Product planning makes sure new products are added, old ones removed, and questionable ones

changed. Product planners work closely with market researchers to develop **sales forecasts** for each product. A sales forecast is a guess or prediction about how many products will be sold in the future, Figure 16–5. The information learned with a survey or market research experiment is very important to product planning and sales forecasts. If many consumers say on a survey that they want a new or different type of product, the company should try to make and sell that product. It is important for companies to plan for the future and try to beat the competition to the market. Product planners work closely with research and development. When consumers say they want a certain type of product, that information must be passed on to R&D so prototypes can be made and tested.

FIGURE 16–5 Market researchers develop sales forecasts of how many products may be sold in the future. *(Courtesy of W.W. Grainger, Inc.)*

FIGURE 16–4 Surveys are used to determine who will buy a product and why. *(Courtesy of Field Facts)*

Advertising

Advertising makes consumers aware of the products they can buy. This is usually done in a way that tries to interest the consumer in buying the product. Advertisers can use a number of media to send their message. We have all seen advertising on television and in magazines and newspapers. Companies who want to sell products buy time on television and space in magazines and newspapers. This, in turn, helps to pay for the programs and publications themselves. The most important part of advertising is good communication.

Some of the advertising methods used by manufacturers include:

- television
- radio
- newspapers
- magazines (general magazines for all people and special magazines just for children, women, men, or other groups)
- billboards
- direct mail
- free samples

Probably no manufacturer uses all of the methods. They try to pick those that would most likely reach their potential consumers, Figure 16–6. For example, a toy manufacturer might use Saturday-morning television to advertise its products. A company that manufactures sports equipment might advertise in special sports magazines. The advertising method should reach the potential consumers. Products needed by everybody, like food and clothes, are often advertised in the **mass media**. Mass media includes television, radio,

newspapers, and general magazines. Whatever method is used to advertise a product, it should attract the attention of consumers, Figure 16–7. Making consumers interested in buying the product is the goal of advertising.

FIGURE 16–6 Advertising is aimed at the potential consumers of a product. *(Reproduced with permission, © PepsiCo, Inc., 1989)*

FIGURE 16–7 Good advertising will represent the product in a way that will attract the attention of consumers. *(Courtesy of Klein Bicycle Corporation)*

One type of advertising that most products use is packaging, Figure 16–8. The package in which the product is sold is an effective form of advertising. Often, consumers can identify a certain brand or product just by looking at the package. Packaging must protect the product; at the same time, attracting consumer attention and advertising can also be achieved with packaging.

Selling

All activities in marketing can be called selling. It is important to remember that some workers will actually visit potential consumers and "make the sell." Salespeople make personal contact with consumers and take orders for the product. Selling often occurs after the product has been manufactured and advertising is going on. It has been found that nothing works as well in

FIGURE 16–8 Good packaging will protect and advertise the product.

Marketing for Youth: Not Child's Play

Marketing for preteens can no longer be considered child's play. The money made by marketing products for young people has increased over the years. This has become a fact of business that will continue well into the 1990s. Households with older parents spend more money on their children. And young people have been spending more money in recent years.

The company that has been the pioneer of advertising directed at young people for the longest time is McDonald's. The fast-food giant has seen children as the key to their long-term success for years. Their marketing approach has been based on three ideas:

1. Advertising to children through such devices as fun-theme meal kits and parties

2. Using advertising to get families to go to their restaurants

3. Viewing children as the grown-up consumers of tomorrow

It is the idea of viewing children as future consumers that makes the most sense. The U.S. Army, Navy, Marines, and Air Force have begun advertising in magazines such as *Boy's Life* and *Scholastic*. This way, they can appeal to young people before they begin to make life decisions. Planting seeds in the minds of young people may be the key to long-term success for any product or service.

As marketers become more aware that children are the consumers of tomorrow, more and more advertising will focus on them.

selling as the personal touch. Many customers make their final product choice based on how they feel toward the salesperson, Figure 16–9.

It is important that a manufacturing company select and train good salespeople. Companies view the following as the role of salespeople:

1. Salespeople represent the goal of all the planning and effort that went into making the product. If the salespeople fail to sell the product, then the whole company has failed.

2. Salespeople represent the company. If consumers don't like the salespeople, they may not like the company either.

3. Salespeople usually work alone and must make decisions on their own. They must have good decision-making skills. They must also understand the company and its products.

4. The job of selling can be expensive. Salespeople often travel from consumer to consumer.

FIGURE 16–9 Consumers often base their purchase decisions on how they feel toward the salesperson. (Courtesy of Walgreen Company)

Product Service

Product service is important to the long-term success of any company. The job of product service is to make sure the product works correctly and

that the consumer knows how to use it. Product service helps ensure repeat business. If consumers are happy with the way a product works, they may buy another in the future, Figure 16–10. Some companies use salespeople for service, while others have a special service staff. For more complex products, service is even more important. Special industrial equipment requires close attention and service, Figure 16–11.

FIGURE 16–10 Product service makes sure the product works correctly. (Courtesy of Radio Shack, a Division of Tandy Corporation)

FIGURE 16–11 Complex products like this industrial machine require close attention and service. (Courtesy of Simmons Machine Tool Corporation)

Summary

Marketing is a very important part of manufacturing. The goal of marketing is to sell the product. There are five major marketing processes: market research, product planning, advertising, selling, and product service. Market researchers study consumers and other manufacturing companies. They try to decide what products consumers will buy with surveys and other studies. They also look at what products other manufacturers are planning to sell. Product planning involves making forecasts of future sales and working with research and development to plan for new products. Advertising makes consumers aware of the product. It also tries to get consumers to buy the product. Selling is the job of the salespeople. These may be the most important people in the company. If they cannot sell the product, the company may go out of business. Product service makes sure the product works as the consumer expects. Each part of marketing is important to the company goal of making and selling products to consumers.

DISCUSSION QUESTIONS

1. What are the five major marketing processes?

2. Why is market research an important part of the manufacturing system?

3. Why would a company conduct market research before it manufactures any products?

4. Why is the role of the salesperson an important one for the marketing of a product?

5. What is product service, and why is it an important part of marketing?

6. Which part of marketing interests you most? Explain why.

7. Have you ever been asked to complete a survey? If so, describe it.

CHAPTER ACTIVITIES

 ## Surveying the Market

OBJECTIVE

It has become more and more important for a manufacturing company to know who their customers are. Making products is very expensive. Before a company decides to make a product they usually make sure there are consumers who will buy it.

In this activity, you will design a market survey to answer specific questions about products your class might manufacture.

EQUIPMENT AND SUPPLIES

1. Sample market survey (from teacher)
2. Product idea or prototype that may be mass produced
3. Paper and pencil
4. Typewriter or computer word processing system
5. Copying or duplicating machine

PROCEDURE

1. Break into small groups. Each group will develop a survey. Later, the class will combine the best parts of each survey into a final survey.
2. Review the sample market survey given by your teacher. Read the questions on the survey to get an idea of what is asked on a survey.
3. Think of the questions you would want to ask consumers about your product design idea or prototype. Some of the questions you will want to answer might be:
   a. How many consumers will buy the product?
   b. How much are they willing to pay?
   c. What do they want the product to look like?
   d. Who is the typical buyer?
   e. How can we reach more of the typical buyers?
   f. What would make our product more desirable?
   g. What changes would consumers like to see in our product idea?
4. For each question, decide on multiple choice selections which will give you the exact single answer you need. For example, suppose you need to know what price people will pay for a banana slicer. You could ask, "How much would you pay for this slicer?" Or you could ask, "If you bought this slicer, how much would you be willing to pay? $3–$5? $6–$10? $11–$20?" The second question limits the type of answer you will get.
5. Do not ask personal questions. Also, do not ask for the person's name to be written on the survey.
6. When each group has developed their survey, get the entire class together. Your teacher will lead a discussion to pick the best survey questions.
7. Compile one survey for the class. If available, use a computer word processing system to make the survey.

8. Think about putting a picture or drawing of the product on the survey.

9. Duplicate copies of the survey and distribute them in the cafeteria, study hall, or other classes.

10. Collect the surveys on a given date. Study the results of the survey and decide how to sell your product to the consumers.

MATH AND SCIENCE CONNECTIONS

Survey results can be reported as totals, such as "15 of 30 people want x, y, and z in their products." They can also be reported as percentages; "50 percent of people want x, y, and z in their products." Both of the statements (15 of 30 and 50%) report the same information. To calculate percentages, use this formula:

$$\frac{\text{NUMBER STATING A CERTAIN PREFERENCE}}{\text{TOTAL PEOPLE WHO WERE ASKED}} \times 100 = \text{PERCENTAGE}$$

RELATED QUESTIONS

1. Why do manufacturing companies conduct surveys?

2. Which type of question is better on a survey: open-ended, where any answer can be written down, or multiple-choice questions? Why?

3. Why do you think you should not ask personal questions or have people write their names on the survey?

4. Is there a better way to find out what consumers think of your product ideas than using a survey?

 Package Design

OBJECTIVE

Most products need some kind of packaging. Package design is a key element of marketing. Manufacturers know that attractive packaging can serve to advertise the product.

In this activity, you will design a package for one of these products:

- Brite White toothpaste
- Johnson's Double Fudge Cookies
- X-TRA Long Life Light Bulbs (they come with a five year guarantee)

EQUIPMENT AND SUPPLIES

1. Paper, cardboard, or index card material (various sizes)

2. Scissors, ruler, color markers

3. Clear tape, glue, or glue stick

4. Drafting tools or computer graphics system

5. Illustration board or foam board

6. Transfer lettering

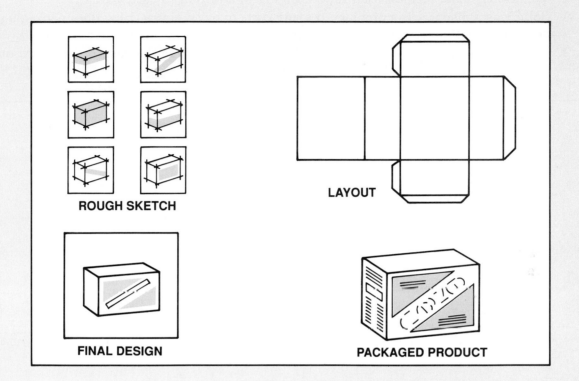

ROUGH SKETCH

LAYOUT

FINAL DESIGN

PACKAGED PRODUCT

PROCEDURE

1. Select one of the products to package.
2. Bring in several packages from home that might be used for your product. Take the packages apart carefully to see how they were made.
3. Make a list of the information that should be included on the package.
4. Make several rough sketches of the design of the package.
5. Choose your best idea and make a final sketch. Your final design should make the consumer want to buy the product.
6. Use drafting equipment or a computer system to create a full-size layout on paper. Cut, fold, and assemble. Check for accuracy and make changes as needed.
7. Use light cardboard or index card material for the final package. Add color and graphic images. Use transfer lettering for words on the package.
8. Make a display of your rough sketches, final design, and the finished package on a large sheet of illustration board or foam board.
9. Share your package design ideas with the class.

RELATED QUESTIONS

1. What are several important purposes of packaging?
2. What other materials are used for packaging in addition to cardboard?
3. Solid waste disposal is a problem in many parts of our country. How does packaging add to the problem? Can you make your package from recycled materials?
4. Find the cost of your package. Use $.03 per square inch as the cost of cardboard or index card material.

CHAPTER 17

Money and Manufacturing

OBJECTIVES

After completing this chapter, you will be able to:

- Describe the importance of money in manufacturing.
- Identify the four responsibilities of the finance department.
- Describe how the finance department determines how much money is needed for a company.
- Explain the three methods manufacturing companies use to obtain money.
- Calculate a break-even point for a product.
- List the steps performed to calculate the pay for a worker.
- Describe the various reports created by the finance department.

KEY TERMS

Assets	EBIT	Profit
Balance sheet	Fixed costs	Purchase order
Bill of materials	Gross pay	Purchasing
Bonds	Income statement	Stock
Bottom line	Interest	Unit costs
Break-even analysis	Liabilities	Variable costs
Budget	Loan	Vendor
Dividends	Net pay	

When a manufacturing company decides to go into business, it soon learns the importance of having enough money. Money is a very important input in any system of technology, Figure 17–1. In manufacturing, money pays for all the inputs needed to design, produce, and sell a pro-duct. Money pays for workers' salaries, tools and machines, utility company energy, and informa-tion, Figure 17–2A–D. Manufacturing companies even need money to get more money. After all, "It takes money to make money."

The goal of manufacturing is to produce pro-

FIGURE 17–1 Money is a very important input for manufacturing. *(Photo by Paul Meyers)*

FIGURE 17–2A Workers must be paid a salary. *(Courtesy of NASA)*

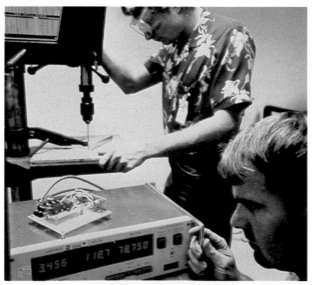

FIGURE 17–2B Money is used to buy tools and machines. Here are a low-cost drill press and an expensive piece of electronic test equipment. *(Courtesy of NASA)*

FIGURE 17–2C Materials to make products, like these rolls of aluminum used to make cans, must be purchased. *(Courtesy of Reynolds Metals Company)*

FIGURE 17–2D A manufacturing plant like this one can cost millions to build and run. *(Courtesy of Toyota Motor Manufacturing, USA, Inc.)*

ducts and make a profit. A **profit** is the money left over after a product is sold and all the expenses are paid. This chapter deals with the importance of money in manufacturing.

The Finance Department

The finance department is responsible for four activities:

1. Determining how much money is needed.
2. Obtaining money.
3. Spending money.
4. Creating reports of income and expenses, Figure 17–3.

Determining How Much Money Is Needed

One of the purposes of market research is to create sales forecasts. After sales forecasts are made, budget forecasts can be made. A **budget** is a plan or guess of the amount of money that will be needed. Each department will need a certain amount of money to do their job. The workers in the finance department must create the budgets for all the other departments. There are four major budgets: the production budget, sales budget, general budget, and master budget, Figure 17–4. The production and sales budgets are for those two departments; the general budget is for all the other departments. And the master budget combines the production, sales, and general budgets into one document.

One important part of making budgets for manufacturing is determining how much it will cost to make the product. A **bill of materials**, which lists every part of a product, their descriptions, and their costs, is used for this purpose, Figure 17–5. Management must know how much

FIGURE 17–3 **Workers in the finance department are responsible for four activities related to money.** *(Courtesy of Union Pacific Corporation)*

money each department needs to design, make, and sell the product.

Obtaining Money

After the budgets have been made, the next step is to obtain the needed money. A new company

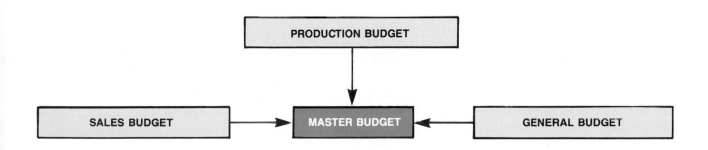

FIGURE 17–4 **A master budget combines the production budget, the sales budget, and the general budget.**

XYZ MANUFACTURING CO.

PRODUCT CODE		X-125-A			MODEL	08
QUANTITY		450	BILL OF MATERIALS		DRAWING	125
PART NO.	QTY.	PART NAME	DESCRIPTION		COST	
					UNIT	TOTAL
1	1	Base	walnut, $1\,1/8 \times 4\,1/4 \times 10\,3/8$		1.64	1.64
2	2	Sides	walnut, $5/8 \times 3\,3/8 \times 7\,7/8$		.48	.96
3	1	Case Top	walnut, $1/2 \times 4 \times 10\,3/8$		.62	.62
4	1	Middle Top	walnut, $1/2 \times 3\,5/8 \times 9\,3/8$		.50	.50
5	1	Cap	walnut, $1/2 \times 3 \times 8\,1/2$		.35	.35
6	4	Frame	walnut, $1/2 \times 1/2 \times 8$		.06	.24
7	2	Retainer	pine, $1/2 \times 1/2 \times 7$		.02	.04
8	1	Handle	#H-750, brass finish		.75	.75
9	1	Clock Face	#547CF, painted, Clock Co. Intl.		1.25	1.25
10	1	Movement	#629MV, battery, Clock Co. Intl.		2.50	2.50
11	1	Glass	$1/8 \times 8\,3/16 \times 7\,7/8$, City Glass		.28	.28
					TOTAL	$9.13

APPROVED BY:_____   DATE:_____

FIGURE 17–5 A bill of materials is a very important part of making budgets. Listed on the bill of materials are all the parts needed to make one product, the cost of each part, and the total cost for the product.

can get money from three sources: loans, bonds, and stocks, Figure 17–6. Banks **loan** money to companies. For the right to borrow the money, the company pays the bank interest. **Interest** is extra money paid above the amount of the loan. Instead of borrowing money from a bank, a company can issue **bonds**. Bonds are like a loan, but a group of people gives the money instead of a bank. Companies also pay interest on bonds. Obtaining money with a loan or bonds is merely borrowing. Stocks are very different, Figure 17–7. When a company sells stock, it is actually selling a part of the company. **Stock** gives people partial ownership of the company. Anyone with money can buy stock. As a part-owner, the stockholder has the right to participate in management decisions, Figure 17–8. People buy stock in the hope that the company will make a profit. Stockholders receive dividends. **Dividends** are money paid to stockholders out of the company's profits.

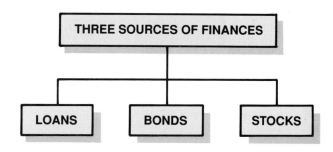

FIGURE 17–6 The three major sources of money for manufacturing companies are loans, bonds, and stocks.

Spending Money

Once a company has decided to make a product, the necessary inputs must be bought. Workers must be paid; tools, materials, and supplies must

FIGURE 17–7 Most large companies obtain money by selling stock. A share of stock gives the stockholder partial ownership in the company. *(Reprinted with permission from Harms,* MANUFACTURING TECHNOLOGY ACTIVITIES, *© 1988 by Delmar Publishers Inc.)*

FIGURE 17–9 Workers in the finance department are responsible for spending money. *(Permission granted by Westvago Corporation/Photography by Jeffrey Sranita)*

FIGURE 17–8 Stockholders are the real owners. Owning shares of stock gives stockholders a vote in the management of the company. *(Courtesy of Fleet/Norstar Financial Group)*

be bought; and rent on buildings must be paid. Interest on bank loans and bonds must also be paid. A large amount of money must be spent to run a manufacturing company. Finance workers are responsible for spending the money, Figure 17–9. Two important activities of spending money are purchasing materials and paying workers.

Purchasing Materials

Purchasing means to obtain something in exchange for money. Finance workers try to purchase (buy) the best quality materials from the lowest cost **vendor** (supplier or seller). When certain materials are needed, a finance worker will send a **purchase order** to the lowest cost vendor, Figure 17–10. The purchase order gives the vendor permission to send the materials. It is like a contract between the vendor and the company. The purchase order tells the vendor exactly what materials are needed, how many are needed, and when they should be delivered to the company. When the materials are sent to the company by the vendor, the finance worker checks the quality

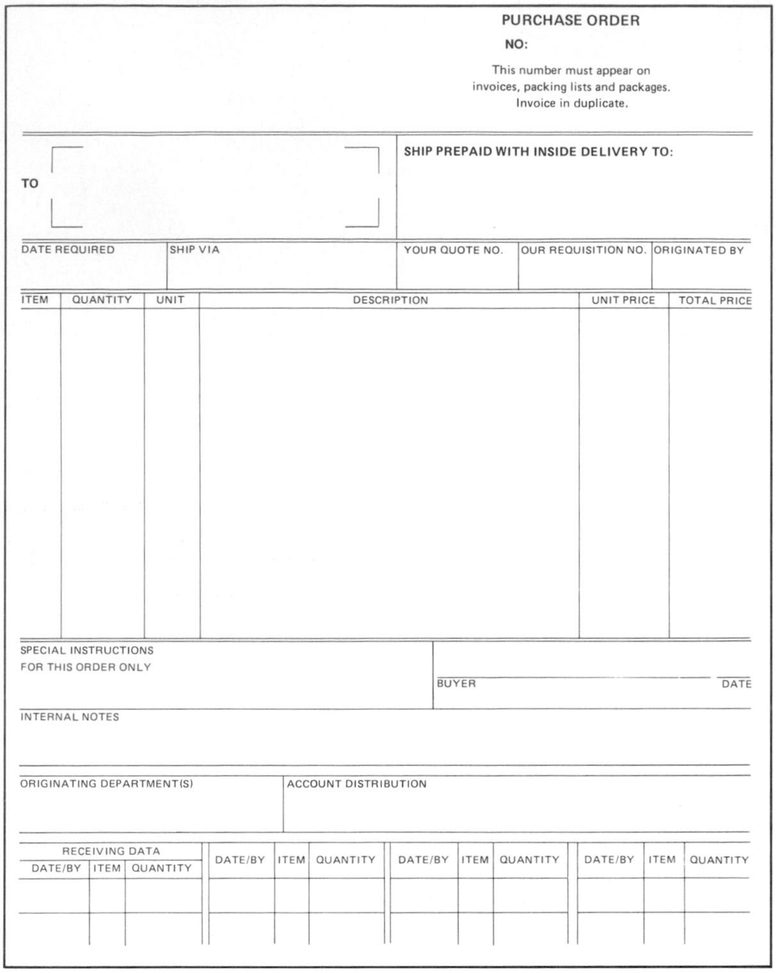

PURCHASE ORDER

NO:

This number must appear on
invoices, packing lists and packages.
Invoice in duplicate.

	SHIP PREPAID WITH INSIDE DELIVERY TO:
TO	

DATE REQUIRED	SHIP VIA	YOUR QUOTE NO.	OUR REQUISITION NO.	ORIGINATED BY

ITEM	QUANTITY	UNIT	DESCRIPTION	UNIT PRICE	TOTAL PRICE

SPECIAL INSTRUCTIONS
FOR THIS ORDER ONLY

	BUYER	DATE

INTERNAL NOTES

ORIGINATING DEPARTMENT(S) | ACCOUNT DISTRIBUTION

RECEIVING DATA			DATE/BY	ITEM	QUANTITY	DATE/BY	ITEM	QUANTITY	DATE/BY	ITEM	QUANTITY
DATE/BY	ITEM	QUANTITY									

FIGURE 17–10 A purchase order gives the vendor permission to send the needed materials to the manufacturing company.

and quantity of the materials, Figure 17–11. As with every job in manufacturing, purchasing and finance workers must do their jobs correctly if the company is to be successful and profitable.

Paying Workers

People work to make money. Payroll is often one of the largest costs for a manufacturing company. Each worker must be paid for the number of hours worked. Large manufacturing companies employ thousands of workers. Company payrolls can be in the millions of dollars per year. The hours an employee works are kept on timecards. Timecard machines record the starting and stopping time for each worker.

FIGURE 17–11 An important part of buying materials is checking their quality and quantity. *(Courtesy of Rockwell International Corporation/Ted Horowitz)*

Figure 17–12 shows the process used to calculate a worker's pay. The process includes the following steps:

1. **Calculate gross pay. Gross pay** is equal to the hourly wage rate multiplied by the number of hours worked.
2. **Deduct Social Security.** Social Security deductions are required by law. In 1989, workers paid approximately 7.51% of their gross pay, up to an annual maximum, to Social Security. In our example, gross pay was multiplied by 0.751 (7.51%). This amount was subtracted from the gross pay.
3. **Deduct federal tax.** Income tax deductions are also required by law. The Internal Revenue Service (IRS) creates tables that employers use to determine the federal tax, Figure 17–13. The worker in our example earned gross wages of $400. Federal tax on this amount was $44. This amount was subtracted from gross pay.
4. **Deduct state tax.** States have varying tax rates. The state tax rate used in our example is 2%. Multiply gross pay times 0.02 (2%), and subtract from gross wages.
5. **Deduct local tax.** Local taxes also vary by region. In our example, 1% was used.
6. **Deduct unemployment compensation.** Unemployment compensation insurance is deducted from the worker's pay in case of future layoff from the job. Workers who contribute and are unemployed because of layoffs collect money from the government. In our example, unemployment compensation is equal to 0.1%.
7. **Other deductions.** Deductions may also be made from a worker's pay for insurances, retirement funds, union dues, or other items. These deductions are based on agreements created between unions and companies.
8. After all the deductions have been subtracted from the worker's gross pay, the **net pay** is left. This is the amount of money the worker will receive in a paycheck.

Notice the amount of money deducted from the worker's pay in our example. The company must pay this money to the proper groups, such as the IRS, state and local tax collectors, and insurance companies.

CALCULATING A WORKER'S PAY

PROCESS	FORMULA	CALCULATION	TOTAL
GROSS PAY			
Gross pay	Hourly wage rate × hours worked	$10 × 40	$400.00
DEDUCTIONS			
Social security	Gross pay × 7.51%* (up to annual maximum)	$400 × 7.51%	− 30.04
Federal tax	Taken from IRS tax tables*	NA	− 44.00
State tax	Gross pay × 2% (or applicable state rate)	$400 × 2%	− 8.00
Local tax	Gross pay × 1% (or local rate)	$400 × 1%	− 4.00
Unemployment Compensation Insurance	Gross pay × .1% (or applicable rate)	$400 × .1%	− 0.40
Other deductions (Health insurance, savings plan, union dues, etc.)	Based on contract agreement	NA	− 7.50
*1989 rates. Rates may change each year.		NET PAY	$306.06

FIGURE 17–12 Quite a few deductions come out of a worker's gross pay. The net pay is the amount of money workers receive in their paychecks.

SINGLE Persons–WEEKLY Payroll Period
(For Wages Paid After December 1988)

And the wages are–		And the number of withholding allowances claimed is–										
At least	But less than	0	1	2	3	4	5	6	7	8	9	10
		The amount of income tax to be withheld shall be–										
$540	$550	$100	$90	$79	$68	$57	$50	$44	$38	$32	$27	$21
550	560	103	92	82	71	60	51	45	40	34	28	22
560	570	106	95	84	74	63	53	47	41	35	30	24
570	580	109	98	87	76	66	55	48	43	37	31	25
580	590	112	101	90	79	68	58	50	44	38	33	27
590	600	114	104	93	82	71	60	51	46	40	34	28
600	610	117	106	96	85	74	63	53	47	41	36	30
610	620	120	109	98	88	77	66	55	49	43	37	31
620	630	123	112	101	90	80	69	58	50	44	39	33
630	640	126	115	104	93	82	72	61	52	46	40	34
640	650	128	118	107	96	85	74	64	53	47	42	36
650	660	131	120	110	99	88	77	66	56	49	43	37
660	670	134	123	112	102	91	80	69	59	50	45	39
670	680	137	126	115	104	94	83	72	61	52	46	40
680	690	140	129	118	107	96	86	75	64	53	48	42
690	700	142	132	121	110	99	88	78	67	56	49	43
700	710	145	134	124	113	102	91	80	70	59	51	45
710	720	148	137	126	116	105	94	83	73	62	52	46
720	730	151	140	129	118	108	97	86	75	65	54	48
730	740	154	143	132	121	110	100	89	78	67	57	49
740	750	156	146	135	124	113	102	92	81	70	59	51
750	760	159	148	138	127	116	105	94	84	73	62	52
760	770	162	151	140	130	119	108	97	87	76	65	54
770	780	165	154	143	132	122	111	100	89	79	68	57
780	790	8	157	46	35	24	114	103	92	81	71	60
8			0			7	116	06	8			63
							1	8				6

FIGURE 17–13 Federal income tax deductions are required by law. The amount of the deduction is found in tables available from the Internal Revenue Service.

Fixed and Variable Costs

A manufacturing company may have millions of dollars in costs. There are two types of costs in manufacturing: fixed costs and variable costs. **Fixed costs** stay the same, or fixed, for a company for a certain time period. No matter how many products the company makes or sells, fixed costs remain the same, Figure 17–14. Fixed costs include money spent on buildings, machines, and furniture. **Variable costs** will vary,—that is, go up or down—with the number of products made. Variable costs include the money spent to make the final product. The materials in the product and wages for the production workers are examples of variable costs. Figure 17–15 shows the relationship between variable costs and the number of products made.

Break-Even Analysis

Company managers want to know how many products must be sold to turn a profit. The pro-

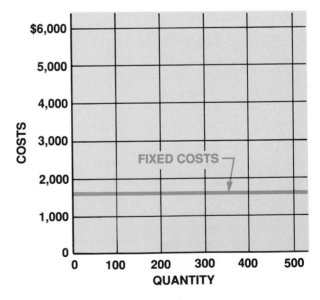

FIGURE 17–14 Fixed costs stay the same (remain fixed) no matter how many products are made. This graph shows total fixed costs for a company to be around $1700, whether they make 100 or 500 products.

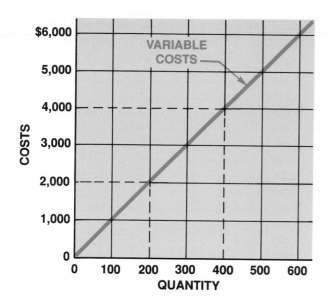

FIGURE 17–15 Variable costs go up or down (vary) with the quantity (number) of products manufactured. As this graph shows, when the quantity doubles, variable costs also double. At 100 products, costs are $1000. At 200 products, costs are $2000.

cess used to find this number is called a **break-even analysis**. At the break-even point, the company makes no profit and has no loss. There are three ways to calculate a break-even point: the math formula method, the graph method, and the computer method.

Math Formula Method. An example of using the math formula method is shown in Figure 17–16. XYZ Manufacturing Company has fixed costs of $750 and variable costs of $2 per unit. With a selling price of $4, the company would have to sell 375 units to break even — no profits and no losses. If they made or sold fewer than 375 products, they would lose money. If they sold more than 375 products, they would make a profit.

$$\text{BREAK-EVEN POINT} = \frac{F}{P - V}$$

where F = Fixed Costs
P = Selling Price
V = Variable Costs

$$\text{BREAK-EVEN POINT} = \frac{\$750}{\$4 - \$2}$$

XYZ MANUFACTURING CO.

FIXED COSTS = $750
VARIABLE COSTS = $2/unit
SELLING PRICE = $4

BREAK-EVEN POINT = 375 units

FIGURE 17–16 One way to calculate a break-even point is with the math formula method.

Graph Method. Use of the graph method of break-even analysis for XYZ Manufacturing is shown in Figure 17–17. Starting with a piece of graph paper, follow these steps:

1. Draw the horizontal and vertical axes, Figure 17–17A.
2. Mark the number of products along the horizontal axis. XYZ Manufacturing plans to make 600 products, Figure 17–17B.
3. Along the vertical axis, mark the dollar amounts for income and expenses, Figure 17–17C. For XYZ Manufacturing, if 600 products were sold at $4 each, the total income would be $2400.
4. Plot the fixed costs line. For XYZ Manufacturing, fixed costs were $750, Figure 17–17D.
5. Plot the total costs line. Total costs are added to fixed costs, Figure 17–17E.
6. Plot the sales income line. For a $4 selling price, sales income would run from $0 to $2400, Figure 17–17F.
7. Find the break-even point. The break-even point is located where the total cost line and sales income line cross. By dropping down to the horizontal axis, the break-even quantity can be found to be 375. As we found with the math formula, XYZ Manufacturing must make and sell 375 units to break even, Figure 17–18.

Notice the shaded areas below and above the break-even point in Figure 17–18. Below the break-even point, the company would lose money. This area is called the loss zone. The area above the break-even point is called the profit zone.

Computer Method. As with many other processes in manufacturing, people are using computers to perform the break-even analysis. Computers are often used to reduce the time needed to calculate the break-even point. Figure 17–19 shows a printout of a computerized break-even analysis for Small Manufacturing Company. Workers input the fixed costs, variable costs, and suggested selling price. The computer finds the break-even point. You can see on the printout that at a selling price of $49.99, the company must make and sell 626 products to break even. More powerful computer software can turn this data into a printed graph. Computers are faster and more accurate than using the math or graph methods.

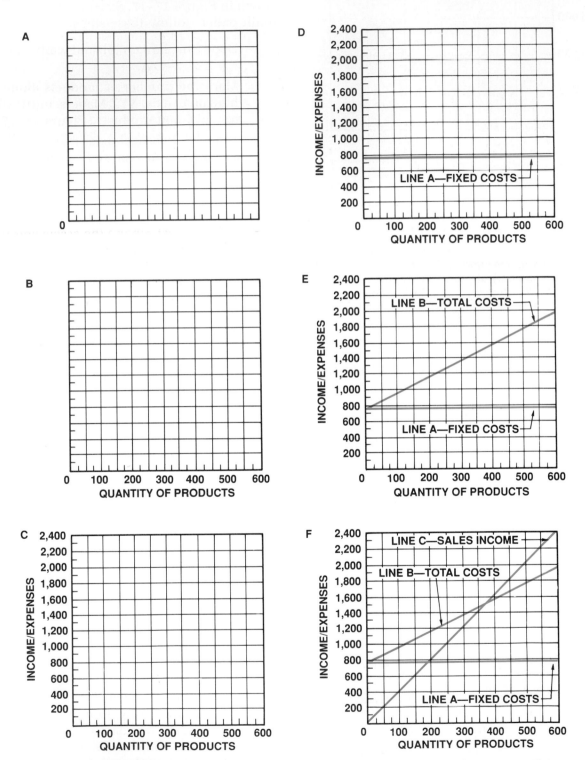

FIGURE 17–17 How to calculate a break-even point using the graph method: (A) Draw the horizontal and vertical axes; (B) mark the quantity of products along the horizontal axis; (C) mark income and expenses along the vertical axis; (D) plot the fixed costs line; (E) plot the total costs line; (F) plot the sales income line.

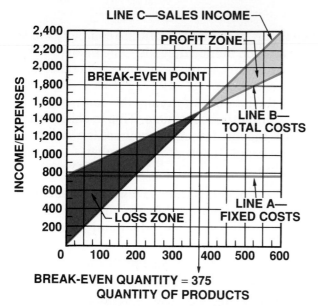

FIGURE 17–18 The point where the total costs line (line B) and the sales income line (line C) cross can be used to find the break-even point.

Creating Reports of Income and Expenses

The fourth job of the finance department is creating reports of income and expenses. Two of the more common reports are the balance sheet and the income statement. With these two reports, the finance department can show company managers and stockholders how well the company is doing with its money. The balance sheet and income statement are prepared at regular times, such as monthly, quarterly, or yearly. Let's look at the basic parts of the balance sheet and income statement.

Creating a Balance Sheet

A **balance sheet** reports the assets and liabilities of the company. **Assets** are the things the

```
!=====================================================================!
!                   SMALL MANUFACTURING COMPANY                       !
!PRODUCT: #374-XYZ                          DATE: Oct. 26, 19XX        !
!=====================================================================!
!            VARIABLE COSTS  --  PRODUCTION BUDGET                    !
!=====================================================================!
!                      (per unit costs)                               !
!      Production Dept. Wages:      21.50                              !
!           Product Materials:       4.63     DIRECTIONS:  Input costs !
!         Production Supplies:       1.60     for items as requested.  !
!                   Packaging:       1.25     Totals will be calculated!
!       Purchased Components:        3.60     automatically by program.!
!          Other Product Costs:       .15                              !
!       Other Production Costs:      0.00                              !
!                                ---------                             !
!   TOTAL PRODUCTION EXPENSES:      32.73/unit                        !
!=====================================================================!
!            FIXED COSTS  --  SELLING BUDGET                          !
!=====================================================================!
!                      (total costs)                                  !
!       Advertising Materials:     300.00                             !
!           Sales Commissions:     400.00                             !
!              Shipping Costs:     275.00                             !
!        Other Marketing Costs:    100.00                             !
!                                ---------                             !
!    TOTAL MARKETING EXPENSES:    1075.00                             !
!=====================================================================!
!           FIXED COSTS  --  GENERAL EXPENSES BUDGET                 !
!=====================================================================!
!                      Wages:    7525.00  (all depts. except production)!
!              Loan Payments:     750.00                              !
!                      Taxes:     425.00                              !
!                 Insurances:     350.00                              !
!             Utilities Costs:    200.00                              !
!                R&D Prototype:    85.00                              !
!   Jigs, Fixtures, & Tools:      320.00                              !
!         Other General Costs:     70.00                              !
!                                ---------                             !
!     TOTAL GENERAL EXPENSES:    9725.00                             !
!=====================================================================!
!                   BREAK-EVEN ANALYSIS                               !
!========!=========!=========!===========!=============================!
! SELLING! FIXED   ! VARIABLE! BREAK-EVEN ! DIRECTIONS:  Input suggested!
!  PRICE ! COSTS   ! COSTS   !   POINT    ! SELLING PRICE. Computer will !
!--------!---------!---------!-----------! automatically calculate the  !
!  49.99 !10800.00 !   32.73 !   625.72   ! number of units needed for   !
!        !         !         !           ! the BREAK-EVEN POINT.        !
!=====================================================================!
```

FIGURE 17–19 This printout is from a computer program used to perform break-even analyses.

company owns. **Liabilities** are what the company owes to other companies and businesses. A balance sheet is shown in Figure 17–20. Assets, such as cash, investments, buildings, and machines, are listed at the top and totaled. The liabilities are also listed and totaled. Below the total liabilities is the total stockholder equity. Total stockholder equity is equal to the value of the stock sold by the company plus any retained earnings. Notice that the total assets ($100,000) for Mighty Manufacturing are balanced by (equal to) total liabilities ($54,900) plus total stockholder equity ($45,100). This is why this sheet is called a balance sheet.

Creating an Income Statement

The **income statement** reports the income and expenses for the company. Income statements are also called profit/loss statements. This is because they report how much profit the company made or how much money they lost. The income statement includes two main parts—income and expenses, Figure 17–21. The income section shows the amount of money the company took in by selling their products. This is called sales income. Income from other sources is also listed. Expenses are listed for a number of items, including:

1. Cost of goods sold—This is the cost to run the production department—It includes materials in the product and production worker wages.
2. Gross profits—Gross profits are equal to total income minus the cost of goods sold.
3. Operating expenses—Included here are the selling expenses, general expenses, and any other expenses the company has.

MIGHTY MANUFACTURING COMPANY
Balance Sheet
Year Ending December 31, 1989

ASSETS

Cash	$ 7,000
Accounts receivable	5,000
Investments	11,700
Product inventory	25,000
Land and buildings	24,000
Machines and equipment	27,300
Total Assets	$ 100,000

LIABILITIES

Credits	8,800
Bank loans	6,700
Bond payments	32,900
Taxes	6,500
Total Liabilities	$ 54,900

Stockholder equity

stock ($5 × 3,000 shares)	$ 15,000
Retained earnings	30,100
Total stockholder equity	$ 45,100
Total liabilities and stockholder equity	$ 100,000

FIGURE 17–20 A balance sheet includes assets and liabilities. The stockholder equity is equal to assets minus liabilities.

```
                    MIGHTY MANUFACTURING COMPANY
                             Income Statement
                      Year Ending December 31, 1989


  INCOME
     Sales income ..................................................................... $ 120,000
     Other income ....................................................................     20,000


          Total Income ................................................................ $ 140,000

  EXPENSES
     Cost of goods sold ..............................................................     95,000


          Gross Profits ............................................................... $   45,000

     Operating expenses
          Selling expenses ...........................................................      4,000
          General expenses ..........................................................      7,500
          Other expenses ............................................................      2,700

     Earnings before interest and taxes (EBIT) ..................................... $   30,800

     Interest Expenses
     Bank loans ......................................................................      2,550
     Bonds ...........................................................................      1,000
     Other ...........................................................................        600

     Taxes ...........................................................................     13,635

  NET INCOME ......................................................................... $   13,015
```

FIGURE 17–21 Income statements, also called profit/loss statements, report how much money a company made or lost.

4. Earnings before interest and taxes (EBIT)—**EBIT** is equal to gross profits minus operating expenses.

5. Interest expenses — Interest the company pays to banks, bond holders, and other groups is listed here.

6. Taxes—Just like workers, the company must pay taxes.

7. Net income—Subtracting interest expenses and taxes from EBIT equals the net income. The net income is called the **bottom line**. The bottom line shows how much money the company made.

Calculating Dividend Payments

After the balance sheet and income statement are made, the dividends can be calculated. The amount of dividends paid to stockholders is found using the income statement and balance sheet. Dividends per share equals net profits (found on the income statement) divided by the number of shares issued (found on the balance sheet). Figure 17–22 shows how Mighty Manufacturing Company would figure its dividends based on its income statement and balance sheet.

$$\text{DIVIDENDS PER SHARE} = \frac{\text{NET PROFITS}}{\text{NUMBER OF SHARES}}$$

$$\text{DIVIDENDS PER SHARE} = \frac{\$13,015}{3,000}$$

$$\text{DIVIDENDS PER SHARE} = \$4.34$$

FIGURE 17–22 The amount of dividends per share paid to stockholders is equal to net profits divided by the number of shares of stock.

Why Do Manufactured Products Cost So Much?

How much profit do companies really make? What are the actual costs of manufacturing a product? Let's look at a simple product and see exactly what costs are involved. A mantel clock made by C&C Clock Corporation sells for $49.99, Figure 1. Cathie and Chris were the original owners of C&C. Before deciding to go into business, they made clocks as a hobby. They did all the market research, made and sold the clocks, and handled their own money. Once they decided to manufacture clocks for a living, they formed a corporation.

First, the company needed a building and the necessary equipment. Cathie and Chris found a completely equipped shop they

FIGURE 1. C&C Clock Corporation sells this mantel clock for $49.99. How much does it cost C&C to make this clock? *(Courtesy of Mason & Sullivan)*

FIXED COSTS (total costs)		VARIABLE COSTS (per unit costs)	
SELLING BUDGET		**PRODUCTION BUDGET**	
Advertising	$ 350		
Shipping	150	Labor	$21.50**
Sales Commissions	1,750		
TOTAL:	$2,250	Materials	
		■ wood	$ 4.35
GENERAL EXPENSES		■ glass	.28
Wages	$6,350*	■ glue	.25
Loan Payment	750	■ fasteners	.45
Taxes	425	■ finish	.15
Insurances	350	■ hardware	.75
Utilities	270	■ movement	2.50
Jigs & Fixtures	250	■ face	1.25
Special Tools	85	■ package	1.25
R&D Prototype	70	TOTAL:	$32.73/unit
TOTAL:	$8,550		

*Includes engineering, management, finances, human resources, and marketing personnel
**10 days × 8 hrs × 28 workers × $9.60/hr

	FIXED COSTS	VARIABLE COSTS	TOTAL COSTS
TOTALS	$10,800	$32,730	$43,530

FIGURE 2. Here are C&C's fixed and variable costs to make 1000 mantel clocks. What is the largest cost for C&C?

could buy for $70,000. They did not have that much money, so they decided to sell stock in the company. The C&C Clock Corporation sold 1000 shares of stock at $62 per share for a total of $62,000. Cathie and Chris bought 501 shares of stock so they could still control ownership of the company. This money plus an $8000 loan from a local bank gave C&C the money needed to start making clocks.

C&C was very successful in manufacturing and selling the mantel clocks. To help us understand why the clocks cost $49.99, Figure 2, provided by C&C, shows us the actual costs to manufacture 1000 clocks. Notice that the production expenses are given in **unit costs**. These are the costs to manufacture each unit or one clock. By examining the fixed and variable costs, you can see every expense involved in making the clocks.

Figure 3 shows the income and profit from the sale of 1000 clocks. Notice the profit on each unit. Only $6.46 of the $49.99 selling price is actual profit for C&C. Notice also that ten percent of the profits were held back for future plant improvements. Based on the 1000 shares of stock sold, stockholders would receive a dividend of $5.81 per share.

INCOME & PROFIT STATEMENT

	PER UNIT	1,000 UNITS
Income	$ 49.99	$ 49,990
Costs	– 43.53	– 43,530
Gross Profit	6.46	$ 6,460
10% Retained	– .65	– 646
Net Profit	$ 5.81	$ 5,814

$$\frac{\text{Net Profits}}{\text{Shares of Stock}} \quad \frac{\$5,814}{1,000} = \$5.81 \text{ Dividend per share}$$

FIGURE 3. C&C's income and profit statement shows a net profit of $5.81 per unit after all costs are paid and ten percent is retained. Stockholders received a dividend of $5.81 per share. Would you buy stock in C&C? Why or why not?

The profit on each unit may seem small. The company had to spend more than $43 on each clock to make only $6.46. In some manufacturing businesses, the profit may be less than a dollar per unit. However, if you sell five million units, that can add up to some nice profits.

Maybe now you can better understand why manufactured products cost so much. There are many expenses that the company must pay in order to produce and sell products.

Summary

The finance department is responsible for four tasks: (1) determining how much money is needed, (2) obtaining money, (3) spending money, and (4) creating reports on money. Determining how much money is needed is done by creating budgets for the various departments in the company. A bill of materials is also important when determining how much money is needed. Manufacturing companies can obtain money through bank loans, issuing bonds, or selling stock. Purchasing materials and paying workers are two important spending activities. Another important part of spending money is performing a break-even analysis. Fixed costs and variable costs are needed to calculate the break-even point. Two important reports created by the finance department are the balance sheet and income statement.

DISCUSSION QUESTIONS

1. Why is money important in manufacturing?
2. What are the four responsibilities of the finance department?
3. How is the bill of materials used to help the finance department determine how much money is needed?
4. What are the three methods manufacturing companies can use to obtain money?
5. What is a break-even analysis? What are the three different types of break-even analyses?
6. What deductions are subtracted from a worker's gross pay?
7. What are two reports created by the finance department? How are these reports used to help calculate dividends per share?

CHAPTER ACTIVITIES

 ## Obtaining Money by Selling Stock

OBJECTIVE

One of the most popular ways to obtain money for a manufacturing corporation is by selling stock. Stock sales allow a company to obtain funds that can be used to purchase the inputs needed for manufacturing.

In this activity, you will sell stock shares in a newly formed manufacturing corporation.

EQUIPMENT AND SUPPLIES

1. Stock certificates (from your teacher)
2. Stockholder's ledger (from your teacher)

PROCEDURE

1. Form a manufacturing corporation with the members of your class.
2. Choose a product and set a selling price.
3. Determine how much money your company needs to make and sell products.
4. Decide how many shares of stock you will sell.
5. Set the selling price for the stock.
6. Be sure to inform investors of the risks and rewards of owning stock in your company.
7. Sell stock certificates to interested investors.
8. On the stockholder's ledger, keep a list of the investors, the number of shares they purchase, and their total investment.
9. Deposit money from stock sales into a bank account.
10. You now have the money you need to start making products.

MATH AND SCIENCE CONNECTIONS

Stock is sold at a certain price for each share, or price per share. To calculate the price per share, use this formula:

$$\text{PRICE PER SHARE} = \frac{\text{TOTAL VALUE OF ALL SHARES OF STOCK}}{\text{TOTAL NUMBER OF SHARES}}$$

RELATED QUESTIONS

1. Why do manufacturing companies sell stock?
2. What do we call the people who buy stock? Why would a person buy stock in a manufacturing company?
3. What are the risks and rewards of owning stock?
4. A company wants to sell stock for $15 per share. How many shares would it need to sell to raise $3000?

 ## Calculating A Break-Even Point

OBJECTIVE

The break-even analysis is an important part of manufacturing. Before a company spends any money on production, a break-even analysis must be performed.

In this activity, you and the members of your class will select a product and calculate a break-even point.

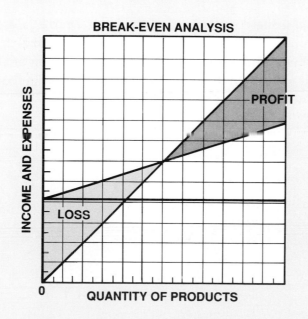

BREAK-EVEN ANALYSIS

EQUIPMENT AND SUPPLIES

1. Product made by members of the class
2. Worksheet for fixed and variable costs (from your teacher)
3. Suggested selling price for the product
4. Depending on which method of calculation you use, you will need the following:
   a. calculator for math method
   b. graph paper and drawing instruments for graph method
   c. computer, software program, and printer for computer method

PROCEDURE

1. Your teacher may divide the class into groups. Each group may use a different break-even analysis method.
2. Complete the worksheet for fixed costs and variable costs.
3. Decide the suggested selling price for the product.
4. Calculate the break-even point for your product.
5. Compare results among your three groups.
6. Change the selling price and see what happens to the break-even point.

MATH AND SCIENCE CONNECTIONS

A break-even point can be calculated with three methods. The math formula method uses the following formula:

$$\text{BREAK-EVEN POINT} = \frac{\text{FIXED COSTS}}{\text{PRICE} - \text{VARIABLE COSTS}}$$

RELATED QUESTIONS

1. Why do manufacturing companies perform break-even analyses?
2. What three factors are needed to perform a break-even analysis?
3. What three methods can be used to perform a break-even analysis?
4. What is the break-even point for a company that has $100 in fixed costs, $8 in variable costs, and a product selling price of $12.99?

SECTION FIVE

CONSTRUCTION SYSTEMS

(Courtesy of Pat Dineen/Photo by Shirley Nelson)

CHAPTER 18

Architectural Design and Drawing

OBJECTIVES

After completing this chapter, you will be able to:

- Identify the factors to consider when designing a structure.
- Describe the different types of drawings included in a set of architectural working drawings.
- Identify approved symbols and abbreviations used on architectural drawings.
- Describe why CAD is used for architectural design and drawing.
- Measure scale drawings with an architect's scale.
- Design and draw plans for a construction project.

KEY TERMS

Architect	Elevations	Section views
Contour lines	Form	Topographical drawings
Detail drawings	Function	Zoning boards
Drafters	Plan views	

Before a structure can be built, it must be designed and plans must be drawn. This phase of construction is called architectural design and drawing. The main person responsible for designing the appearance of a structure is an **architect**. Engineers design the strength of the structure and any mechanical systems, like heating or air conditioning. **Drafters** turn the architect's and engineer's design ideas into working drawings and plans. The number of architects, engineers, and drafters involved often depends on the size of the structure being designed.

Design Considerations

Designing a structure can be a difficult and time consuming task. The first step is for the architect and engineer to meet with the client (the owner of the structure). Together, they often discuss the following design considerations:

- What the structure will be used for (its function)
- What the structure should look like (its form)
- How much space is required, today and tomorrow

■ The condition of the land where the structure will be built

■ How much money the client has available for the structure

■ Community regulations that must be followed

■ Any special requirements the client may have

One of the most important considerations in any design is **function**. Function means how the structure will be used. Bridges serve the function of providing smooth, safe, and quick travel over waterways, Figure 18–1. Homes serve the function of providing safety and protection from the weather. All structures serve some function.

Another important consideration is form. **Form** is the size and shape of a structure. Bridges should be high enough above the water to let boats pass. Also, the roadway on a bridge must be large enough for the trucks and other vehi-

cles that pass over it. Sometimes architects design the form of a structure to be a work of art, Figure 18–2.

Cost is always an important design consideration. The size of the structure, the materials used, and special features all affect the final cost. Architects and engineers must often find a balance between cost and the other considerations.

Community regulations are also an important consideration. Most communities have **zoning boards** to study the needs of the community for new structures. Zoning board members consider the size of the community, resources on hand, and the location and condition of existing structures. All new structures must meet the regulations developed by the zoning board.

FIGURE 18–2 The form of this auditorium is one of its most distinctive features. (Courtesy of New York State Department of Commerce)

From Design to Drawings

As the architect and engineer develop their designs for a new structure, they often make rough sketches. They also make notes related to the design considerations. As changes are made in the design, new, more detailed sketches are drawn. Figure 18–3 shows an architect's first rough sketch for a house. Several changes and improvements will be made before this drawing can be used to build the home. Once the client approves the detailed sketches, the process of preparing architectural working drawings can begin.

FIGURE 18–1 The function of this bridge is to permit safe, smooth traffic flow. (Courtesy of the Eaton Corporation)

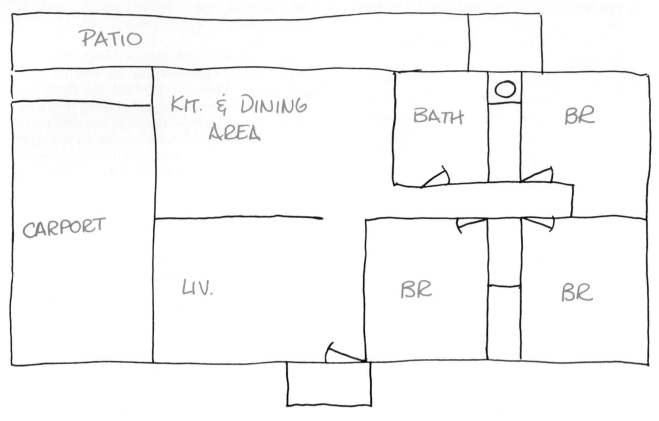

FIGURE 18–3 Rough sketch for a home done by an architect.

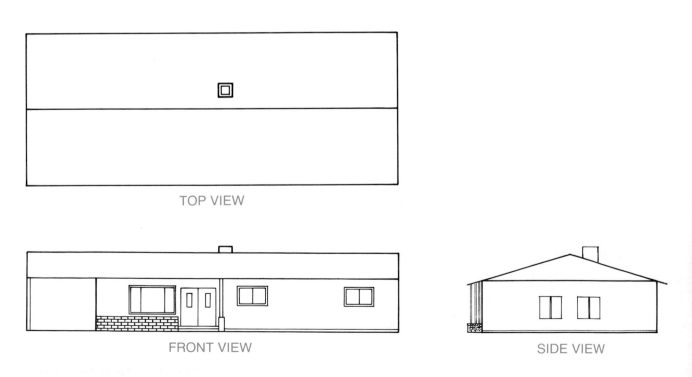

FIGURE 18–4 Multi-view drawing for a home.

Architectural Working Drawings

Working drawings are a set of drawings and plans that include all the information needed to build a structure. Drafters make working drawings. Both multi-view drawings, Figure 18–4, and pictorial drawings, Figure 18–5, are used in architectural working drawings. (See Chapter 2 for a discussion of multi-view and pictorial drawings.) Four types of architectural working drawings are done: elevations, section views, plan views, and detail drawings.

Elevations

Elevations are drawings of a structure as it is seen from its different sides. The front of a building is drawn as the front elevation. The side views are called the right-side and left-side elevations. The side opposite the front is called a rear elevation, Figure 18–6.

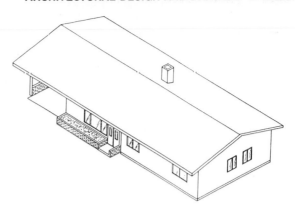

FIGURE 18–5 Pictorial drawing of a home.

Section Views

Section views are similar to elevations. Instead of showing the outside of the structure, they show the inside features. Section views are drawn by imagining that half the structure has

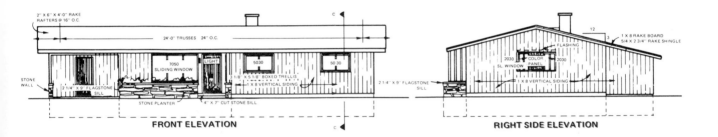

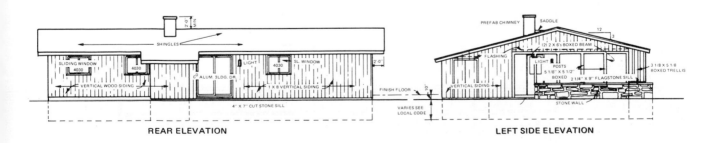

FIGURE 18–6 Front, rear, and side elevations from a set of working drawings.

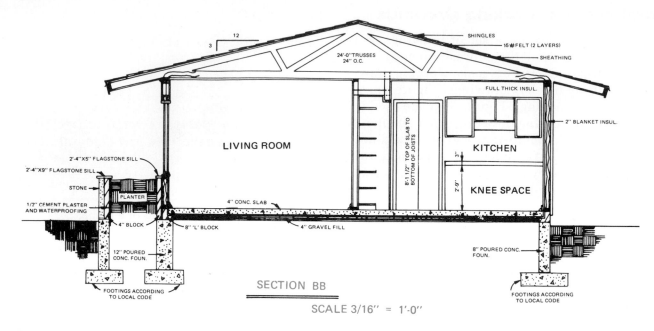

FIGURE 18–7 Section view of a house. *(Courtesy of Home Planners, Inc.)*

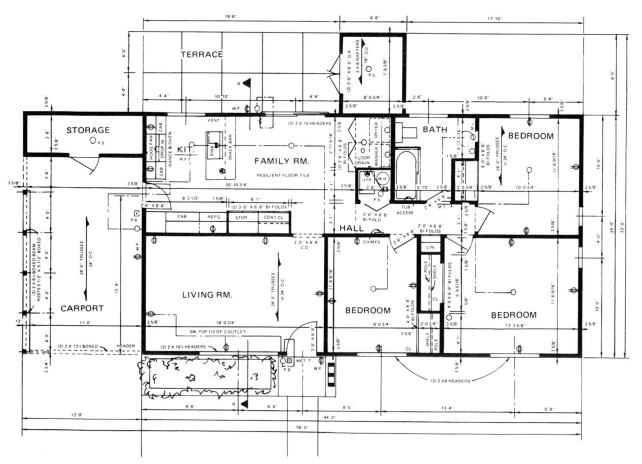

FIGURE 18–8 Floor plan. *(Courtesy of Home Planners, Inc.)*

been cut away, Figure 18–7. Notice that the section view includes the foundation.

Plan Views

Plan views are drawings that look down from above the structure. One common type of plan view is the floor plan, Figure 18–8. Drafters make floor plans by imagining that the top half of the building is cut away so they can see inside. The finished drawing shows the size and location of all the rooms, walls, doors, and windows. A separate floor plan is drawn for the basement and each floor.

Detail Drawings

In most complex structures, the elevations, section views, and plan views do not provide enough information. In those cases, detail drawings are made. **Detail drawings** provide construction workers with specific details on how certain parts of the structure should be built, Figure 18–9.

Structural and Civil Drawings

For heavy construction projects like bridges and highways, structural and civil drawings are made. Structural drawings are used for bridges and skyscrapers, Figure 18–10. They include information about the size and kind of material to be used, the location of parts, and how the parts are to be fastened. Notice that plan, elevation, and section views are used in the structural drawing. Detail drawings can also be included.

Civil drawings are made for highways and any other project involving moving earth. **Topographical drawings** are used in these types of projects. They show the rise and fall of the land. Surveyors measure the elevation (height) and location of certain points on the land, Figure 18–11. The surveyor's notes are then made into topographical drawings, Figure 18–12. The lines on a topographical drawing are called **contour lines**. The numbers on the contour lines show

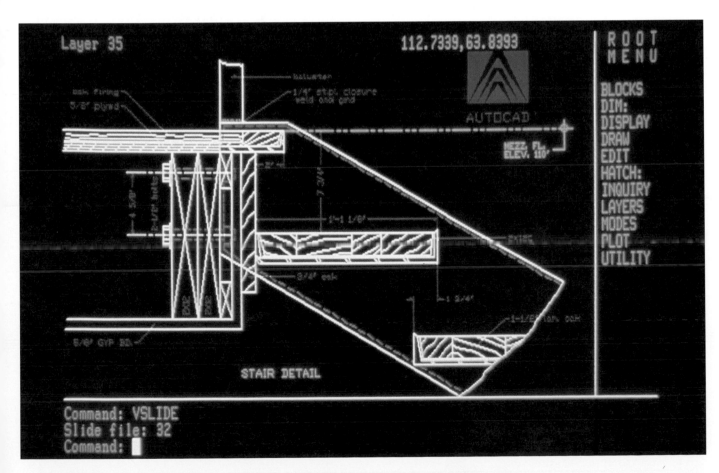

FIGURE 18–9 Detail drawing for one part of a structure. *(Courtesy of Autodesk, Inc.)*

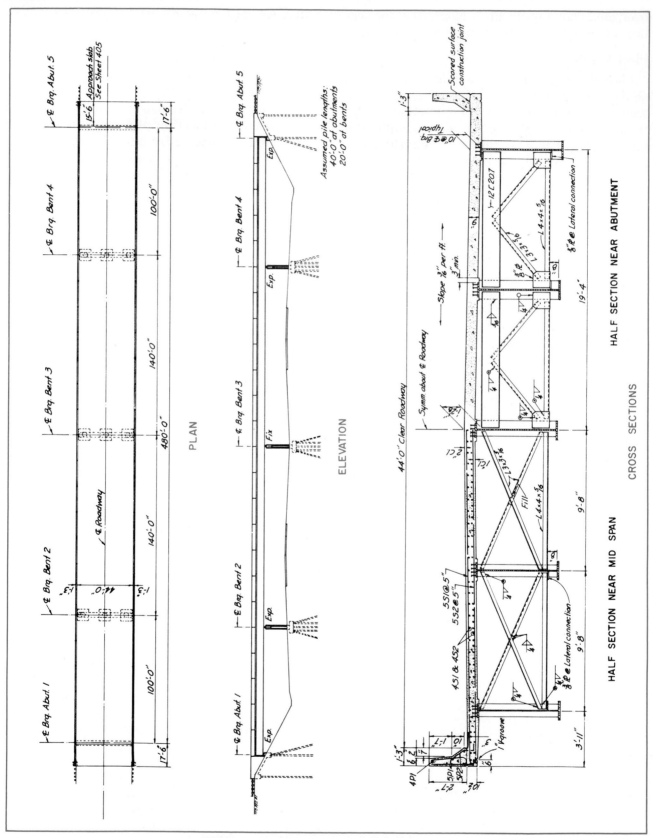

FIGURE 18–10 Structural drawings for a bridge.

the elevation of the line above sea level. The distance between each line in this drawing is two feet. You should be able to calculate the difference in elevation between points A and B as six feet (B equals 286 feet and A equals 280 feet).

Architectural Symbols and Abbreviations

Making accurate architectural working drawings takes many hours of work. To save time in drawing, certain symbols and abbreviations have been approved by the American National Standards Institute (ANSI), Figure 18–13. These symbols and abbreviations are used on drawings instead of drawing the complete details. This makes the drawing process more efficient.

Computer-Aided Drafting (CAD)

CAD is also used to reduce the time involved in making architectural working drawings. The

FIGURE 18–11 Land surveyors measure the elevation and size of land. *(Courtesy of the Lietz Company, copyright © 1986)*

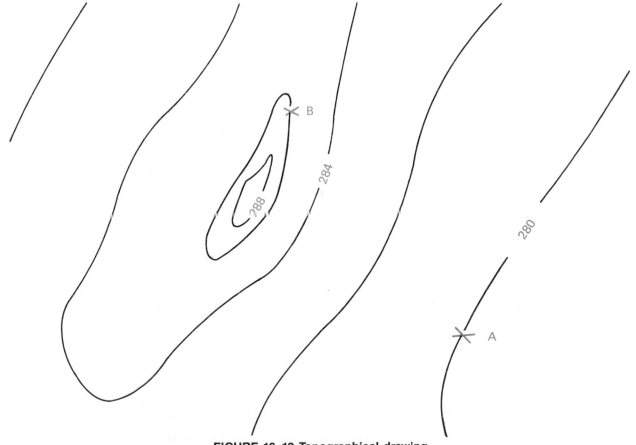

FIGURE 18–12 Topographical drawing.

PLAN AND SECTION INDICATIONS

ELEVATION INDICATIONS

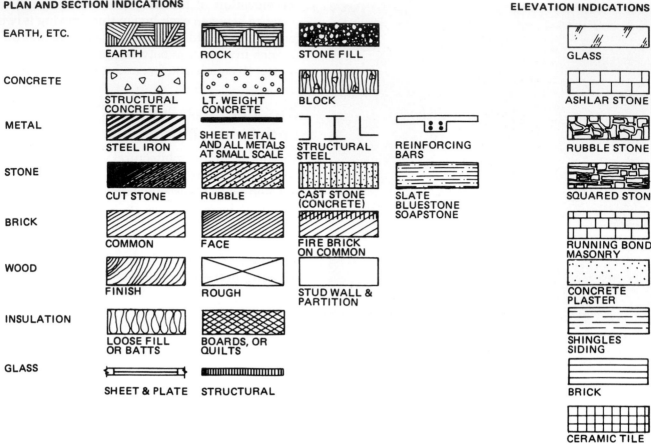

EARTH, ETC.
EARTH | ROCK | STONE FILL
GLASS

CONCRETE
STRUCTURAL CONCRETE | LT. WEIGHT CONCRETE | BLOCK
ASHLAR STONE

METAL
STEEL IRON | SHEET METAL AND ALL METALS AT SMALL SCALE | STRUCTURAL STEEL | REINFORCING BARS
RUBBLE STONE

STONE
CUT STONE | RUBBLE | CAST STONE (CONCRETE) | SLATE BLUESTONE SOAPSTONE
SQUARED STONE

BRICK
COMMON | FACE | FIRE BRICK ON COMMON
RUNNING BOND MASONRY

WOOD
FINISH | ROUGH | STUD WALL & PARTITION
CONCRETE PLASTER

INSULATION
LOOSE FILL OR BATTS | BOARDS, OR QUILTS
SHINGLES SIDING

GLASS
SHEET & PLATE | STRUCTURAL
BRICK

CERAMIC TILE

STACK BOND MASONRY

PLUMBING SYMBOLS

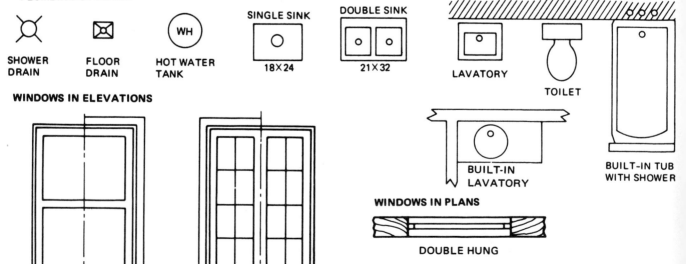

SHOWER DRAIN | FLOOR DRAIN | HOT WATER TANK | SINGLE SINK 18×24 | DOUBLE SINK 21×32 | LAVATORY | TOILET | BUILT-IN TUB WITH SHOWER

BUILT-IN LAVATORY

WINDOWS IN ELEVATIONS

WINDOWS IN PLANS

DOUBLE HUNG | CASEMENT

DOUBLE HUNG

CASEMENT

FIGURE 18–13 Architectural symbols and abbreviations.

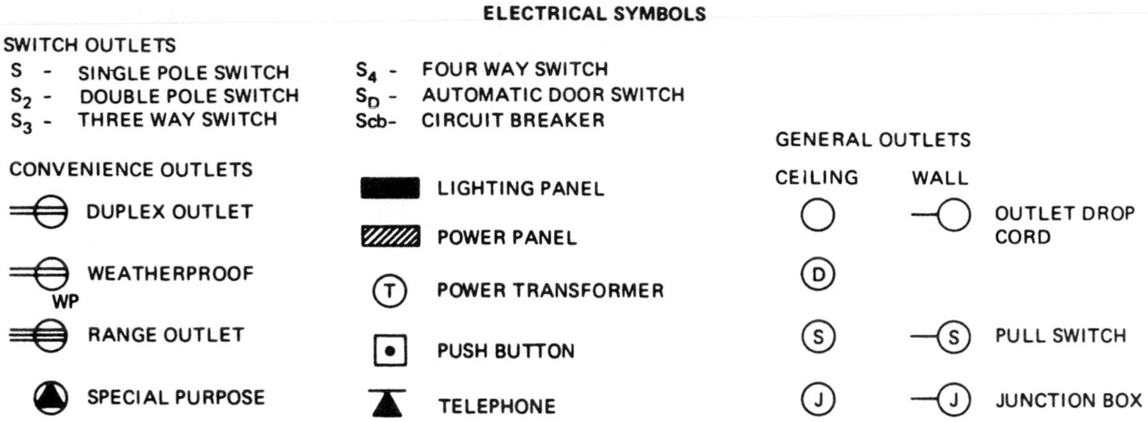

ELECTRICAL SYMBOLS

SWITCH OUTLETS

S -	SINGLE POLE SWITCH	S_4 -	FOUR WAY SWITCH
S_2 -	DOUBLE POLE SWITCH	S_D -	AUTOMATIC DOOR SWITCH
S_3 -	THREE WAY SWITCH	Scb-	CIRCUIT BREAKER

CONVENIENCE OUTLETS

DUPLEX OUTLET

WEATHERPROOF
WP

RANGE OUTLET

SPECIAL PURPOSE

LIGHTING PANEL

POWER PANEL

POWER TRANSFORMER

PUSH BUTTON

TELEPHONE

GENERAL OUTLETS

CEILING   WALL

OUTLET DROP CORD

PULL SWITCH

JUNCTION BOX

ABBREVIATIONS USED ON WORKING DRAWINGS

AWG	American Wire Gauge	GL	Glass
B	Bathroom	HB	Hose Bibb
BR	Bedroom	C	Hundred
BD	Board	INS	Insulation
BM	Board Measure	INT	Interior
BTU	British Thermal Unit	KD	Kiln Dried
BLDG	Building	K	Kitchen
CLG	Ceiling	LAV	Lavatory
C to C	Center to Center	LR	Living Room
CL or C̶L	Centerline	MLDG	Molding
CLO	Closet	OC	On Center
COL	Column	REF	Refrigerator
CONC	Concrete	R	Riser
CFM	Cubic feet per minute	RM	Room
CU YD	Cubic Yard	SPEC	Specification
DR	Dining Room	STD	Standard
ENT	Entrance	M	Thousand
EXT	Exterior	T & G	Tongue and Groove
FIN	Finish	UNFIN	Unfinished
FL	Floor	WC	Water Closet
FTG	Footing	WH	Water Heater
FDN	Foundation	WP	Waterproof
GA	Gauge	WD	Wood

FIGURE 18–13 (Continued).

basic components of a CAD system include a computer, input devices, output devices, and software, Figure 18–14. Each of these components was described in Chapter 2.

CAD has many advantages for architectural drawing and design. ANSI-approved architectural symbols and abbreviations can be stored in the CAD computer and called up when needed. Elevations, floor plans, sections, and other architectural drawings can be changed without having to be completely redrawn. Drawings can be done in layers, so that a basic design outline can be used for several different projects, Figures 18–15 and 18–16. By using a CAD function called zoom, a design can be shown as close up or as far away as desired, Figures 18–17 and 18–18.

Scales

It is not possible to make architectural drawings actual size, so they are drawn to scale. The

FIGURE 18–14 Microcomputer CAD system. *(Courtesy of Hewlett-Packard Company)*

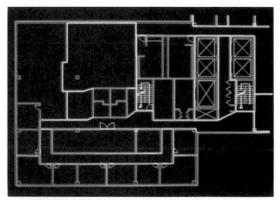

FIGURE 18–15 A basic floor plan. *(Courtesy of Intergraph Corporation)*

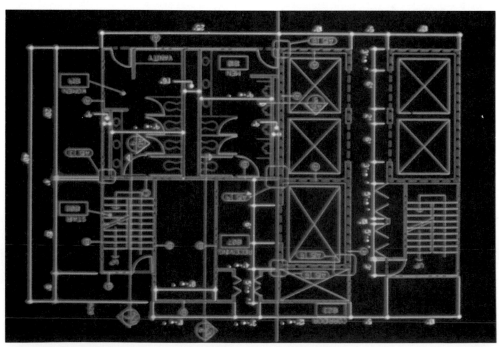

FIGURE 18–16 Dimensions and plumbing symbols are added to the floor plan in another layer. *(Courtesy of Intergraph Corporation)*

FIGURE 18–17 Interior view of an office created on CAD. *(Courtesy of Intergraph Corporation)*

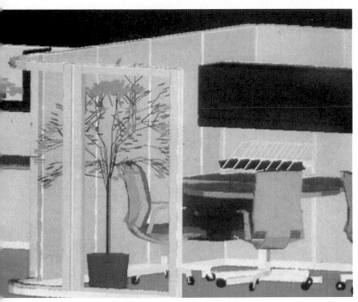

FIGURE 18–18 Zooming magnifies the office so that all detail can be seen. *(Courtesy of Intergraph Corporation)*

dimensions of all parts are reduced to a size that can be drawn on a sheet of paper, Figure 18–19. For example, most floor plans for homes are drawn 1/48th of actual size. At this scale, 1/4 inch on the drawing represents 1 foot on the actual building. This scale is written ¼″ = 1′-0″.

To make drawings to scale, an architect's scale is used, Figure 18–20. Eleven different scales are found on the architect's triangular scale:

Full scale

⅛″	= 1′-0″	¼″	= 1′-0″	⅜″	= 1′-0″
¾″	= 1′-0″	½″	= 1′-0″	1″	= 1′-0″
1½″	= 1′-0″	3″	= 1′-0″		
³⁄₃₂″	= 1′-0″	³⁄₁₆″	= 1′-0″		

FIGURE 18–20 Triangular architect's scale. *(Courtesy of Koh-I-Noor Rapidograph, Inc.)*

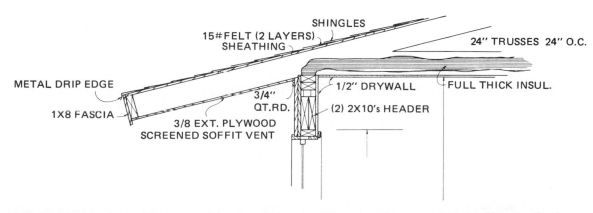

FIGURE 18–19 This detail drawing is done in a ½″ = 1′-0″ scale. *(Courtesy of Home Planners, Inc.)*

Applying Mathematics: Reading the Architect's Scale

To read the triangular architect's scale, turn it to the ¼-inch scale, Figure 1. There are twelve lines marked on the scale from the zero towards the ¼ mark on the left. So, each line equals one inch on the ¼″ = 1′-0″ scale.

The ⅛-inch scale is combined with the ¼-inch scale and on the opposite end, Figure 2. There are six lines from zero to the ⅛ mark. So, each line equals two inches on the ⅛″ = 1′-0″ scale.

The same technique can be used on each of the faces of the architect's scale. Each time, remember that the divisions on the end of the scale are equal to one foot.

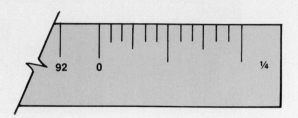

FIGURE 1. Close-up view of the ¼-inch scale.

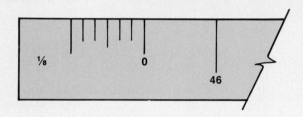

FIGURE 2. Close-up view of the ⅛-inch scale.

Two scales are combined on each face, except for the full scale. The scales can be combined because one is twice as large as the other. For example, the ⅛″ and ¼″ scales are combined and placed on opposite ends of the scale.

Modeling

Models are often built by architects and engineers as they design a structure. The models are built to scale from paper, cardboard, wood, clay, or plastic. Many models have all the details of the finished structure. Architects build models to test the appearance of the structure. Engineers build models to test the strength and operation of the structure. Modeling, like symbols and abbreviations, CAD, and scales, can be used to reduce the time needed to design a structure.

Summary

Before structures are built, they must be designed and plans must be drawn. This phase of the project is called architectural design and drawing. The architect and engineer design the structure after considering factors such as function, form, costs, and zoning board regulations. Architectural working drawings, including elevations, section views, plan views, and detail drawings, are done by drafters after the design is created. For heavy construction projects, structural and civil drawings are made. The process of making architectural working drawings is improved by the use of approved symbols and abbreviations, CAD, and scales.

DISCUSSION QUESTIONS

1. What are some of the factors that must be considered when designing a structure?
2. Which design considerations do you think are most important? Why?
3. What are the four different types of drawings included in a set of architectural working drawings? What are the differences among the four types?
4. What are structural and civil drawings? When are they used?
5. Why are symbols and abbreviations used on working drawings? What organization approves the symbols and abbreviations?
6. Why is CAD used for architectural drawing and design?
7. What does it mean to draw a structure "to scale"? What are some of the scales used in architectural working drawings?

CHAPTER ACTIVITIES

 ## Home Floor Plan

OBJECTIVE

In this chapter, you read about plan views in architectural design and drawing. One type of plan view is a floor plan. In this activity, you will sketch a floor plan for your home.

PROBLEM

Sketch a floor plan for one floor of your home, Figure 1.

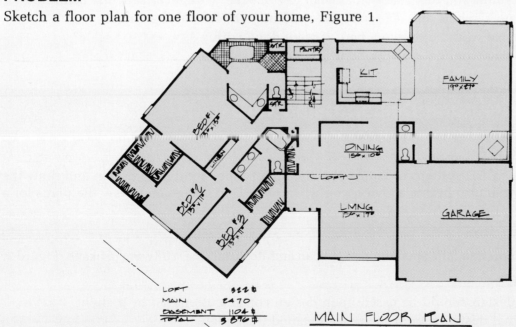

FIGURE 1. Floor plan. *(Reprinted from* ARCHITECTURAL DRAFTING AND DESIGN *by Jeffries and Madsen, © 1986 by Delmar Publishers Inc.)*

GUIDELINES

1. Sketch your floor plan on grid paper.
2. Use a scale, such as ¼″ = 1′-0″, that matches the size of the grid on your paper.
3. Use the approved symbols and abbreviations described in this chapter.
4. Include the roof size and overall dimensions.
5. Be sure to include all doors, windows, and built-in kitchen and bathroom fixtures.

EQUIPMENT AND SUPPLIES

1. Grid paper (¼″ or ⅛″)
2. Tape measure
3. Drawing/sketching pencils
4. Symbols/abbreviations chart

PROCEDURE

1. Get a friend to help you. Pick one floor in your home.
2. Measure the size of each room and the location of all walls, doors, and windows.
3. Just as architects do, make a rough sketch as you measure your home floor plan.
4. Locate all electrical and plumbing systems on your sketch.
5. Once you have all the measurements and notes on your rough sketch, do a final sketch. Use the proper symbols and include the required dimensions.

MATH AND SCIENCE CONNECTIONS

Drawings done with a ¼″ = 1′-0″ scale are $\frac{1}{48}$ size. That means the structure is 48 times as large as the drawing. This is so because there are forty-eight ¼″ measurements in each foot. (Each inch has four ¼″ measurements and there are twelve inches per foot. So, 4 x 12 = 48.)

RELATED QUESTIONS

1. How would you describe a floor plan to someone who never saw one before?
2. Why are symbols and abbreviations used in architectural drawings?
3. What size would a drawing be if it were done with a ½″ = 1′-0″ scale?
4. What size would a drawing be if it were done with a ⅛″ = 1′-0″ scale?

 # Architectural Design Using CAD

OBJECTIVE

Architects use CAD systems to reduce the amount of time required to design and draw the plans for a structure. In this activity, you will use CAD to design and draw the plans for a structure.

PROBLEM

Design the floor plan for a small home using an architectural CAD software package, Figure 2.

GUIDELINES

1. Your design should be based upon design considerations set by a client.
2. The final design should include a detailed floor plan.
3. Also included should be an elevation view of one room.

EQUIPMENT AND SUPPLIES

1. Grid paper and pencils
2. Architectural CAD software
3. CAD data disks
4. Microcomputer system
5. Computer printer

PROCEDURE

1. Locate a client (teacher, student, principal, relative, etc.) who will work with you on this project. Describe the activity and ask the client to prepare for a design interview. In the interview, ask questions related to the following design considerations:
   a. Style—what are their tastes and interests?
   b. Furniture—what furniture would they like?
   c. Special Needs and Wants—needs must be included, wants may be included. Needs and wants many be colors desired, number of bathrooms, floor coverings, special features, etc.
   d. Price—what is the client willing to spend?
   e. Time Schedule—when does the client want the finished design?
2. Make several rough sketches of your design ideas on grid paper.
3. Pick your best ideas and include them in one or two CAD drawings of the floor plan.
4. Print out your drawings and prepare a presentation for the client.
5. Meet with the client to get feedback on your drawings.

MATH AND SCIENCE CONNECTIONS

CAD systems save money by reducing the amount of time needed to make and remake drawings. If the client asked for several changes in your design, do you think it would be faster to do them by hand, or by computer? Why?

RELATED QUESTIONS

1. What are the advantages of using CAD to make architectural drawings? What are the disadvantages?
2. What were four important design considerations in this activity?
3. What people are involved in the design and drawing of a structure?
4. Would you like to be an architect? Why or why not?

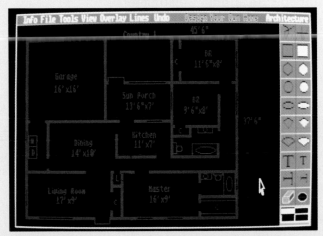

FIGURE 2. Architectural design done with a CAD program. *(Courtesy of Abracadata, Ltd.)*

CHAPTER 19

Construction Specifications and Contracts

OBJECTIVES

After completing this chapter, you will be able to:

- Describe the purpose of writing specifications for a construction project and identify the divisions in specifications.

- Explain how CAD, computers, and word processing simplify specification writing.

- Describe the process of estimating, bidding, and contracting work for a construction project.

- Discuss the roles of contractors, subcontractors, and other people in managing a construction project.

- Explain the process of financing construction projects with loans.

KEY TERMS

Bar chart

Bid

Construction specifications

Contract

Contractor

CSI Format

Estimate

General contractor

Mortgage

Subcontractor

Working drawings contain as much information as possible about the materials to be used and the size and location of parts for a structure. However, it is impossible to include all the necessary information. **Construction specifications**, commonly called specs, are written documents that give detailed information not shown on the working drawings, Figure 19–1. Specs describe the amount, quality, and type of materials to be used and the methods for putting them together.

Specification writing requires a great deal of knowledge about construction materials, practices, and regulations. On smaller projects, the specs are brief and may be written by the architect. On larger projects, specification writers with

272

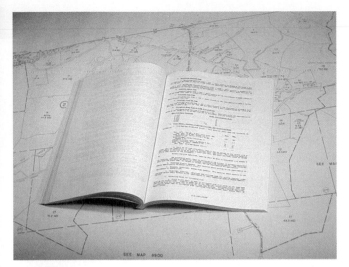

FIGURE 19–1 Construction specifications are written instructions.

a background in architecture and engineering perform this task.

The accepted standard for specification writing is the Construction Specification Institute's **(CSI) Format**. This format uses sixteen divisions:

1. Division 1 — General Requirements
2. Division 2 — Site Work
3. Division 3 — Concrete
4. Division 4 — Masonry
5. Division 5 — Metals
6. Division 6 — Wood and Plastics
7. Division 7 — Thermal and Moisture Protection
8. Division 8 — Doors and Windows
9. Division 9 — Finishes
10. Division 10 — Specialities
11. Division 11 — Equipment
12. Division 12 — Furnishings
13. Division 13 — Special Construction
14. Division 14 — Conveying Systems
15. Division 15 — Mechanical
16. Division 16 — Electrical

Division 1 — General Requirements is an overall division describing contracts, schedules and temporary utilities. Divisions 2 through 16 deal with the actual construction of the structure. When specs are written, the divisions are arranged as nearly as possible in the order of work.

CAD, Computers and Specification Writing

Writing specifications for a large project can take many hours of work. The use of CAD and computer word processing systems has made the job a little easier, Figure 19–2. Some powerful architectural CAD programs automatically calculate the number of parts needed. Word processors store written specs for certain materials and parts. Once stored, the specs can be called up from the computer's memory for the next project. This can be a real time saver.

FIGURE 19–2 Computers are used for CAD and word processing to make specification writing easier. *(Courtesy of Bechtel Group, Inc.)*

Bids, Estimates, and Contracts

When the working drawings and specifications are complete, a contractor is selected. A **contractor** plans the building of a project according to a written agreement. On large projects, contractors normally bid on the job.

A **bid** is an offer to build a structure for a certain amount of money and within a certain period of time. In order to prepare a bid, a contractor reviews the working drawings and specifications.

After reviewing the drawings and specs, the contractor makes cost estimates. An **estimate** is a calculation that is carefully made. It includes three factors: direct costs, overhead costs, and profit, Figure 19–3. The labor of workers and the materials used to build the structure are direct costs. Overhead costs include insurance, office costs, and equipment maintenance. Profit is the amount of money the contractor expects to make after paying for direct costs and overhead costs.

After a contractor determines the price for the project, a bid is submitted. Often several contractors will bid on one project. The contractor who submits the lowest bid, and follows the plans and specs, is awarded a contract for the job. A **contract** is a legal agreement between the contractor and the people paying for the structure. Contracts usually include the following:

Completion Schedule. A date for the completion of the project is specified. For large projects, dates are given for completion of various parts of the project.

Payment Schedule. On smaller structures, payment in full is made after the job is complete.

For larger structures, partial payments are made as the work progresses. One method is for the contractor to receive partial payment for the work done each month. Another typical payment schedule for a house may be as follows:

20% when the foundation is complete
30% when the structure is enclosed
30% when the utilities are installed
20% when the house is completed

Responsibilities. The owner is responsible for having the property surveyed. The architect may be responsible for the overall management of the project. The contractor is responsible for building the structure.

Insurance. Certain kinds of insurance are required during construction. Contractors usually have liability insurance, and the owner usually has property insurance. Insurance protects them from being sued for accidents on the job site.

Termination. Contracts can be terminated (ended) if one party fails to follow the contract.

Contractors and Subcontractors

Most contracts make the contractor responsible for directing the building processes from start to finish. The contractor must supply the workers, equipment, and materials needed to complete the structure. In many cases, this is not always possible and the contractor may hire **subcontractors**. Subcontractors hire workers who specialize in certain jobs, such as carpentry, plumbing, electrical, or other trades, Figure 19–4. When subcontractors are used, the first contractor is called a **general contractor**. On large jobs, one general contractor may use several subcontractors to complete the work. Both the general contractor and the subcontractors must be able to manage their workers, equipment, and materials. General contractors must be able to manage all phases of a construction project, including the work done by individual subcontractors. Most general contractors start out as subcontractors.

Managing a Construction Project

Contractors must understand their role in the management of a construction project. Figure 19–5 shows the various people involved in man-

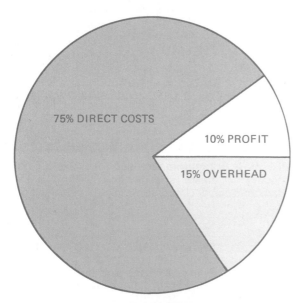

FIGURE 19–3 On one construction project 75% of the cost was for direct expenses; 15% was for overhead; and 10% was profit.

FIGURE 19–4 The painter is a subcontractor on a house-building job. *(Courtesy of Richard T. Kreh, Sr.)*

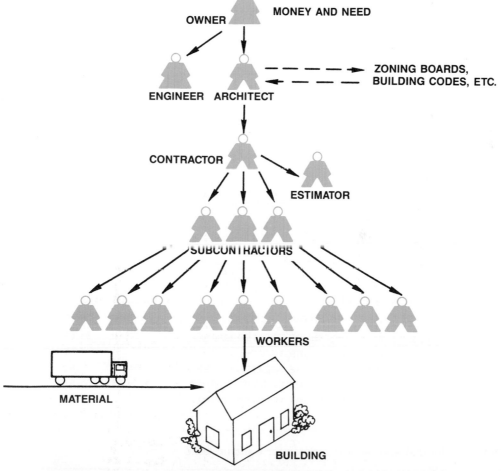

FIGURE 19–5 The management structure for a construction project.

Applying Mathematics: Estimating

The person who makes the estimates for a project is called an estimator. A good estimator must be able to use math. He must also have experience in construction technology. In Chapter 4, you read about "ball park" estimates. Although they are important, ball park estimates are never used for bids. More precise estimates must be made. Let's look at how an estimator calculates costs for materials and labor.

Estimating Material Costs

For materials like wood, plywood, and steel, the estimator simply counts the number of pieces needed for the structure. A picnic table made from 2 x 2, 2 x 4, and 2 x 6 lumber will serve as a simple example, Figure 1. The number of lumber pieces needed is found on the plans. With the costs per piece, an estimate of the total building costs can be made. Experienced estimators know that most 2x lumber comes in eight-foot lengths. They also know that there may be mistakes as the table is built. This information is figured in the calculation. The estimator may add a certain amount, such as 15%, for these possibilities. The estimator will also add for materials like screws and bolts, sandpaper, and the finish.

Counting also works for building parts like doors, windows, sinks, and electrical outlets. All this information can be found on the working drawings and specifications. Once the total number is known, the estimator simply multiplies by the cost per piece to find the total costs, Figure 2.

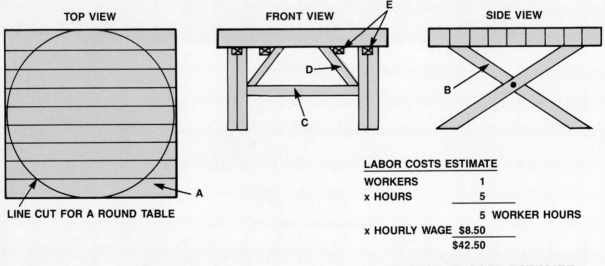

LABOR COSTS ESTIMATE

WORKERS	1
x HOURS	5

5 WORKER HOURS

x HOURLY WAGE $8.50

$42.50

PICNIC TABLE PARTS LIST

PART	QTY	SIZE	MATERIAL
A TOP	9	2 x 6 x 48	PRESSURE TREATED LUMBER
B LEG	4	2 x 4 x 40	PRESSURE TREATED LUMBER
C CROSS	1	2 x 4 x 29¾	PRESSURE TREATED LUMBER
D BRACE	2	2 x 2 x 10½	PRESSURE TREATED LUMBER
E CLEAT	4	2 x 2 x 29¼	PRESSURE TREATED LUMBER

MATERIALS COST ESTIMATE

COST/PIECE	TOTAL
$1.69	$15.21
1.35	5.40
.89	.89
.25	.50
1.02	4.08
TOTAL COSTS:	$26.08

FIGURE 1. Material and labor estimates for a picnic table.

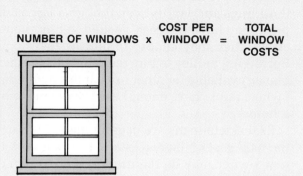

NUMBER OF WINDOWS x $\dfrac{\text{COST PER}}{\text{WINDOW}}$ = $\dfrac{\text{TOTAL}}{\substack{\text{WINDOW} \\ \text{COSTS}}}$

18 WINDOWS   x $85.00   = $1530.00

FIGURE 2. Cost estimate for windows in a house.

Concrete is sold by volume, not by the piece. The volume measure used is a cubic yard (3ft x 3ft x 3ft). Figure 3 shows the calculations for a concrete patio that measures four inches thick by fifteen feet square. The total costs are found by multiplying the total cubic yards of concrete needed by the cost per cubic yard. Once again, a small amount may be added, just in case.

Estimating Labor Costs

Estimating labor costs requires an understanding of three factors:

- The number of workers needed
- The number of jobs they must perform
- The time required for each job.

Estimators use past experience to do these calculations. Careful notes are kept each time a structure is built. Then, when they estimate a similar job in the future, they refer back to the notes.

For our picnic table, an estimator may know that one experienced worker could do all the measuring, cutting, and building in five hours. Figure 1 also shows the labor costs calculations.

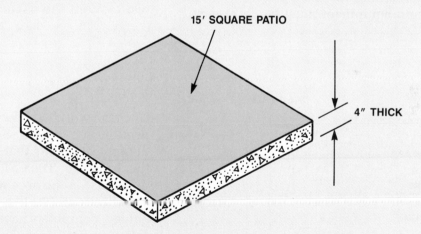

15′ SQUARE PATIO

4″ THICK

$$\text{VOLUME (CUBIC YARDS)} = \frac{\text{THICKNESS (FT.) x WIDTH (FT.) x LENGTH (FT.)}}{27 \text{FT.}^3 / \text{CUBIC YARD}}$$

$$\text{VOLUME (CUBIC YARDS)} = \frac{.33 \text{ FT. x 15FT. x 15FT.}}{27 \text{FT.}^3 / \text{CUBIC YARD}}$$

$$\text{VOLUME (CUBIC YARDS)} = 2.75$$

FIGURE 3. Volume estimates for a patio.

aging and carrying out a construction project. The owner provides the need for the structure and the money to pay. The architect and engineer design the structure to meet the owner's needs, as well as zoning regulations. Drafters turn the design ideas into working drawings. The general contractor is responsible for making sure the structure gets built correctly and on time. Large contracting firms have estimators who bid on projects. Subcontractors are often hired for certain jobs in building the structure. The workers, or tradespeople, actually build the structure.

Also shown on Figure 19-5 is the importance of on-time delivery of materials and equipment. This requires good planning and scheduling. On small projects, contractors rely on their past experience as a guide for scheduling. For large projects, more planning is required. One technique used to schedule materials and equipment is the **bar chart**, Figure 19-6. The bar chart lists the jobs to be done down the left side. To the right of each job is a horizontal bar. The bars show the time it should take to complete each job. With this type of information, the manager knows when materials and equipment must be delivered.

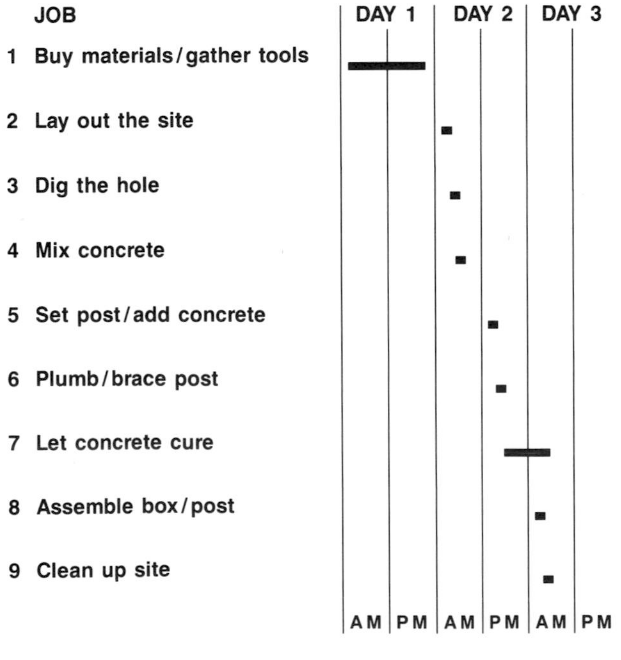

Financing Construction Projects

Most owners do not pay cash for their real estate (buildings and land). Even though a large company may have the cash needed to buy the real estate, they may choose to borrow the money. Borrowing money to buy real estate leaves more money available for other uses. Individual homeowners rarely have enough cash available to buy a home.

Real estate loans are usually made for part of the total cost of the real estate. For example, the new owner may be required to pay 20 percent of the total selling price in cash. The lender, usually a bank, loans the remaining 80 percent to the borrower. The borrower agrees to make payments (usually monthly) to the lender. These payments are made up of a percentage of the loan and interest. Interest is extra money paid back to the lender for the right to borrow money.

Lenders require some form of guarantee that the loan payments will be made. The most common guarantee is called a **mortgage**. A mortgage permits the lender to assume ownership of the real estate if the payments are not made.

Summary

Once the working drawings are complete, specifications are written. Specifications provide all the project details that may not be included on the drawings. The CSI Format, with its sixteen divisions, is the accepted standard for writing specifications. Just like many other areas of technology, computers have been applied to specification writing to make the job easier and more efficient.

Contractors compete for building contracts by placing bids for the job. They must make estimates of the costs for workers, materials, equipment, and profits in order to create a bid. On larger projects, general contractors hire several subcontractors to do the actual work.

The future owners of a new structure provide the need for the structure and the money to pay. They hire the other people involved in the project. Many owners finance their construction projects with loans.

JOB	DAY 1	DAY 2	DAY 3
1 Buy materials/gather tools	▬▬		
2 Lay out the site		▪	
3 Dig the hole		▪	
4 Mix concrete		▪	
5 Set post/add concrete		▪	
6 Plumb/brace post		▪	
7 Let concrete cure		▬▬▬	
8 Assemble box/post			▪
9 Clean up site			▪
	AM PM	AM PM	AM PM

FIGURE 19-6 Bar chart for a mailbox installation project.

DISCUSSION QUESTIONS

1. What is the purpose of writing specifications?
2. What is the name of the organization that created the accepted format for construction specifications?
3. How are computers being used to make specification writing more efficient?
4. What is the process used by owners to hire contractors? Be sure to explain bids and estimates.
5. What are general contractors and subcontractors?
6. What is the management structure for a construction project? Draw an illustration to show the people involved.
7. How do most owners pay for new construction projects?

CHAPTER ACTIVITIES

Estimating Construction Costs

OBJECTIVE
Labor, equipment, and materials are three cost items that must be estimated in construction. In this activity, you will estimate these costs for a simple construction project, Figure 1.

PROBLEM
Given the plans for a structure, estimate the construction costs.

GUIDELINES
1. The structure will be an A-Frame Club House (or another structure picked by your teacher).
2. You will estimate costs for labor, equipment, and materials.
3. You will use actual material prices from lumberyards (or prices supplied by your teacher).
4. You will assume a construction crew of five workers who get paid $8.50 per hour.

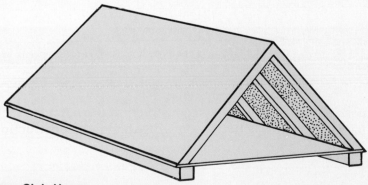

FIGURE 1. A-Frame Club House

EQUIPMENT AND SUPPLIES

1. Plans for the A-Frame Club House (or other structure)
2. Estimating Construction Costs Worksheet (from your teacher)
3. Telephone book
4. Calculator

PROCEDURE

1. Obtain the worksheet and structure plans from your teacher.
2. With the help of your teacher, list the tools and equipment needed to build the structure.
3. Ask your teacher for a tool and equipment rental price. Estimate the total equipment costs.
4. Call a local lumberyard for material prices. Estimate the total materials costs.
5. Work with your teacher to determine the labor costs. Remember, you have five workers who are paid $8.50 per hour. Estimate total labor costs.
6. Add the equipment, material, and labor costs to get an estimate of the total costs.

MATH AND SCIENCE CONNECTIONS

Estimators use simple math every day. Yet, if they make a mistake, they could cost their company thousands of dollars. Estimators always check their work to make sure their estimates are accurate.

Expressed as a mathematical formula, the total costs for a structure would be:

$$
\begin{array}{ll}
\text{Equipment Costs} & = \text{Equipment Used} \times \text{Rent/Purchase Price} \\
+ \text{Material Costs} & = \text{Materials Used} \times \text{Cost per Piece} \\
+ \text{Labor Costs} & = \text{Workers} \times \text{Hours Worked} \times \text{Hourly Wage} \\
\hline
= \text{Total Costs} &
\end{array}
$$

RELATED QUESTIONS

1. What type of math calculations did you perform in this activity?
2. Why are accurate estimates necessary?
3. How would this activity have been different if the structure were a concrete dam?
4. Are labor and materials a direct cost or an overhead cost?

 # Scheduling a Construction Project

OBJECTIVE

Scheduling is an important part of managing a construction project. One technique used to schedule a job is the bar chart. In this activity, you will schedule a construction project using a bar chart.

PROBLEM

Schedule the jobs required to construct a structure, Figure 1.

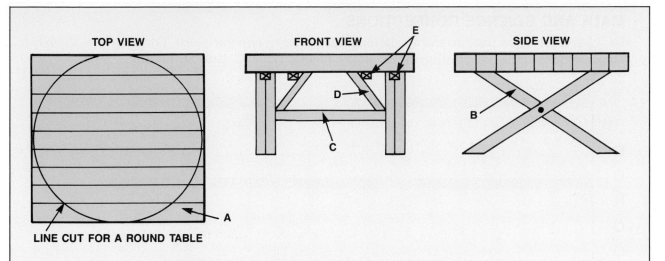

TOP VIEW   FRONT VIEW   SIDE VIEW

LINE CUT FOR A ROUND TABLE

PICNIC TABLE PARTS LIST

PART	QTY	SIZE	MATERIAL
A TOP	9	2x6x48	PRESSURE TREATED LUMBER
B LEG	4	2x4x40	PRESSURE TREATED LUMBER
C CROSS	1	2x4x29¾	PRESSURE TREATED LUMBER
D BRACE	2	2x2x10½	PRESSURE TREATED LUMBER
E CLEAT	4	2x2x29¼	PRESSURE TREATED LUMBER

FIGURE 1. Picnic table plans

GUIDELINES

1. The structure will be the picnic table described in this chapter (or another structure picked by your teacher).
2. You may assume unlimited access to tools, materials, and people.
3. Your bar chart should be drawn on grid paper.

EQUIPMENT AND SUPPLIES

1. Plans for the Picnic Table (or other structure)
2. Scheduling a Construction Project Worksheet (from your teacher)
3. Felt markers

PROCEDURE

1. Obtain the worksheet and structure plans from your teacher.
2. In small work groups, make a list of the jobs that must be performed to make the structure.
3. List the jobs, in the order they should be done, down the left side of the worksheet.
4. Across the top of the grid and to the right of your list, label each block so it is equal to a certain unit of time, such as 15 minutes, hours, or half hours.
5. Make an estimate of the time required to do each job. Draw horizontal bars with a felt marker to show the time estimate for each job.

MATH AND SCIENCE CONNECTIONS

Bar charts are often used to show relationships between two variables. For example, the bar chart below shows the enrollment for four schools, A, B, C, and D, Figure 2.

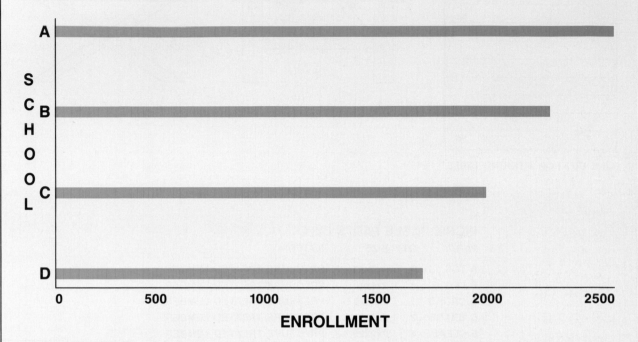

FIGURE 2. Bar charts can be used to compare the enrollments in four schools.

RELATED QUESTIONS

1. What was the hardest part of this activity? What was the easiest part?
2. Can you think of any other uses for bar charts?
3. Who makes bar charts in construction?
4. How are bar charts used in construction to help manage a project?

The Substructure: Site Work and Foundations

KEY TERMS

Batter boards	Frost line	Setback
Bedrock	Percolation test	Slab-on-grade
Building codes	Pier	Solar orientation
Building permit	Pile	Substructure
Excavation	Plot plan	Superstructure
Footing	Screeding	

Building Codes and Plot Plans

Building codes are laws that control construction. One thing building codes control is where a structure can be placed on the building site. In some communities, this is based on the setback of neighboring buildings. **Setback** is the distance that a building is from the street. Codes of this type ensure that a new home is not built closer to the street than other homes.

Building codes usually require a plot plan to be included with the working drawings, Figure 20–1. A **plot plan** is a drawing that shows the location of a structure on the site. It shows the location of the street and trees. It also shows the boundaries and size of the plot (the land on which the structure will be built).

283

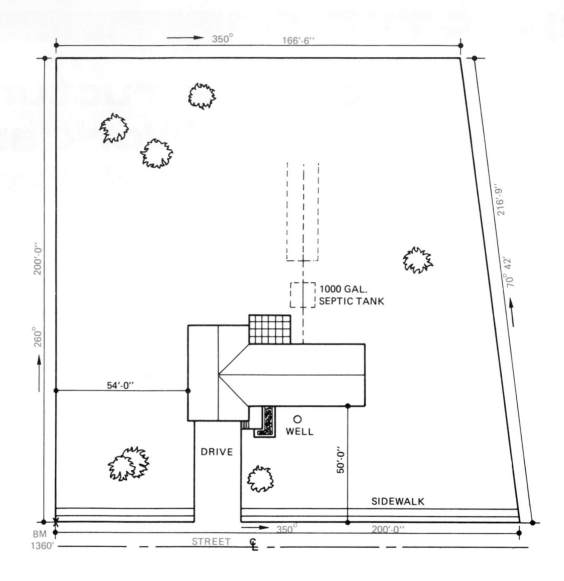

FIGURE 20–1 A plot plan.

Architects or surveyors draw the plot plan on large construction projects. For residential construction a general contractor often prepares a plot plan. Plot plans must be submitted to local agencies to obtain a building permit. A **building permit** gives the builder permission to construct a structure.

Plot Plan and Energy Efficiency

The location of a building in relation to its site is important to energy efficiency. The location of the sun, the direction of winds, and the position of trees and other buildings all affect energy efficiency. The position of a building in relation to these factors is called **solar orientation**.

Trees are very important to solar orientation. Deciduous trees (trees that lose their leaves) can be used to control solar energy striking a building. In the winter, the solar energy can pass through the branches and enter the windows, Figure 20–2. This solar energy can partially heat the home. In the summer, the leaves shade the building and protect it from the high heat, Figure 20–3.

In certain locations, the wind must also be considered, Figure 20–4. Land next to large bodies of water is often windy. In the summer, cool air

FIGURE 20–2 The winter sun passes through deciduous trees.

FIGURE 20–3 The summer sun is blocked by deciduous trees.

blows off the water to cool a house during the day. At night, warm air moves past the house from the land. In the winter, the high winds can cause a house to lose heat faster than normal. Trees can be grown close to a home to block the negative effects of winds. Houses built close to one another can also block the wind.

Site Work

Laying Out Building Lines

After a building permit is obtained, site work can begin. The first step is to stake (lay) out the position of the structure. Stakes are driven into the ground where the corners of the building will be.

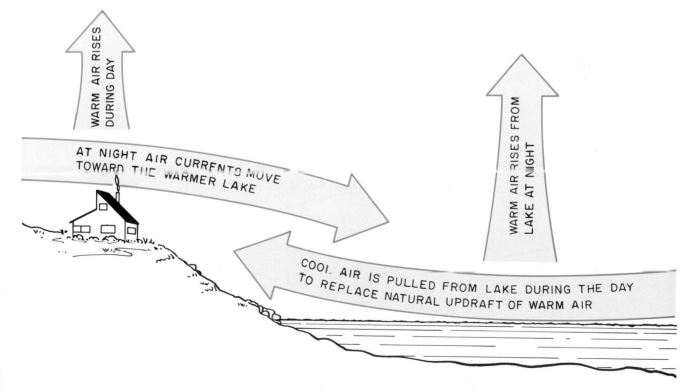

FIGURE 20–4 The location of a building can affect the heating and cooling loads.

The front of the building is usually laid out first. Figure 20–5 shows how the location of the front of a building is found by measuring back from the street. Notice how the stakes are located at the corners.

The sides of the building are found by laying out lines at right angles to the front. Right angles can be measured by the 6-8-10 method explained in Chapter 8. A builder's level or transit can also be used, Figure 20–6. If a builder's level is used, a target rod must be held over the opposite front corner stake. A transit can be aimed directly at the stake, so a target rod is not needed. When the level or transit is in position, the telescope is turned 90 degrees. This gives the location of one side of the building. The same technique is used to find the other side.

When both side lines are known, the back corners are found by measuring from the front corners. If the building has an irregular shape, it is divided into several smaller rectangles, Figure 20–7.

It is important to make sure the rectangles are square. If the lines are not correct, the building may not be built correctly. Measuring the diagonals is one method to check the squareness of the rectangle.

Erecting Batter Boards

The land where the stakes are placed will be excavated (cut out). **Batter boards** are erected at each corner several feet outside the building

FIGURE 20–6 This worker is laying out the building lines on a site.

lines, Figure 20–8. They help workers find the corners during excavation. Strings are attached to the batter boards and stretched between each corner. A plumb bob is used to make sure the strings cross at the exact corner of the building. Saw kerfs are cut in the batter board to mark the position of the lines in case the lines get cut or moved.

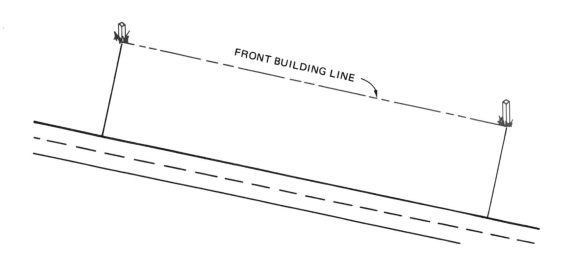

FIGURE 20–5 The first step in laying out a building is to drive stakes for the front building line.

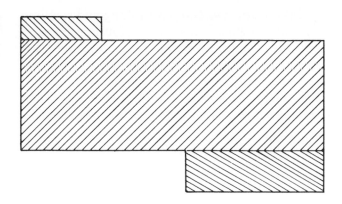

FIGURE 20–7 Irregularly shaped buildings are divided into smaller rectangles. Then each part is laid out separately.

FIGURE 20–8 Batter boards are erected at each corner.

Excavating

Even small buildings have great weight. The weight of the building must be supported by the earth directly beneath it. Because of this it is important to consider the soil on which the building will rest.

Most soil contains some moisture. In many parts of the country, this moisture freezes during the winter. As the ground freezes, it swells just as water does. When the ground thaws, it shrinks back. This swelling and shrinking can cause problems for buildings. To overcome these problems, the base of buildings must be placed below the frost line. The **frost line** is the depth to which the earth freezes. To place the base below the frost line, the earth must be excavated, Figure 20–9. **Excavation** involves digging the soil with heavy earthworking equipment.

Excavating is usually done by a subcontractor who specializes in this work. Heavy equipment,

FIGURE 20–9 Earthmoving equipment is used for excavating a site.

like backhoes, bulldozers, and dump trucks are used. Excavators must understand the engineering characteristics of soil.

Types of Soil

Early in the design of a structure, soil surveys are obtained of the site, Figure 20–10. Soil surveys can be obtained from the Soil Conservation Service of the United States Department of Agriculture. They provide engineers with information about the characteristics of the soil.

There are many types of soil, including sand, gravel, loam, silt, and clay. On a construction site, these soils can be mixed together. Soil experts identify soils with certain tests. A sieve test is one such test. Soil is passed through a series of different sized sieves (screens). The amount of soil that passes through each size is recorded. When compared to a chart like that shown in Figure 20–11 the soil can be accurately named.

Engineering Qualities of Soil

Four important engineering qualities of soil include shrinking and swelling, drainage, depth to bedrock, and compaction. Each of these is discussed below.

FIGURE 20–10 Soil surveys are prepared by soil scientists. *(Courtesy of the UDSA Soil Conservation Service)*

Shrinking and Swelling. Shrinking and swelling soil can cause cracks in the foundation of a building, Figure 20–12. Soils with a high clay content can have large amounts of shrinking and swelling. Soil engineers check shrinking and swelling by measuring a sample of wet soil. Then they measure it again after is has been dried.

Drainage. Some soils allow water to drain off more readily than others. Poor drainage can cause several problems. Water can get under the foundation and raise a building. It can also push against a foundation wall and cause cracks. Engineers check drainage with a **percolation test**. Water is poured into a hole dug in the ground. The amount of time it takes for the ground to absorb the water relates to the drainage. Percolation tests are usually required by inspectors when a new septic system is planned.

Depth to Bedrock. The earth's hard rock crust is called **bedrock**. The bedrock is covered with a thin layer of soil. This soil layer varies from a few inches to hundreds of feet deep. Smaller buildings can be built on solid soils. Heavy buildings, like skyscrapers and dams, must be built on bedrock. Designers of these structures must know the depth to bedrock on the site.

One simple method to determine the depth to bedrock is the sounding rod method. Long sections of steel rod are driven into the ground. When the rod strikes bedrock, it makes a ringing sound. Several soundings are made on each site to make sure the rod has not hit a boulder.

Compaction. Compaction is the squeezing or compressing of soil. It is also called settling. Heavy structures cause the soil under them to settle. Uneven settling can cause a structure to break apart.

Some soils compact more than others. Engineers drill holes in the ground to find out what type of soil is under a new building site. For many construction jobs, the soil is compacted with heavy equipment before the structure is built.

Foundation Design

The foundation is the base upon which a structure rests. Since it is located below the structure, the foundation is also called the **substructure**.

ANALYSIS OF SOIL GRADATION

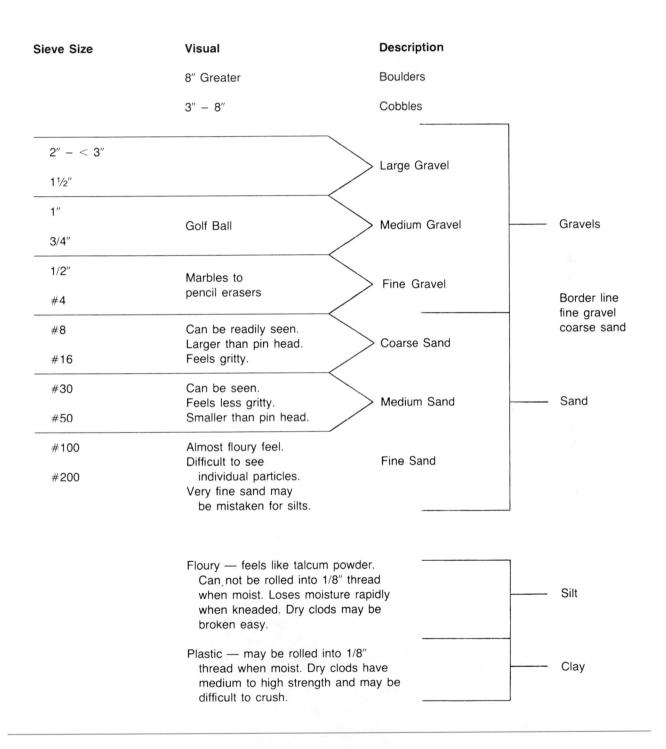

Sieve Size	Visual	Description
	8″ Greater	Boulders
	3″ – 8″	Cobbles
2″ – < 3″ 1½″		Large Gravel
1″ 3/4″	Golf Ball	Medium Gravel
1/2″ #4	Marbles to pencil erasers	Fine Gravel
#8 #16	Can be readily seen. Larger than pin head. Feels gritty.	Coarse Sand
#30 #50	Can be seen. Feels less gritty. Smaller than pin head.	Medium Sand

#100
#200
	Almost floury feel.
	Difficult to see
	  individual particles.
	Very fine sand may
	  be mistaken for silts.
		Fine Sand

Gravels

Border line
fine gravel
coarse sand

Sand

Floury — feels like talcum powder.
Can not be rolled into 1/8″ thread
when moist. Loses moisture rapidly
when kneaded. Dry clods may be
broken easy.

Silt

Plastic — may be rolled into 1/8″
thread when moist. Dry clods have
medium to high strength and may be
difficult to crush.

Clay

FIGURE 20–11 Soils are named according to their grain size.

FIGURE 20–12 Shrinking and swelling of the soil can cause cracks in the foundation of a structure. *(Courtesy of Paul E. Meyers)*

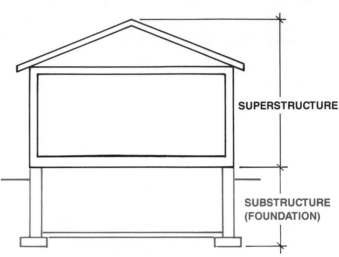

FIGURE 20–13 Buildings have a substructure (foundation) and a superstructure.

The part of the structure above the ground is called the **superstructure**, Figure 20–13.

Structures are very heavy. When placed on soil they tend to settle and compact the soil. The foundation spreads out the weight of the structure evenly over the soil. This reduces the possibilities of uneven settling or compaction. The foundation is built first and provides a strong and stable base for the rest of the structure. Without a strong, rigid foundation, the superstructure would flex and crack. The foundation can also be used to enclose a basement. A basement can be a useful space if enough headroom is provided and the area is dry. There are three main types of foundations: slab-on-grade, wall, and pier or pile.

Slab-on-Grade Foundations

In **slab-on-grade** construction, the first floor of the building serves as the foundation, Figure 20–14. A concrete slab is placed directly on the grade (ground level) of the site. Slab-on-grade

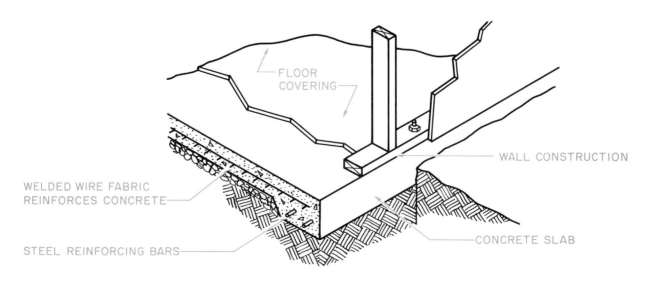

FIGURE 20–14 Parts of a slab-on-grade foundation.

construction is popular in warmer climates for three main reasons:

1. The ground does not freeze, so there is no concern for the action of the freezing and thawing;
2. The cost of excavation is saved;
3. Warmer climates often have excess ground water that makes the soil less stable.

Constructing a slab is simpler than other types of foundations. There are several steps that must be taken before the concrete is placed. First, forms are built by carpenters on the site to contain the concrete around the outside edge of the slab. The forms also control the height and levelness of the top of the slab. Next, any pipes that will run under or through the slab are positioned before the concrete is placed. Large-diameter pipes are normally installed as sleeves. Piping is installed in these sleeves later, Figure 20–15. Finally, wire mesh is put in place to reinforce the finished slab. The actual placement and finishing of the concrete is the quickest part of the job.

Wall Foundations

In areas where the earth's surface freezes in the winter, the foundation must reach below the frost line. Usually, wall foundations are used. A wall foundation has two parts: the footing and the wall.

The **footing** is a strip of concrete placed around the perimeter of the structure. It is the base upon which the wall part of the foundation is supported, Figure 20–16. The footing is located below the frost line. It spreads the weight of the structure above it over a larger area than the foundation wall would alone.

Architects design the footing based on building code regulations. Figure 20–17 shows a foundation plan for a home. The dashed lines on each side of the foundation represent the footing.

Footing Design. Footings are usually made of concrete. Sometimes carpenters build forms for the footing similar to the slab-on-grade foundation. Other times, concrete for the footing is poured directly in an accurately excavated ditch. After the concrete is placed in the form or ditch, it is leveled, Figure 20–18. This is called **screeding**. A screed is a straightedge used to push excess concrete into low spots. Screeding produces a rough surface that provides for better bonding of the mortar in which concrete blocks are set.

Foundation Wall Design. Foundation walls are usually made of concrete or concrete blocks. When concrete is used, carpenters erect forms of plywood or metal, Figure 20–19. Concrete is cast (placed) between the forms to shape the walls. After the concrete is cast and cured, the

FIGURE 20–15 Piping sleeves are placed under the slab area before the concrete is placed. *(Photo by Larry Jeffus)*

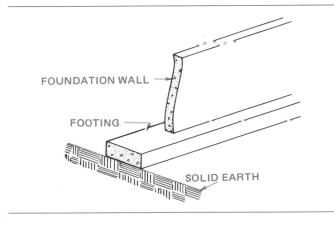

FOUNDATION WALL

FOOTING

SOLID EARTH

FIGURE 20–16 The parts of a wall foundation. The foundation wall can be solid concrete or concrete blocks.

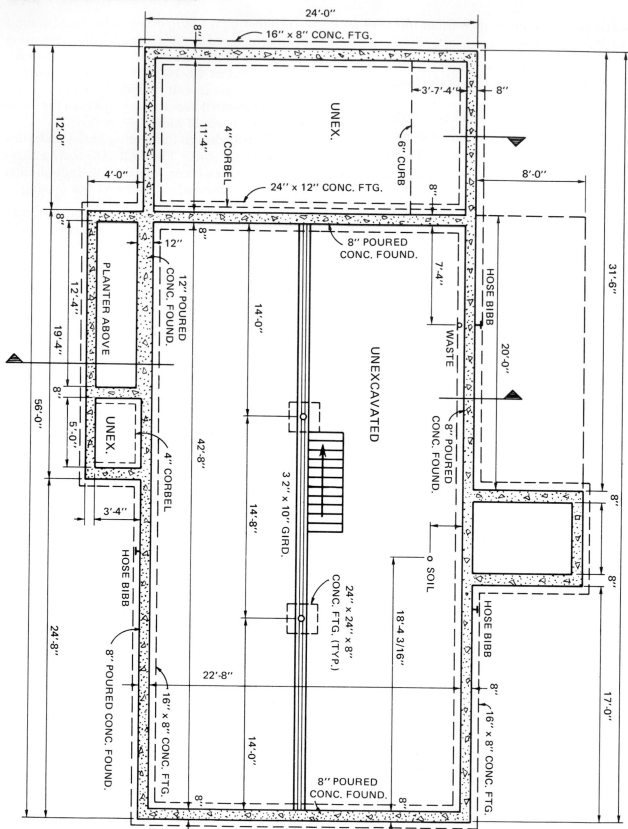

FIGURE 20–17 Foundation plan. The dashed lines show the location of the footing. *(Courtesy of Home Planners, Inc.)*

FIGURE 20–18 These workers are screeding off the concrete in a footing. Their screed rides on the wooden forms.

FIGURE 20–19 Forms for concrete foundation walls. *(Courtesy of Portland Cement Association)*

forms are stripped (removed), cleaned, and stored for use on the next job.

Concrete blocks are also popular. Each block is laid by hand by a mason. Mortar (bonding ce-

ment) is placed between each block. This is more time consuming than casting a solid concrete wall. Although block foundation walls take longer to build, the materials are less expensive.

Laying a Concrete Block Foundation. Before any mortar is mixed or blocks are laid, the masons mark the location of the wall on the footing with

Wooden Foundation Walls

Cast concrete or concrete blocks are two common materials for foundation walls. However, a type of specially treated wood foundation is gaining popularity, Figure 1. The wood for these foundations is treated with preservatives. Pressure is used to completely soak the wood in preservatives, which prevent the wood from decaying. The wooden foundation walls are also protected by covering them with a thin sheet of plastic during construction. Wood is much lighter than concrete or concrete blocks, so settling is reduced.

FIGURE 1. Wooden foundations are gaining popularity. *(Courtesy of the American Wood Council)*

a chalk line. The mortar is mixed and a full bed of mortar is spread along the footing for a length of several blocks. The blocks are set in the mortar bed. As each block is laid, it is checked for levelness and plumbness with a spirit level, Figure 20–20.

The corners are constructed first. As each course (level) of blocks is laid, a length of mason's line is stretched between the tops of the corner blocks, Figure 20–21. The masons use this line as a guide in setting the blocks to the right height. Plumb is still checked with a level. Enough courses are laid to produce the height needed for a crawl space or basement.

Foundation blocks can be laid in a stacked bond or a running bond, Figure 20–22. When the mortar is just hard enough so that it leaves a thumbprint when pressed, the joints are tooled. The most common jointing tool is the convex jointer. Jointing produces a smooth, attractive mortar joint, Figure 20–23.

FIGURE 20–21 These masons are laying blocks to a line. *(Photo by Richard T. Kreh, Sr.)*

FIGURE 20–20 A mason checks blocks for plumbness. *(Photo by Richard T. Kreh, Sr.)*

FIGURE 20–22 (A) stacked bond, (B) running bond.

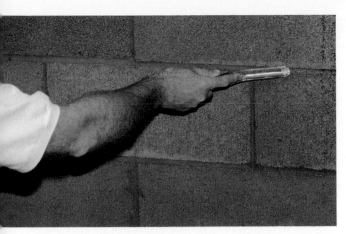

FIGURE 20–23 Before the mortar is completely set, the mason smooths the joint with a jointer.

Pile and Pier Foundations

Pile and pier foundations are used in heavy construction projects like bridges and skyscrapers. **Piles** are columns of steel or wood that are placed under a structure. They are driven deep into the ground by a pile driver, Figure 20–24. **Piers**

are usually made from cast concrete. A deep hole is dug in the ground and concrete is poured in, Figure 20–25. Both piles and piers are placed deep enough to place the foundation on bedrock.

Summary

The substructure (foundation) of any structure is built first. Its location on the site is identified with a plot plan. Building codes must be followed and building permits must be obtained before the site work begins. Site work involves laying out building lines, erecting batter boards, excavating, and constructing foundations. The location of a structure on the site can affect energy efficiency.

The type of soil on the site is important to the design of the foundation. The different types of soil include sand, gravel, loam, silt, and clay. Engineers must study the qualities and characteristics of these soils to design strong, durable foundations.

The three main types of foundations are slab-on-grade, wall, and pile or pier. Slab-on-grade foundations are often used in warm, wet climates. Wall foundations are used where the ground freezes in the winter. Piles and piers are used in heavy construction.

FIGURE 20–24 A piledriver drops a huge weight on the pile to drive it into the ground. *(Courtesy of Northwest Engineering)*

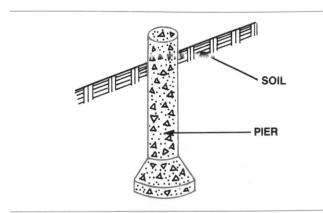

SOIL

PIER

FIGURE 20–25 Piers are made of concrete cast into a deep hole.

DISCUSSION QUESTIONS

1. What is the relationship between a plot plan, building codes, building permits, and energy efficiency?
2. After a building permit has been obtained, what are the steps involved in constructing a foundation?
3. Why are sites excavated so the foundation can be placed below the frost line?
4. What are the five types of soil?
5. What are the four engineering qualities of soil? How does each quality affect foundation design?
6. What are the key features of a slab-on-grade foundation? Use sketches for your explanation.
7. What are the key features of a wall foundation? Use sketches for your explanation.
8. What are the key features of a pile or pier foundation? Use sketches for your explanation.

CHAPTER ACTIVITIES

 Soil Testing

OBJECTIVE

Before a foundation is designed, the soil on the site is tested. Soil engineers measure percolation, soil shrinking and swelling, grain size, and compaction. In this activity, you will test the soil in an area specified by your teacher.

EQUIPMENT AND SUPPLIES

1. Shovel
2. 5-gallon bucket
3. Soil samples of clay, sand, gravel
4. Measuring tape or rule
5. Soil sieves

PROCEDURE

Percolation

1. Dig a hole 12 inches in diameter and 30 inches deep, Figure 1. Pour five gallons of water into the hole and record the time it takes for the water to completely drain. This is the percolation time for that area.
2. Do percolation tests in areas with different soil types. Compare the percolation times.

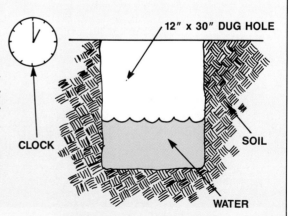

FIGURE 1. Percolation test

Shrinking and Swelling

3. Obtain several different soil samples in clear plastic cups.
4. Wet down each sample. Mark the depth of each sample on the side of the cup.
5. Freeze the samples and check the amount of swelling.
6. Dry the soil samples placing them in a warm location. After they are dry, measure the depth of the soil, Figure 2.

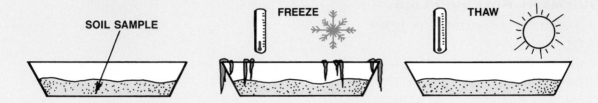

FIGURE 2. Shrinking and swelling test

Compaction

7. Obtain several different soil samples in small trays. Measure the depth of each sample.
8. Place a weight, such as a brick, on each sample. Remove the brick and measure the depth of each sample, Figure 3.

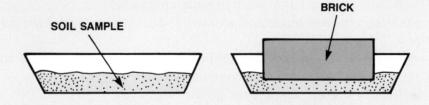

FIGURE 3. Compaction test

MATH AND SCIENCE CONNECTIONS

Although we may not think of it, soil is an important natural resource. Poor planning during construction can cause excessive soil erosion.

Also, rainwater runoff must be considered when designing a structure. The soil absorbs rainwater. When concrete or asphalt is used for roads, sidewalks, tennis courts, patios, and so on, rainwater can not be absorbed. This can cause excess runoff and flooding problems in low-lying areas.

RELATED QUESTIONS

1. How are soil percolation times important to construction? What type of soil had the "best" percolation times?
2. How are shrinking and swelling characteristics important to construction? Which soil had the best shrinking and swelling characteristics?
3. How is soil compaction important to construction? Which soil had the best compaction characteristics?
4. Who conducts soil tests? What U.S. government agency provides information on soil characteristics?

Footing Design

OBJECTIVE

Engineers design footings for structures. They must know how each type of footing will settle on a particular type of soil. In this activity, you will experiment with the importance of footing design.

EQUIPMENT AND SUPPLIES

2–3 containers measuring 1′ x 1′ x 1′

2–3 types of soil

2–3 pieces of lumber 2″ x 6″ x 10″

2–3 pieces of lumber 2″ x 4″ x 10″

2–3 concrete blocks 8″ x 8″ x 16″

Hammer and nails

Safety glasses

PROCEDURE

1. Fill each container with a different type of soil.
2. Wearing your safety glasses, nail the 2 x 6 x 10's to the 2 x 4 x 10's to make T shapes, Figure 1.
3. Place a T with the 2 x 4 x 10 on edge in each container.
4. Carefully place the concrete block on top of the T. This simulates the weight of a structure, Figure 2.
5. Measure and compare the depth to which the T settles in each type of soil.
6. Predict what would happen if the Ts were turned over and the block were placed on the edge of the 2 x 4 x 10. Check your prediction.

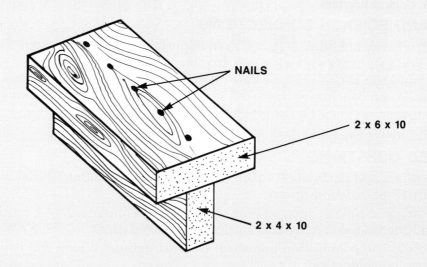

FIGURE 1. 2 x 4 "T" design

MATH AND SCIENCE CONNECTIONS

Structure loads on soil are often rated in pounds per square foot. The surface area of your T in this activity is very small. A rating of pounds per square inch can be used.

To find the load on each square inch of soil under your T footing, do this:

a. Weigh the block and T.
Add their weights together.

block weight	_____
T weight	+ _____
TOTAL WEIGHT	=

b. Measure the length and width on the edge of the 2 x 4 x 10 in inches.

2 x 4 length	= _____
2 x 4 width	= _____

c. Multiply the length by the width to get an area measure.
Your answer will be square inches.

length x width	= _____

d. Divide the weight of the T and block by the area in square inches.

TOTAL WEIGHT	_____
SQ. INCHES	_____

e. Your answer will be pounds per square inch.

LBS/SQ. INCH	= _____

RELATED QUESTIONS

1. Why is settling or compaction important in construction?
2. Which soil provided the best footing in this activity?
3. What can be done to reduce the settling in the other soils?
4. What was the load on your simulated footing in pounds per square inch?

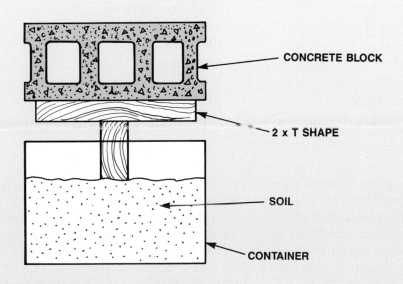

CONCRETE BLOCK

2 x T SHAPE

SOIL

CONTAINER

FIGURE 2. Loading the simulated foundation

CHAPTER 21

Floor and Wall Systems

OBJECTIVES

After completing this chapter, you will be able to:

■ Identify and describe the parts of a floor framing system.

■ Identify and describe the parts of a wall framing system.

■ Explain how wood, metal, concrete, and masonry are used in floor and wall systems.

■ Differentiate between bearing wall and skeleton frame construction.

■ List and compare the different types of loads that structures must support.

KEY TERMS

Bearing-wall construction	Joist	Subfloor
Box beam	Load-bearing wall	Tilt-up construction
Dead load	Live load	Trimmer
Deck	Partition	Underlayment
Girder	Sheathing	Wind bracing
Header	Sill	
Jack studs	Skeleton-frame construction	

Preparing the site and building the foundation are important first steps in constructing any structure. Site work and the foundation provide the base upon which the rest of the structure is built. When the foundation is complete, the next step is to build the first floor and its walls. Both walls and floors are built by framing. Framing involves combining a number of smaller pieces to produce a larger, strong unit. This chapter focuses on framing floors and walls for buildings.

Floor Systems

Any floor system must perform two basic functions. It must provide the necessary support for any loads placed on it, and it must provide a **deck** (flat surface). The main support for the first floor deck is the foundation wall. The floor is framed from a number of beams. The beams, plus any columns or posts placed under the floor inside the foundation wall, also support the floor. The deck can be made of lumber, plywood, metal, or concrete. Architects design the layout of the floor. Structural engineers make sure the floor design can support its loads, Figure 21–1.

Floor Framing Design

The design of the floor framing consists of the sill, girders and posts, joists, and the subfloor, Figure 21–2. Each part is important to the strength of the floor system. Let's look at each part.

The Sill

The **sill** is the wood member directly on top of the foundation. It provides a base for fastening the floor frame to the foundation. The sill is usually made from 2″ x 6″ or 2″ x 8″ lumber. It is attached to the foundation with anchor bolts which have been inserted into the top of the foundation wall.

Girders and Posts

Usually buildings are too wide for continuous pieces of lumber to span the full width. The floor framing is supported by one or more **girders** running the length of the building. A girder is a large steel I-beam or built-up lumber, Figure 21–3. It is supported at regular intervals by posts or columns.

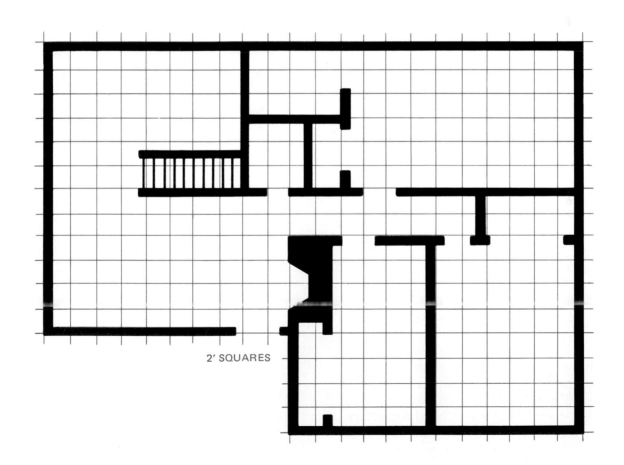

2′ SQUARES

FIGURE 21–1 Architects design floor plans and engineers make sure the design can support its loads.

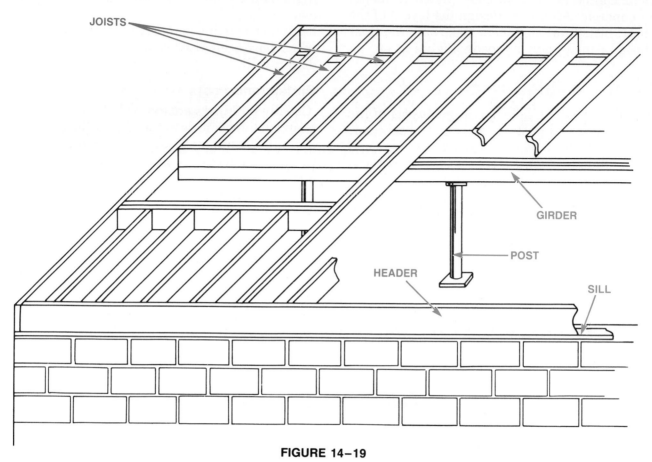

FIGURE 14–19

FIGURE 21–2 The main parts of a floor frame include the sill, girders and posts, and the joists. The subfloor will be added next.

Box Beams

Many girders are made by nailing together several pieces of 2″ x 8″ or 2″ x 10″ lumber face to face. Recently, many buildings have been constructed with **box beams**, Figure 1. A box beam is made up of a lumber frame covered with plywood. The lumber used usually measures 2″ x 4″. The plywood covering makes the box beam rigid and strong. Box beams are lighter in weight and use less material than normal girders. That makes them more cost efficient in construction.

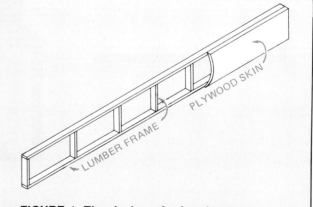

FIGURE 1. The design of a box beam.

A

B

FIGURE 21–3 Girders can be made from (A) steel I-beams or (B) built-up wood beams.

Joists: Light Construction

Joists are closely spaced beams that rest on the foundation walls or girders. In light, residential construction, wood joists are used most often. The inner ends of the joists either overlap the girder or butt against it. When the joists butt against the girder, they can be supported by metal joist hangers, Figure 21–4. The outer ends rest on the sill and are nailed to the header. The **header** is a band of lumber that runs around the top of the sill and ties together the joists (see Figure 21–2).

Floor joists are usually spaced either 16 inches apart or 24 inches apart. Wood floor joists used in houses are usually 8, 10, or 12 inches deep. Where walls, stairs, or other extra weight will be located above, the joists are doubled for added strength.

FIGURE 21–4 Metal hangers used to attach joists to a girder.

Joists: Heavy Construction

In heavy and commercial construction projects, steel and concrete are often used for joists. The three main types of joists used in heavy construction are the open-web steel, structural steel, and concrete beam.

Open-Web Steel Joists. In metal-frame buildings, open-web steel joists are used, Figure 21–5A. These joists consist of a top chord, bottom chord, and diagonal web members, Figure 21–5B. They

A

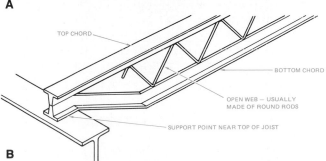

TOP CHORD

BOTTOM CHORD

OPEN WEB — USUALLY MADE OF ROUND RODS

SUPPORT POINT NEAR TOP OF JOIST

B

FIGURE 21–5 Open-web steel joists (A) are used in metal framing. Their parts are shown in (B). *(Photo courtesy of Cargill, Wilson, & Acree, Inc.)*

can span fairly wide areas and can support a heavier load than wood joists. Open-web steel joists are either riveted, bolted, or welded in place.

Structural Steel Joists. On skyscrapers and bridges, structural steel joists are used for the floor or deck framing, Figure 21–6. Structural steel beams, such as I-beams, can support much greater loads than any other type of joists. They can also be spaced much farther apart. Structural steel does, however, have some disadvantages. One is its weight. Structural steel beams are much heavier than wood or open-web joists. A crane is needed to raise them into place.

Concrete Beams. Concrete joists can be either site-cast or precast. Site casting refers to placing concrete in forms at the construction site, Figure 21–7. This is time-consuming, but there is no

FIGURE 21–7 These cement masons are casting a concrete beam on the site. *(Courtesy of Symons Corporation)*

FIGURE 21–6 Structural steel floor joists are used in this skyscraper.

limit to the size of beam that can be made. Constructing with precast concrete beams is quicker. Concrete is placed in reusable forms at a precast yard. When the concrete has cured, the beams are delivered to the construction site ready to use. Workers position the beam and fasten the concrete members with steel fittings which were cast into the members, Figure 21–8. Precast concrete beams must be small enough to fit on trucks for delivery to the site.

Subflooring

The final part of the floor is the subfloor. The **subfloor** is the first layer of material applied over the floor joists, Figure 21–9. Once the subfloor is in place, carpenters and other construction personnel have a smooth platform on which to work. Most subflooring used in light construc-

FIGURE 21–8 This worker is directing the placement of a precast concrete beam. *(Courtesy of Portland Cement Association)*

FIGURE 21–9 Carpenters install subflooring over the floor joists. *(Courtesy of American Plywood Association)*

tion is ½″ or ⅝″ plywood. Nails and glue are used to fasten the plywood to the joists. **Underlayment** (a second layer of plywood) is laid over the subfloor. Carpet or floor tile is placed over the underlayment.

Steel Decking. In steel structures sheet steel provides a rapid means of installing a deck or floor. Usually, a ribbed sheet steel is used. This is covered with a thin layer of concrete to produce the floor.

Reinforced Concrete Slabs

The simplest floor to construct is a reinforced concrete slab. Reinforced concrete slabs are similar to the slab-on-grade foundation. Concrete slabs can be used for the floors in structures that are several stories high.

Reinforced concrete is not very strong as a floor material above the first floor. In order to produce a stronger reinforced concrete floor, ribs are often included, Figure 1A. The ribs greatly increase the strength of the concrete. Even stronger floors are created when a waffle slab is used, Figure 1B.

FIGURE 1. (A) The ribs greatly improve the strength of this concrete slab. (B) Even stronger slabs are created with this waffle slab. *(Photo B courtesy of Portland Cement Association)*

Wall Systems

After the first floor is in place, the walls can be constructed. The walls in a house serve many purposes. Some walls support the weight of the structure above them. These are called **load-bearing walls**. Other walls, called **partitions**, only support their own weight. They divide the house into separate rooms. Outside walls protect the inside of the house from weather. They must also create a pleasing appearance inside and outside.

Walls can be constructed of masonry, concrete, or with wood or metal frames. In home construction most walls are of the wood-frame type. Steel and concrete are used for larger structures.

Wall-Framing Members

The basic parts of a wall frame are the top and bottom plates and the studs, Figure 21–10. The plates are the horizontal members at the top and bottom of the wall. They hold the ends of the studs in place. They also provide a surface to fasten the wall in place on the floor framing. In most walls, the plates and studs are made of 2″ x 4″ lumber. The wall frame also provides a surface on which to nail the inside and outside wall coverings. Where two outside walls meet at a corner, an extra stud must provide this nailing surface.

One or more of the studs must be cut off for window and door openings. These are called rough openings. The weight normally supported by the removed studs must be transferred to other studs. A header transfers this load to the sides of the opening, Figure 21–11. To support the extra weight in a rough opening, a second stud is used. This stud is called a **trimmer**. The trimmer supports the header and strengthens the full stud.

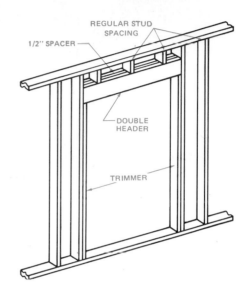

FIGURE 21–11 Headers and trimmers help support wall loads around a doorway.

Window rough openings do not extend to the floor. At the bottom of the window rough opening is a horizontal member called a sill. The sill extends from one stud to the other, Figure 21–12. Extra studs, called **jack studs**, are added below the sill for extra strength.

Sheathing

Sheathing is the first layer of wall covering on the outside of the walls. It provides a surface on which to nail the finished siding. Boards can be used, but the most common sheathing materials are sheets of plywood, fiberboard, and rigid plastic foam. Plywood makes the entire structure more rigid. Fiberboard and rigid foam are not as strong. If they are used, plywood sheathing is added at the corners of the building for strength, Figure 21–13.

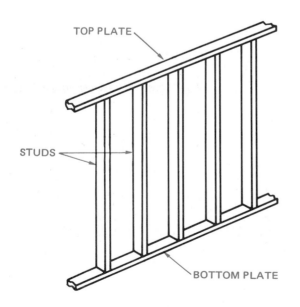

FIGURE 21–10 Studs and plates are the basic parts of a wall frame.

After the studs are cut, all the materials are laid out on the subfloor and nailed together. Many carpenters prefer to nail the wall sheathing on before the wall is tilted up into place. The carpenter must make sure the wall is square before the sheathing is applied. This can be checked by measuring the diagonals. The sheathing panels make the wall frame rigid. Once the wall framing is completely assembled, it is tilted up and slid into position, Figure 21–14. The bottom plate is nailed to the floor. The wall sections are plumbed and braced to hold them in place.

After all of the framing is in place, a second top plate is nailed in position. The corners of the top member of the top plate are lapped over the bottom members. This helps tie the walls together and makes the structure stronger.

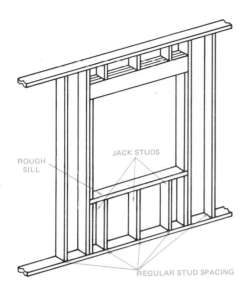

FIGURE 21–12 The sill and jack studs help support loads around a window.

Constructing Walls

The first step in constructing walls is to mark their location on the subfloor. A chalk line is usually used to mark the location of a wall. After the location is marked the top and bottom plates are cut. Next, the locations of all door and window openings are marked on the plates. Then, the location of all studs are marked on the plates at 16- or 24-inch intervals. The location of extra studs for corners is also marked.

FIGURE 21–14 Carpenters are tilting up a completed wall frame assembly.

FIGURE 21–13 Plywood is often used to brace the corners when fiberboard or rigid foam is used. *(Photo by Larry Jeffus)*

Using Lasers to Build Floors and Walls

Many builders now use lasers to lay out the location of floors and walls. Laser light is perfectly straight. It can be used in place of a chalk line. The laser shown in Figure 1 creates a 360-degree laser reference line. The light source rotates to project the laser reference line. The reference line can be used to mark the location of floors and walls. The location of walls is projected on the floor and ceiling at the same time. Lasers of this type can be used for almost any leveling or plumbing job in construction.

SAFETY NOTE: Some lasers produce powerful light beams that can be harmful to humans. Some lasers used in manufacturing can even cut materials. The lasers used in construction are low power and can not cut materials. However, you should **never** look directly into the light beam of a laser.

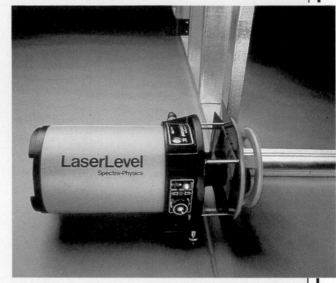

FIGURE 1. Lasers are used to help carpenters build walls and floors. *(Courtesy of Spectra-Physics)*

Special Wall-Framing Considerations

As the carpenters build the walls for a building, they must allow for features that will be added later. Openings sometimes have to be framed for heating, ventilating, plumbing, and air-conditioning (HVAC), Figure 21–15. If the design calls for 4-inch pipes, thicker walls are framed with 2″ x 6″ lumber. Of course, this must all be planned ahead of time.

Energy Efficiency of Walls. In recent years construction systems have been developed to conserve energy. Developments are intended to reduce the loss of heat in winter or the gain of heat in summer.

When wall framing is done with 2 x 4s spaced 16 inches on centers, a fairly large amount of solid wood is exposed to the surface of the wall. Wood conducts heat out of the building. Only the space between the solid wood framing can be filled with insulation. By using 2 x 6 studs

FIGURE 21–15 Openings must be framed into walls for heating, ventilating, plumbing, and air conditioning systems. *(Photo by Richard T. Kreh, Sr.)*

placed 24 inches on centers, the area of solid wood exposed to the surface of the wall is reduced by 20 percent. This also allows for two inches more insulation, Figure 21–16.

Using rigid plastic foam board sheathing can add as much insulating value as that found within a conventional 2 x 4 wall frame. The insulating properties of the sheathing can be further improved by an aluminum foil surface. The aluminum foil surface reflects radiant heat so that it cannot penetrate the wall.

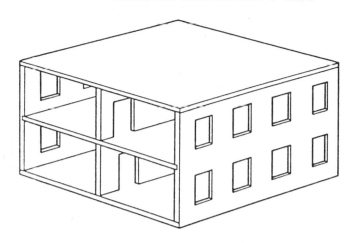

FIGURE 21–16 Insulation is added between wall studs for improved energy efficiency.

Bearing-Wall Construction versus Skeleton-Frame Construction

The types of construction used in buildings can be classified as either bearing-wall or skeleton-frame. The basic difference between the two types is in the way the building loads are supported. In **bearing-wall construction**, solid walls support the structure, Figure 21–17A. In **skeleton-frame construction**, the walls are made of individual members with space between them, Figure 21–17B. The frame looks like a skeleton. The wood frame walls described above are an example of skeleton-frame construction. Masonry

FIGURE 21–17 With bearing-wall construction (top), the walls support the structure. In skeleton-frame construction (bottom), individual members are combined to make a frame. *(Photo courtesy of Niagara Mohawk Power Corporation)*

and concrete walls are examples of bearing-wall construction.

Bearing Walls

Masonry. Masonry units, such as bricks or blocks, are widely used in bearing-wall construction up to a few stories high, Figure 21–18. For very tall structures, masonry bearing walls are seldom used. To support the load of a skyscraper, masonry walls would have to be very thick at their base.

Concrete. Precast concrete panels are very common materials for bearing-wall construction, Figure 21–19. Reinforced concrete provides excellent strength. The precast method allows the walls to be assembled quickly after they are delivered to the site.

To eliminate the need for transportation, heavy panels can be cast at the construction site. Forms are made for the edges of the panels and for the window and door openings. Reinforced steel is placed in the forms and tied or welded, then the concrete is placed. When the concrete has cured, the panels are tilted up into place, Figure 21–20. This is called **tilt-up construction.**

Skeleton-Frame Construction

Structural Steel. Skyscrapers can be constructed with reinforced concrete. Steel provides better strength with much less weight than concrete. This makes it a common material for frames of tall structures. Steel frames can also be put up quickly. This saves time and money.

The individual pieces of steel are lifted into position with a crane. Until the middle of this century, nearly all connections in structural steel were made with rivets. Today, welding and bolts

FIGURE 21-19 Precast concrete panels used for bearing-wall construction. *(Courtesy of Portland Cement Association)*

have taken the place of rivets. It is common for both methods to be used on one construction job.

Welding has become an important trade in the construction industry. Welding is done by wel-

FIGURE 21–18 Masonry is used in bearing-wall construction.

FIGURE 21–20 Tilting up a concrete wall panel. *(Courtesy of Portland Cement Association)*

ders or ironworkers. One of the most important aspects of their jobs is that they work in high places under dangerous conditions, Figure 21–21.

Vertical Loads. Structural engineers are responsible for making sure the structure will stand up under certain types of loads. There are two types of vertical loads that must be supported by walls — dead loads and live loads. The **dead load** is the weight of all the parts of the structure. For example, the dead load includes the weight of the foundation, walls, roof, windows, and all other parts of the structure, Figure 21–22. **Live loads** include the weights of people, furniture, and rain and snow. Live loads can be removed, but dead loads are a permanent part of the structure.

Lateral Loads. In taller buildings the wind applies large forces to the sides of the building. In some areas, earthquakes also apply forces sideways to the building. These horizontal forces are called lateral loads.

Lateral loads are resisted by wind bracing or shear walls. **Wind bracing** refers to diagonal

FIGURE 21–22 Dead load includes the weight of all the materials used to build a structure.

members added to skeleton-frame construction. Wind bracing forms triangular shapes on the frame, Figure 21–6. This prevents the building frame from flexing in that direction.

Summary

After the foundation is completed, floors and walls can be constructed. Both floors and walls are framed. The floor framing system is composed of the sill, girders and posts, joists, and the subfloor. Wall framing systems are composed of the top and bottom plates, studs, headers, sills, and sheathing. Wood is the primary floor and wall framing material in light construction. Steel and concrete are used primarily in heavy construction.

Wall systems can be described as bearing wall or skeleton frame. Bearing walls are solid, while skeleton frames are made from a number of individual parts with space between them. Structures, including floors and walls, must support vertical and lateral loads.

FIGURE 21–21 Ironworkers must be able to work in high places. *(Courtesy of Cargill, Wilson, & Acree, Inc.)*

DISCUSSION QUESTIONS

1. What are the different names given to the parts used to frame a floor?
2. What are the different names given to the parts used to frame a wall?
3. What are some important considerations when building floors and walls?
4. How are wood, metal, concrete, and masonry used in floor and wall systems?
5. What is the difference between bearing-wall and skeleton-frame construction?
6. What are the different types of vertical and lateral loads that structures must resist?
7. How are lasers used in building walls and floors?

CHAPTER ACTIVITIES

 ## Testing Model Box Beam Girders

OBJECTIVE

Girders are used to support floor joists. They can be made by nailing together several pieces of large lumber. Box beams are improved girders. They are made by nailing together several pieces of smaller lumber in a frame. The frame is then covered with a plywood skin. In this activity, you will make and test model box beam girders.

EQUIPMENT AND SUPPLIES

Grid paper
Wax paper
Modeling knife
Chip board or drawing paper
Safety glasses
Sheet of cardboard

$\frac{1}{8}$″ x $\frac{1}{8}$″ balsa wood
Hot glue or white glue
2 hand screw clamps
Bucket and sand
Bathroom scale
Supply of straight pins

PROCEDURE

1. Sketch the plans for the scale model box beam, Figure 1, on grid paper. Count the number of squares on the plans and transfer the size to your grid paper.

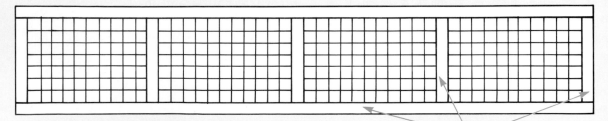

FIGURE 1. Plans for a model box beam.

BALSA WOOD FRAME

2. Tape your finished sketch to the sheet of cardboard. Tape the wax paper over the sketch. This will prevent the glue from fastening your box beam to your sketch.

3. Place the pieces of balsa wood over the parts on your sketch. Cut the balsa wood to the exact sizes needed.

4. After all of the pieces are cut to length, glue them together over the sketch and wax paper. Use the straight pins to hold the pieces in place as the glue dries.

5. When the glue is dry, remove the straight pins. Place the box beam on top of the chip board (or drawing paper). The chip board will simulate the plywood skin on your box beam. Trace around the box beam very accurately. Cut out chip board rectangles for both sides of your box beam.

6. Glue the chip board to both sides of the box beam. Secure with straight pins while the glue dries.

7. When the beam is complete, measure its weight. This is its dead load.

8. Test the strength of your box beam. Clamp the beam in an upright position between the clamps. Position the beam between two tables. Place the bucket over the center of your beam. Add sand to the bucket until the box beam fails. Weigh the sand. This is the live load.

9. Calculate the efficiency of your box beam with the following formula:

$$\text{EFFICIENCY} = \frac{\text{LIVE LOAD}}{\text{DEAD LOAD}} \times 100$$

10. Determine who had the most efficient box beam in the class.

MATH AND SCIENCE CONNECTIONS

The efficiency you calculated is a percentage. This number tells you the percentage of its own weight that the box beam was able to support. For example, if your efficiency is 200, your beam was able to support 200% of its own weight.

RELATED QUESTIONS

1. What was a typical efficiency for the box beams tested in your class?
2. Why do you think some beams were more efficient than others?
3. Why are box beams used instead of built-up girders?
4. What is the purpose of a box beam girder in constructing a floor?

 ## Testing Concrete Beams

OBJECTIVE

Concrete is often used as a material for beams, especially for heavy construction projects. Concrete is not really a very good beam material, however; it must often be reinforced. In this activity, you will experiment with different concrete beam reinforcement techniques, Figure 1.

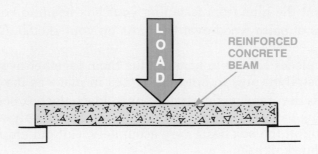

FIGURE 1. Build and test reinforced concrete beams.

EQUIPMENT AND SUPPLIES

Concrete Beam Test Graph (from teacher)
Concrete beam form (get plans from your teacher)
Motor oil or plastic wrap
Concrete mixture (fine aggregate)
Bucket
Concrete mixing tools

Bathroom scale
Reinforcement:
  fishing line, wire, coat hanger
Safety glasses
Water
Sand
Screed (small piece of wood)

PROCEDURE

1. Prepare the concrete beam form. Each person in the class should make his own form. (Follow all safety rules.)
2. Label each form as NONE, LINE, WIRE, or HANGER. These names relate to the type of reinforcement that will be used. Try to mark an equal number of each type.
3. Mix enough concrete to fill all the forms. Ask your teacher for assistance.
4. Line the wood forms with motor oil or plastic wrap. This will make it easy to remove the concrete beams once they have cured.
5. Place the correct reinforcement in each form. The reinforcement should be placed near the bottom of the beam. Design a way to keep the reinforcement from lying on the bottom of the form.
6. Place concrete in each form. Be sure to keep the reinforcement from lying on the bottom. Shake or vibrate each form so all the air bubbles are removed from the concrete.
7. Screed off the top of each concrete beam. Fill in any low spots. Level the concrete with the top of the form.
8. Let the concrete cure for at least 24 hours.
9. When the concrete has cured, test each beam. Make sure you keep the different types of reinforcement in separate groups. Support each beam at its ends. **Make sure the reinforcement is on the bottom of the beam**. Hang a bucket across the center of the beam and add sand or other weights, Figure 2. Keep adding weight until each beam cracks and fails.
10. Calculate the average weight that caused each type of beam to fail. Graph the average weight on the Concrete Beam Test Graph.

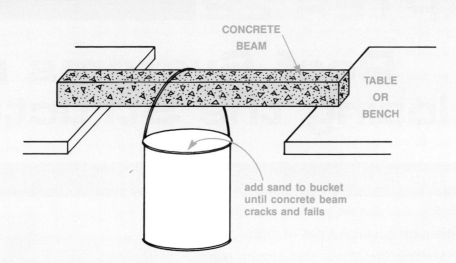

FIGURE 2. Testing a concrete beam.

MATH AND SCIENCE CONNECTIONS

1. It is the chemical reaction between the water and cement in the concrete mixture that makes concrete hard. The strength of concrete is directly related to the ratio of water to cement in the mixture. The lower the water content, the stronger the concrete will be (if all other ingredients are kept constant). There must be just enough water to mix the cement around each grain of sand and aggregate.
2. Graphing data is an important math skill. Graphs are often used to show the relationship between two or more factors quickly and easily.

RELATED QUESTIONS

1. Did all of the reinforced beams hold more weight than beams without reinforcement?
2. Which reinforcement technique worked best? Why?
3. What materials are used to reinforce real concrete beams?
4. Would the reinforcement have worked if it had been placed near the top of the beam instead of near the bottom? Why or why not?

CHAPTER 22

Roof Systems and Enclosing the Structure

OBJECTIVES

After completing this chapter, you will be able to:

- Identify the most common types of roofs.
- List and describe the various roof framing terms.
- Describe the processes and parts involved in framing and completing a roof.
- Describe the purposes of enclosing a structure.
- Describe the various types of doors and windows, siding or masonry veneer, and exterior trim.
- Compare a variety of materials used to finish the outside of residential buildings.

KEY TERMS

Bond	Gusset	Rise
Bottom chord	Hinge gain	Run
Casing	Hip roof	Sash
Collar beam	Jamb	Shed roof
Cornice	Lath	Soffit
Fascia	Mansard roof	Span
Fascia Header	Measuring line	Tail
Frieze	Pitch	Top chord
Gable roof	Pneumatic structure	Web
Gambrel roof	Rafter	
Geodesic dome	Ridge board	

As the floors and walls are constructed, they start to enclose the inner space of a structure. Two more steps are needed to completely enclose the structural frame: adding the roof system and enclosing the structure.

The roof of a house protects the structure and the people inside from rain and snow. Enclosing the structure involves installing windows, doors, and other parts that fill holes in the structural frame.

Roof Systems

Five types of roofs are commonly used in residential construction, Figure 22–1. Each type of roof can be changed slightly for a different style.

Gable Roof

The **gable roof** is one of the most common types used on houses. It has two sloping sides that meet at the ridge (top). The triangle formed

316

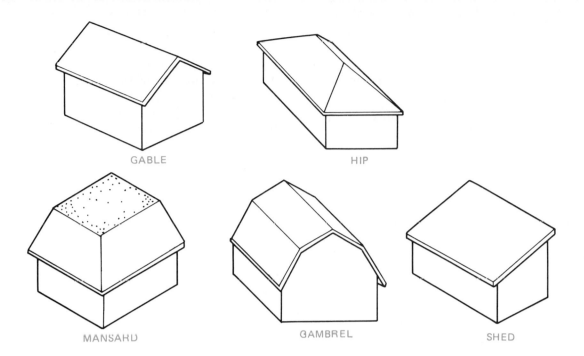

GABLE

HIP

MANSARD

GAMBREL

SHED

FIGURE 22–1 Common roof types.

at the ends of the house between the wall and the roof is called the gable.

Gambrel Roof

The **gambrel roof** is similar to the gable roof. The sides of this roof slope very steeply from the walls. At a point about halfway up the roof, they have a more gradual slope.

Hip Roof

The **hip roof** slopes on all four sides. The hip roof has no exposed wall above the top of the plates. This results in all four sides of the house being equally protected from the weather.

Mansard Roof

The **mansard roof** is similar to the hip roof. The lower half of the roof has a very steep slope and the top half is more gradual. This roof is often used in commercial construction, such as on stores.

Shed Roof

The **shed roof** is a simple sloped roof with no ridge. The shed roof is not as common as the other types. It is used on some modern houses and additions to houses.

Roof Design

The roof is framed just like the floors and walls. The roof framing members that extend from the wall plates to the ridge are called **rafters**. On a shed roof the rafters span the entire structure.

Ceiling Joists

Before carpenters can begin setting rafters, ceiling joists must be installed. These are similar to floor joists. The ceiling joists perform two functions. They provide a surface on which to attach the ceiling. They also prevent the rafters from pushing the walls outward. The size and direction of the ceiling joists are shown on the floor plan. The ceiling joists can be nailed to the top plate or fastened with special metal anchors.

Roof-Framing Terms

The main parts in a roof above the ceiling joists are the rafters, collar beams, and the ridge board,

Figure 22–2. The **ridge board** is nailed between the tops of the rafters. **Collar beams** are placed between every third pair of rafters for extra support. The rafters are placed at the angle of the roof.

There are a number of special roof-framing terms. They are shown on Figure 22–3.

Span is the total width covered by the rafters. This is usually the distance between the outside walls.

Run is the width covered by one rafter. If the roof has the same slope on both sides, the run is one half the span.

Rise is the height from the top of the wall plates to the top of the roof.

Measuring line is an imaginary line along the center of the rafter. This is where the length of the rafter is measured.

Tail is the portion of the rafter that extends from the wall outward to create overhang.

Pitch is the steepness of the roof. This is usually given as the number of inches of rise per foot of run. For example, if the height of the roof changes 5 inches for every 12 inches horizontally, the pitch is referred to as 5 in 12.

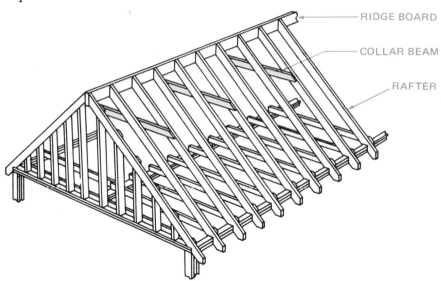

FIGURE 22–2 Roof-framing parts include the rafters, collar beams, and the ridge board.

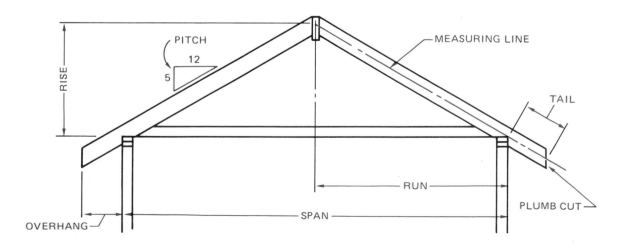

FIGURE 22–3 Roof-framing terms.

Applying Mathematics: Calculating Rafter Lengths

Carpenters use rafter tables to determine the length of the rafters. These tables are available in handbooks. They are also printed on framing squares, Figure 1. Here is how you would find the length of rafters in Figure 2.

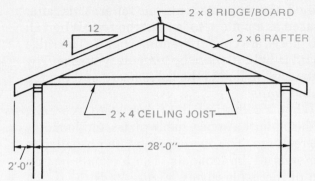

FIGURE 2. Example problem for calculating rafter length.

FIGURE 1. This is a close-up of the rafter tables printed on a framing square.

	18	17	16	15	14	13	12
LENGTH COMMON RAFTERS PER FOOT RUN	21 63	20 81	20	19 21	18 44	17 69	16 97
" HIP OR VALLEY 16 INCHES CENTERS	24 74	24 02	23 32	22 65	22	21 38	20 78
DIFF IN LENGTH OF JACKS 2 FEET	28 7/8	27 3/4	26 11/16	25 5/8	24 9/16	23 9/16	22 5/8
" JACKS	43 1/4	41 5/8	40	38 7/16	36 7/8	35 3/8	33 15/16
SIDE CUT OF JACKS USE	6 11/18	6 15/16	7 3/16	7 1/2	7 13/16	8 1/8	8 1/2
" HIP OR VALLEY	0 1/4	8 1/2	8 3/4	9 1/16	9 3/8	9 5/8	9 7/8

	12	11	10	9	8	7	6	5	4	3	2
	16 97	16 28	15 62	15	14 42	13 89	13 42	13	12 65	12 37	12 16
	20 78	20 22	19 70	19 21	18 76	18 36	18	17 69	17 44	17 23	17 09
	22 5/8	21 11/16	20 13/16	20	19 1/4	18 1/2	17 7/8	17 5/16	16 7/8	16 1/2	16 1/4
	33 15/16	32 9/16	31 1/4	30	28 7/8	27 3/4	28 13/16	26	25 5/16	24 3/4	24 5/16
	8 1/2	8 7/8	9 1/4	9 5/8	10	10 3/8	10 3/4	11 1/16	11 3/8	11 5/8	11 13/16
	9 7/8	0 1/8	10 3/8	10 5/8	10 7/8	11 1/16	11 5/16	11 1/2	11 11/16	11 13/16	11 15/16

PROCEDURE	EXAMPLE
1. Find the number of inches of rise per foot of run on the regular graduations on the square.	4
2. Under this number, find the length of the rafter per foot of run. A space between the numbers indicates a decimal point.	12 65″
3. Multiply the length of the rafter per foot of run by the number of feet of run.	12 65″ x 16 = 202 40″ (round off to 202 ½″)
4. Add the length of the tail and subtract one half the thickness of the ridge board. This is the length of the rafter as measured along the measuring line.	202 ½″ + 18″ − 0 ¾″ ————— 219 75″

Building the Roof Frame

When the length of the rafters has been determined, the carpenter cuts one rafter. This rafter is used as a pattern to lay out the rest. When all the pieces are cut out, framing begins. The ridge board is held in place by temporary braces. The rafters are nailed to the ridge board and the top of the walls.

When a rafter and the ridge board overhang at the ends of a house, gable plates and lookouts are required. The overhanging rafter is nailed to the ends of the lookouts. The lookouts rest on top of the gable plates, Figure 22–4.

The final member of the roof frame is the **fascia header.** This is a piece of lumber the same size as the rafters nailed to the ends of the rafters. The fascia header will support trim to be added later.

Trussed Rafters

Roof trusses are another way to frame a roof. Roof trusses are assembled in a shop. They are then transported to the construction site and set on the walls. Roof trusses reduce building time and make a stronger roof. They are used quite often today in construction.

Figure 22–5 shows the parts of a roof truss. The top members of the truss are called **top chords**. These are similar to rafters. The bottom of the truss is called the **bottom chord**. These are similar to ceiling joists. Braces between the chords are called **webs**. **Gussets** (plywood or metal plates) are nailed over the pieces at each joint. There are many styles of roof trusses. Figure 22–6 shows just a few common designs.

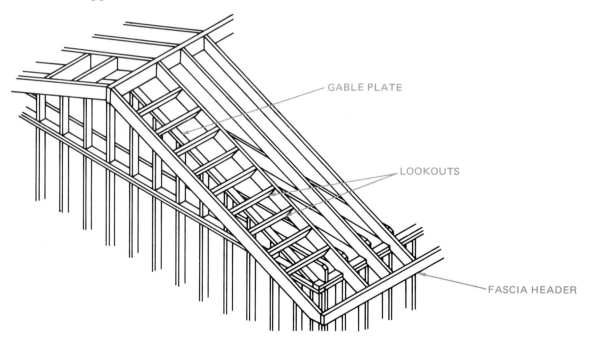

GABLE PLATE

LOOKOUTS

FASCIA HEADER

FIGURE 22–4 Framing for an overhang on the gable end.

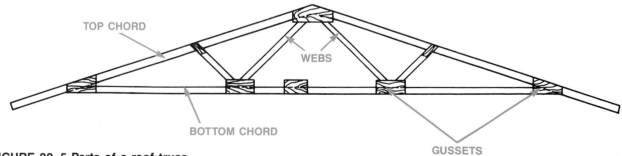

TOP CHORD

WEBS

BOTTOM CHORD

GUSSETS

FIGURE 22–5 Parts of a roof truss.

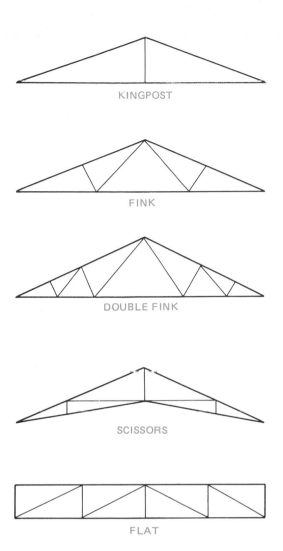

KINGPOST

FINK

DOUBLE FINK

SCISSORS

FLAT

FIGURE 22–6 Some common types of roof trusses.

FIGURE 22–7 Plywood sheathing being installed over the rafters. *(Photo by Larry Jeffus)*

Roof Covering

Roof covering usually involves three steps. The first step is to install sheathing over the rafters, Figure 22–7. This is the same as adding subflooring over floor joists. Underlayment is added next over the sheathing. It protects the sheathing from the weather. Asphalt-saturated felt is the most common roofing underlayment material. Finally, a top layer of shingles is added. Asphalt strip shingles and wood shingles and shakes are the most common types of shingles. Clay tile and metal roofing are also used.

Figure 22–8 shows the arrangement of shingles on a roof. Shingles are applied starting at the eaves and working toward the ridge. Each row of shingles across the roof is called a course. Each course is laid with the bottom edge lined up with the tops of the tabs of the course below. (A tab is the part that is exposed to the weather.) Also, each course overlaps so the cutouts between tabs will not fall over an end joint in the course below. After every few courses the roofers strike a chalk line to be sure that the courses are straight and parallel with the ridge.

At the ridge, the tops of the shingles are lapped over the ridge. Ridge shingles can be purchased precut or made by cutting strip shingles apart at each tab.

Enclosing the Structure

After the roof is installed, the structure is enclosed. Enclosing the structure involves adding trim, doors, windows, and exterior wall coverings. Many of these features serve two purposes. First, they protect the frame and inside of the building. Second, they improve the appearance of the structure. Colors and special designs and details are often added for appearance.

Cornice: Roof Trim

The trim at the edge of the roof makes up the **cornice**. It is usually made of wood or a combination of wood and aluminum. Wood cornices are

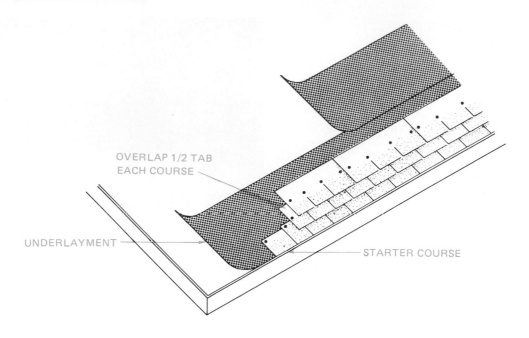

FIGURE 22–8 Arrangement of shingles on a roof.

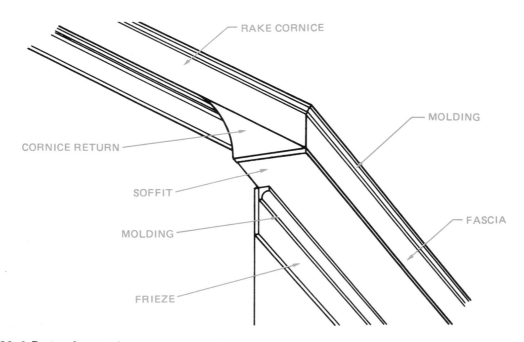

FIGURE 22–9 Parts of a cornice.

made by carpenters. There are several parts to the cornice. The main parts are the **frieze**, **soffit**, and **fascia**, Figure 22–9. The frieze is a horizontal piece against the wall of the building. The soffit is the covering on the underside of the

rafters. The fascia is nailed to the ends of the rafters. Molding can also be used to create a pleasing appearance. The soffit has openings to allow air to flow between the rafters and ventilate the attic.

Walls and Roof All-in-One: Geodesic Domes and Pneumatic Structures

Two interesting innovations in construction technology are geodesic domes and pneumatic structures. Both of these structures have the unique feature of providing walls and a roof in one structure.

Geodesic Domes

Geodesic domes are structures made of a series of triangles placed on the surface of a sphere. The triangle is a very strong and stable shape. Geodesic domes are also very strong and stable. The edges of the triangles on a geodesic dome are called struts. They can be made from wood, aluminum, or steel. Epcot Center in Disney World is a famous geodesic dome. Some families have built geodesic domes for their homes, Figure 1.

Pneumatic Structures

Pneumatic structures have a plastic sheet shell that is supported by air pressure. Plastic sheet material is formed in the desired shape (usually an arch). Then huge fans blow air under the sheet to inflate it. Pneumatic structures are often used to cover tennis courts, swimming pools, or sports practice fields. For large sports stadiums the pneumatic structure is placed on top of a wall around the arena. This raises the pneumatic structure high above the playing field. Several cities have pneumatic structures over their sports stadiums, including the Hoosier Dome in Indianapolis, Pontiac Silverdome in Michigan, Metrodome in Minnesota, and Carrier Dome in Syracuse, New York, Figure 2.

FIGURE 1. Geodesic dome.

FIGURE 2. Pneumatic structure. *(Courtesy of Syracuse University Photo Center)*

The Houston Astrodome

On April 9, 1965, the Houston weather forecast called for rain, possibly thunderstorms, with strong winds. That was the day the Astrodome was opened. The world's first all-weather, multi-purpose stadium is the model by which every sports arena since has been measured.

The cost of the Astrodome was $31.6 million.

The stadium structure covers nine and a half acres of land. With the parking areas (and other facilities), the complex covers 260 acres.

The outside diameter of the Astrodome structure is 710 feet. The stadium's floor is approximately 25 feet below normal ground level.

The roof of the dome has a clear span of 642 feet (twice that of any previous structure). It has a maximum height of 208 feet above the playing field. This would allow the construction of an 18-story building under the roof.

The roof design consists of a steel frame with trussed beams arching upward to meet at the center and braced in a diamond pattern. This steel skeleton supports a roof containing 4,596 cast acrylic sheet skylights (7' 2" x 3' 4"). The roof can withstand hurricane winds of 135 mph and gusts of 165 mph.

The stadium is completely air-conditioned. Smoke and hot air can be drawn out through the top of the roof.

Exterior Doors

Exterior doors give protection and privacy. They are also an important part of any construction design. They may be of the swing type or sliding type. The main entrance is usually a decorative swing-type door. Where the door opens into an entrance hall with no other windows, the door often has side lights (small windows beside the door). A glazed door has a window.

Flush doors have a smooth facing that is glued to a solid or hollow core, Figure 22–10. Panel doors have two or more panels in a framework of rails and stiles, Figure 22–11. Rails are the

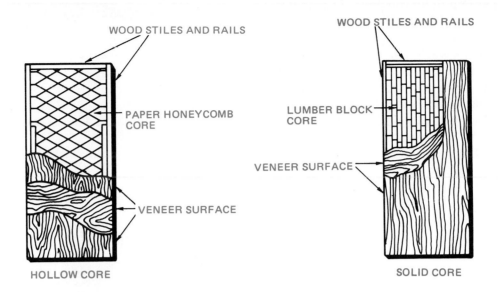

WOOD STILES AND RAILS

PAPER HONEYCOMB CORE

VENEER SURFACE

HOLLOW CORE

WOOD STILES AND RAILS

LUMBER BLOCK CORE

VENEER SURFACE

SOLID CORE

FIGURE 22–10 Construction of a flush door.

RAILS

STILES

PANELS

FIGURE 22-11 Rails and stiles on a panel door. *(Courtesy of Morgan Products, Ltd.)*

horizontal members. Stiles are the vertical members.

Doors are made of wood, aluminum, or steel. Wood has long been the standard material for doors. Metal doors are also becoming popular.

Installing Doors

Doors are hung on hinges in a door frame. The door frame is made up of four main parts and the trim, Figure 22–12. The side **jambs** are cut

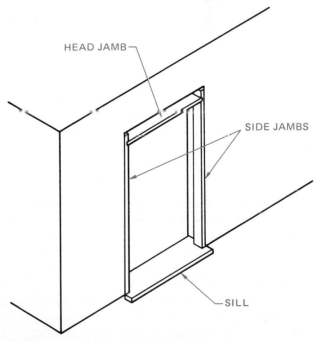

HEAD JAMB

SIDE JAMBS

SILL

FIGURE 22–12 Parts of a door frame.

to the proper length for the door. The head jamb is cut to the necessary width to fit flush on both sides of the wall. The sill is made of either oak or metal to resist wear. The interior and exterior trim is called **casing**.

Carpenters install the individual parts and fit the door. Hanging doors requires skill and careful work. The door jamb and sill are first assembled and fitted into the opening. Next, the door is trimmed with a plane to fit the opening. Then gains are cut for the hinges. A **hinge gain** is a recess made in both the door edge and the jamb that accepts the hinge. The hinges are installed and the door is hung in the jambs. If the jambs do not include a precut doorstop rabbet, stop molding is nailed to the jambs. Finally, a lock set is installed.

Prehung Doors

In some construction, prehung doors are specified. Prehung doors include the door, door frame, and all the necessary hardware in one unit. These units are put in the rough opening, plumbed, and nailed in place.

Windows

Windows are made up of a sash and a frame. The **sash** is the glass and the frame that holds it. The frame of a window holds the sash in the wall.

Windows are named according to the way they open. The simplest kind of window has a fixed sash that does not open. Double-hung windows have two sashes that slide up and down, Figure 22–13. A sliding window is similar to a double-hung window turned on its side. They slide from side to side. Casement windows are hinged on the side and swing out like doors. Awning windows are hinged at the top so the bottoms swing outward.

The size of windows is usually referred to by sash size. The floor plan shows the sash size and location of the windows. The type of window is also given on the floor plan by a symbol.

Installing Windows

Windows are installed by carpenters in much the same way as doors. The complete window unit without the interior casing is slid into the rough openings, Figure 22–14. Shims are placed

FIGURE 22–13 Double-hung windows. *(Courtesy of Anderson Corporation, Bayport, MN 55003)*

under the window sill to level the window. The unit must be perfectly level and square to operate properly. Finishing nails or casing nails are driven through the casing and into the studs. These nails are set below the surface of the wood and covered with plastic wood or putty.

Exterior Wall Coverings

The purpose of exterior wall coverings is to protect the structure from the weather. They also create an attractive appearance. Siding and masonry veneer are two common exterior wall coverings.

Wood Siding

Wood siding can be either vertical or horizontal. Siding must be applied so that water can run down the wall without running behind the siding.

Horizontal wood siding is installed starting at the bottom of the wall. The first piece is leveled and nailed to the wall frame. Aluminum or galvanized nails are used to prevent rusting. The

FIGURE 22–15 Horizontal wood siding is butted against the edge of the corner board.

siding is made in a wide range of styles to look like wood siding, Figure 22–16. The trim for vinyl siding is also made from vinyl of a matching color.

FIGURE 22–14 This carpenter is installing a window. *(Courtesy of Anderson Corp., Bayport, MN 55003)*

second course of siding is applied so that a small amount of the first course is covered.

Wood siding is made in widths from 6 inches to 10 inches. After every few courses, a chalk line is snapped to guide the siding installers.

Siding installers must be accurate when they measure and install siding. Large joints between the ends of two pieces would allow water to seep through. At the corners the siding is either butted snugly against corner boards, Figure 22–15, or covered with metal corners.

Vertical siding is plumbed with a spirit level as it is applied. Tongue-and-groove siding is nailed with finishing nails driven at an angle through the tongue.

Vinyl Siding

Many homes are covered with vinyl siding. Vinyl siding is made from colored plastic. It does not have to be painted like wood siding. Vinyl

FIGURE 22–16 Vinyl siding looks like wood, but it doesn't need painting. *(Courtesy of Reynolds Metals Company)*

The first step in applying horizontal vinyl siding is to nail a starting strip to the bottom of the wall, Figure 22–17. J-channels are nailed around windows and doors. Corners are trimmed with corner posts. The siding is cut to length. Where more length is needed, the pieces are overlapped slightly. The bottom of the first piece is slid up onto the starting strip. Then the top edge is nailed. The top edge of the siding is shaped to hold the bottom of the next piece.

Sheet Siding

Plywood and hardboard are made with special patterns for use as exterior wall coverings. These are usually a vertical pattern, Figure 22–18. They make attractive, weathertight coverings. They also reduce the amount of labor involved. Most sheet sidings are simply face-nailed to the building.

FIGURE 22–18 Several styles of plywood siding. (Courtesy of the American Plywood Association)

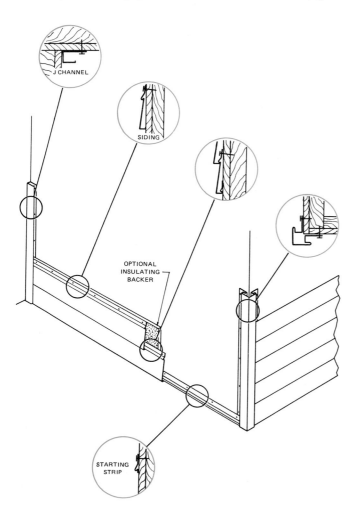

FIGURE 22–17 Details of vinyl siding installation.

Masonry Veneer

Masonry veneer covers a structure with masonry materials, such as brick or stone. Most houses that appear to be made of bricks are actually brick veneer. Masonry veneer is attractive and needs almost no maintenance.

Unlike other kinds of wall covering, masonry veneer requires a special foundation design. The weight of the masonry veneer is more than most soil can support. So the veneer must rest on the foundation, Figure 22–19.

Masonry veneer is built by masons after the building frame is completed. The masonry units are laid in mortar. This is similar to laying block for foundation walls. Architects may specify special arrangements of bricks to create decorative effects, Figure 22–20. The arrangement of bricks to create a pattern is called a **bond**. Masons must know how to lay bricks in a variety of bond patterns.

Wall ties are used to prevent the masonry veneer from pulling away from the building frame. Wall ties are metal devices that are nailed to the wall sheathing. When the masonry units are installed, the wall ties are placed between the mortar joints.

Stucco

Stucco is plaster made with portland cement. It is applied in two or three coats over wire lath. First, the wall sheathing is covered with waterproof building paper. Next the **lath** (usually wire netting) is stapled to the wall, Figure 22–21.

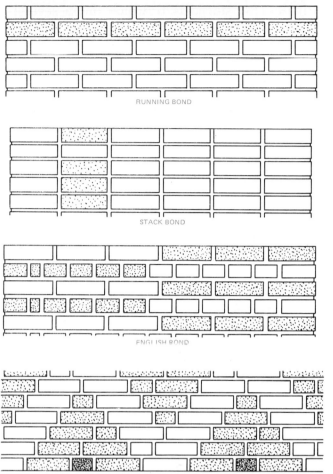

RUNNING BOND

STACK BOND

ENGLISH BOND

GARDEN WALL BOND. SHADED PORTIONS SHOW ANOTHER PATTERN CREATED BY THE BONDING.

FIGURE 22–20 A variety of effects can be created by using different bond patterns.

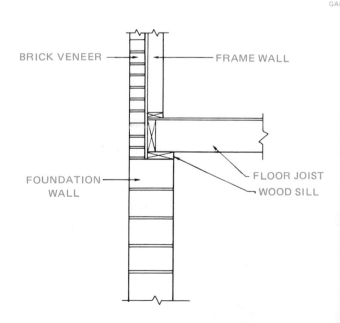

BRICK VENEER

FRAME WALL

FOUNDATION WALL

FLOOR JOIST

WOOD SILL

FIGURE 22–19 For masonry veneer, the foundation wall must be thick enough to provide a base for the bricks.

FIGURE 22–21 Wire lath ready for stucco.

Finally, the stucco is troweled on in several coats. The finish coat is sprayed on, Figure 22–22.

Summary

Five common roof types are gable, hip, mansard, gambrel, and shed. The main parts of a roof frame are the ridge board and the rafters. The width covered by a roof is the span. The run is half the span. The rise is the height of a roof above the walls. Pitch is the steepness of a roof. Roofs are covered with roof sheathing (usually plywood), underlayment, and asphalt or wood shingles.

Enclosing the structure involves adding trim, doors, windows, and exterior wall coverings. Enclosing protects the structure and decorates its appearance. The cornice trims and protects the ends of roof rafters. Exterior doors, made of wood or metal, come in glazed door, flush door, and panel door styles. The trim around a door is called the casing. Windows styles available include fixed sash, double-hung, sliding, casement, and awning. Common exterior wall coverings include wood siding, vinyl siding, sheet siding, masonry veneer, and stucco.

FIGURE 22–22 Spraying on stucco plaster.

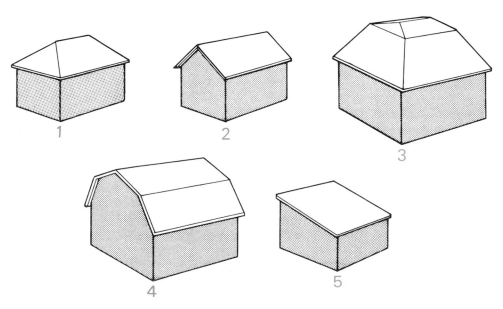

FIGURE 22–23 Can you name each type of roof?

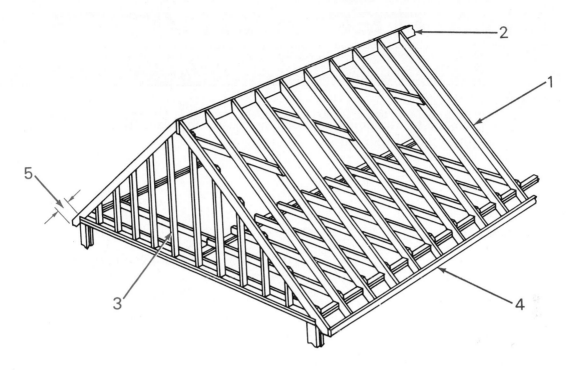

FIGURE 22–24 How many parts can you name?

DISCUSSION QUESTIONS

1. What type of roof is indicated in each drawing in Figure 22-23?
2. How many of the roof frame and roof truss parts can you name in Figure 22-24?
3. Draw an illustration to show the span, run, rise, and pitch of a roof frame.
4. What are the purposes of enclosing a structure?
5. What are the main parts of a cornice?
6. What is a window sash? What is sash size?
7. What exterior wall coverings were described in this chapter?

CHAPTER ACTIVITIES

Testing Model Roof Trusses

OBJECTIVE

Engineers often build and test model structures like trusses. In this activity, you will build and test the strength of model roof trusses, Figure 1.

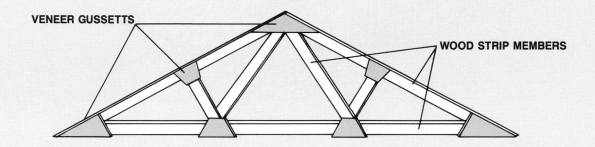

VENEER GUSSETTS

WOOD STRIP MEMBERS

FIGURE 1. Model roof truss.

EQUIPMENT AND SUPPLIES

1. ⅛ " x ¼ " balsa or pine wood strips
2. Wood veneer or chip board
3. White glue, hot glue, or other adhesive
4. Supply of straight pins
5. Grid paper and drawing tools
6. Sheet of cardboard (larger than grid paper)
7. Waxed paper
8. Utility knife
9. Truss-testing device, Figure 2
10. Safety glasses

PROCEDURE

1. Lay out a scale model roof truss on the grid paper. Pick one of the designs in Figure 22-6, or design one of your own.
2. Tape your truss design to the cardboard. Lay the wax paper over the truss design and pin in place.
3. Lay the strips of wood over the wax paper and cut to size with the utility knife. Cut all the pieces needed to make the truss.
4. Pin each piece in place over the wax paper. Then cut and glue gussets over the truss members. Make the gussets from the veneer or chip board.

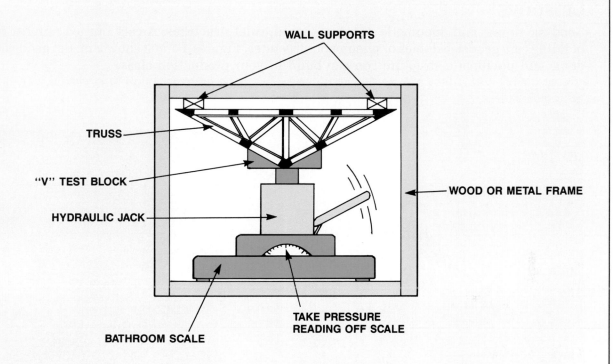

FIGURE 2. Truss-testing device.

5. Make two trusses. When the trusses are dry, glue short pieces of wood between them. This will create a two-truss assembly.
6. Place the truss assembly in the truss tester.
7. Apply pressure to the truss with the jack and listen for cracks. Watch the reading on the scale at the same time.
8. When you hear cracking, remember the reading on the scale.
9. Compare the strength of the various trusses made by your classmates.

MATH AND SCIENCE CONNECTIONS
Weigh the trusses before testing them. Then calculate their strength efficiency by dividing the truss weight by the weight the truss was able to support.

RELATED QUESTIONS
1. What are the advantages of using trusses for roofs?
2. Why do you think engineers test models before building the real thing?
3. Have you ever seen a roof truss before? Where?
4. What types of loads would a roof truss have to support?

Building Geodesic Domes and Pneumatic Structures

OBJECTIVE

Geodesic domes and pneumatic structures are unusual structures. A very limited number of buildings are geodesic domes or pneumatic structures. Figures 1 and 2 show a model geodesic dome and pneumatic structure you can build in your production class.

FIGURE 1. Model geodesic dome.

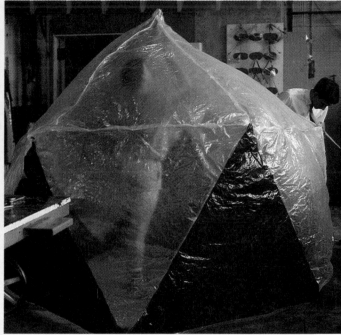

FIGURE 2. Model pneumatic structure.

EQUIPMENT AND SUPPLIES

See these sheets (from your teacher):
Geodesic Dome Building Procedure
Pneumatic Structure Building Procedure

PROCEDURE

1. Decide whether you will build a geodesic dome or pneumatic structure. Groups of students should work together to build one structure.
2. Gather the required materials. Study the building procedure sheets.
3. Follow the procedures to build a geodesic dome or pneumatic structure.

INVESTIGATION

1. Consider doing the following with your geodesic dome and/or pneumatic structure:
   - Inflate a pneumatic structure. Test how long it takes for the shell to fall when the fan is turned off.

■ Test the strength of the geodesic dome and pneumatic structures by placing weights on them.

■ Ask your science teacher how you can measure the air pressure inside a pneumatic structure.

■ Ask your science teacher to explain atmospheric air pressure.

■ Ask your math teacher to explain the concept of a polyhedron.

2. Design ways to build larger, real-life geodesic and pneumatic structures or smaller tabletop models.

3. Design a door system for the structures you built.

MATH AND SCIENCE CONNECTIONS

1. Pneumatic structures use air pressure to support the plastic sheet shell. The normal atmospheric air pressure is 14.7 pounds per square inch at sea level. In order to raise the plastic sheet, a slightly larger air pressure is required.

2. The geodesic dome you built is part of a regular polyhedron. In math terms, a regular polyhedron is a structure that encloses a space with several equal-sized, straight-sided planes where (1) all the struts are of equal length, (2) all angles between the struts are equal degrees, and (3) all corners lie on the surface of a sphere. A cube is a simple regular polyhedron.

RELATED QUESTIONS

1. What do you like most/least about geodesic domes and pneumatic structures?

2. Would you like to live in a geodesic dome home? What problems might you encounter?

3. What are the advantages of geodesic domes and pneumatic structures? Where are they used most often? Why?

4. What does the word pneumatic mean? What does the word geodesic mean? (Look them up in a dictionary.)

CHAPTER 23

Utilities: Plumbing, Electrical, and Climate Control Systems

OBJECTIVES

After completing this chapter, you will be able to:

- Describe the need for various utilities in structures.
- Explain the design and installation of plumbing systems.
- Explain the design and installation of electrical systems.
- Explain the design and installation of heating and cooling systems.

KEY TERMS

Conduit	Grounded	Solder
Ducts	HVAC	Thermostat
Effluent	National Electrical Code	Trap
Finish plumbing	R value	Utilities
Flux	Rough plumbing	Vent
Greenhouse effect	Service panel	

A structure that will be used by people requires a number of services, such as water, sewage, electrical, and heating and cooling. The sources of these services are called **utilities**. This chapter examines three service systems: plumbing, electrical, and climate control. The design of each of these systems is included in the working drawings. Engineers are often responsible for these designs.

Plumbing Systems

Rough and Finish Plumbing

When the basic structure of a house is completed, the plumbing installation begins. This is done before any interior wall covering is applied. Plumbers install only the rough plumbing at this time. **Rough plumbing** includes installation of main supply lines, main sewer lines, and all branch piping, Figure 23–1. Although they are considered part of the finished plumbing, bathtubs are generally installed at this time also. This is so the interior wall covering can be finished next to the tub.

When the interior of the house nears completion, the plumbers return to install the **finish plumbing**. This includes all the fixtures such as sinks, lavatories, and toilets.

Supply and Sewage Plumbing Systems

The plumbing system must perform two basic functions. First, fresh water must be supplied to all points of use, such as sinks, showers, and washing machines. This function is called supply plumbing. Second, waste water must be carried away after use. This function is called sewage plumbing. Plumbing systems can also be used to supply natural gas to a building.

Design of Supply Plumbing

In most communities, water is distributed through a system of water mains (large pipes) under the street. When a new house is built, the water department taps (makes an opening in) the

FIGURE 23–1 Plumbing is roughed in after the framing is done.

main. The supply plumbing from the tap to the house is installed by plumbers.

The main supply pipe entering the house is larger in diameter than the pipes going to each sink or other point of use. The main supply pipe for a house is usually ¾-inch or 1-inch pipe. A water meter is intalled on the supply pipe. It measures the amount of water used. The local

Using Lasers in Plumbing

Installation of plumbing to and from the supply and sewage systems requires accurate alignment of pipes. The traditional method for aligning sections of pipe has been to use wire or string and a level. This method is slow and not always accurate. A laser can be used to do the job quickly and accurately, Figure 1. The laser is positioned at one end of the pipe. If the pipe is to be pitched (slope up or down), the laser is tilted to that angle. The position of each length of pipe is adjusted so the laser beam strikes a target attached to the end of the pipe. The laser is placed inside large-diameter pipes. When the pipe is too small, the laser and target are positioned on top of the pipe.

FIGURE 1. Lasers can be used in plumbing to align pipes. *(Courtesy of Spectra-Physics)*

water company reads the meter to determine the proper water bill. A main water shutoff valve is located near the meter.

From the water meter, the supply pipe runs to the water heater. A tee is installed between the meter and water heater to supply cold water to the house. From this point, smaller pipes are run to each point of use.

When a valve is suddenly closed at a fixture, the water tends to slam into the closed valve. This causes a sudden pressure buildup in the pipes and may cause the pipes to hammer (a sudden shock in the supply piping). To prevent this, an air chamber is installed near each fixture. An air chamber is made up of a short vertical section of pipe that traps air, Figure 23-2. When a valve is suddenly closed, the air chamber acts as a shock absorber to reduce hammer.

Design of Sewage Plumbing

Sewage plumbing removes waste water. To do this, a pipe runs from each point of use to the main sewer. The main sewer carries the **effluent** (fouled water and solid waste) to the municipal sewer or septic system.

Traps. The sewer contains foul-smelling, germ-laden gases. If a pipe were simply run to the sewer, these sewer gases would enter the building. To prevent this, a trap is installed at each point of use. A **trap** is a curved area of plumbing that fills with water to prevent sewer gases from entering the building, Figure 23-3. Not all traps are easily seen. Some fixtures, such as toilets, have built-in traps.

Vents. As the water rushes through a trap it is possible for a siphoning action to be started. To illustrate this siphoning action, a piece of garden

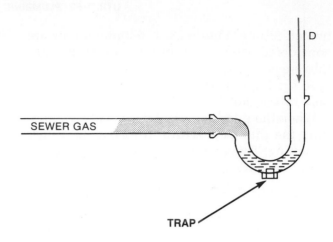

FIGURE 23-3 A trap prevents sewer gas from entering the building.

hose or tubing can be used to draw the water out of an open container. By sucking water through the hose and holding the discharge end at a point lower than the water level in the container, the water will run through the hose without further sucking. This siphoning action can draw the water from the trap. This will leave the sewer open to the inside of the building.

To prevent traps from siphoning, a vent is installed near the trap. The **vent** is an opening that allows air pressure to enter the system and break the suction of the trap, Figure 23-4. Usually all of the fixtures are vented into one main vertical

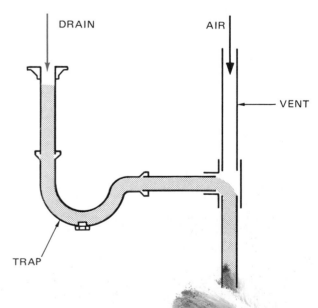

FIGURE 23-4 A vent allows air to enter the system to prevent siphoning.

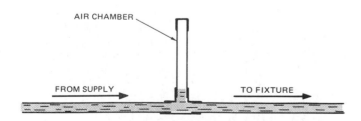

FIGURE 23-2 An air chamber acts like a shock absorber in a plumbing system.

vent pipe, called a stack. The open end of the stack extends to the outside of the house.

Plumbing Materials

The materials most often used for plumbing parts are galvanized iron, copper, plastics, and cast iron. Plumbing parts include pipe and fittings. Pipe is the long straight pieces. Fittings are used to make turns, connect pieces of pipe, and control water flow.

Galvanized iron was once the most widely used material for pipe and fittings. Today, it is not used very much because it corrodes easily. Copper is often used for plumbing today because it resists corrosion. However, it is relatively expensive. Copper pipes and fittings can be threaded or unthreaded for soldered joints.

Solder is a soft metal made by combining tin and lead. Plumbers use a torch to heat copper pipe and fittings and melt solder in the joint. When the copper cools, the solder hardens and seals the joint so it won't leak. A **flux** is used to make the solder flow all around the pieces of pipe that are being connected.

Plastic plumbing materials are very popular today. They are lightweight, noncorrosive, and easily joined. Plastic pipe fittings are joined with solvent cement. The solvent cement softens the plastic and melts together the pieces being joined.

Cast iron is widely used for sewage plumbing because of its strength and resistance to corrosion. It is seldom used for supply plumbing. A common type of cast-iron pipe has an enlarged end into which the straight end of the next piece fits. The joint is then packed with a fiber packing called oakum. Melted lead is poured over the oakum to fill the joint and make it strong, Figure 23–5.

Fittings

A wide assortment of fittings is available. Most fittings are made of the materials of which pipe is made. Some common fittings include couplings, unions, elbows, tees and wyes, cleanouts and valves. Couplings, Figure 23–6, are used to solder together two pipes in a straight line. Unions are similar to couplings, but they have threads so the pieces of pipe can be screwed together. Unions can be unscrewed if needed. Elbows, Figure 23–7, are used to turn the direc-

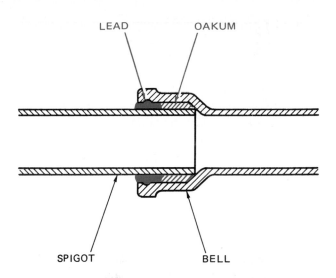

FIGURE 23–5 Bell-and-spigot joints in cast iron pipe are sealed with oakum and molten lead.

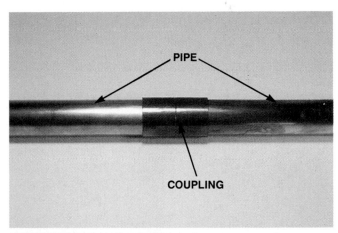

FIGURE 23–6 Couplings are used to permanently connect two pieces of pipe with solder.

FIGURE 23–7 Elbows are used to change the direction of a pipe. This is a 90-degree elbow. Elbows are also available for 45-degree bends.

tion of a pipe either 90 degrees or 45 degrees. Tees and wyes, Figure 23–8, have three openings to allow a second pipe to join the first from the side. Tees have a 90-degree side opening. Wyes have a 45-degree side opening. Cleanouts allow access to sewage plumbing for cleaning. A threaded plug can be removed so a cleanout tool can be run through the line. Valves, Figure 23–9, are used to stop, start, or control the flow of water. The familiar faucets on a sink are a type of valve.

Electrical Systems

Generating and Transmitting Electricity

Electrical power often must be transmitted great distances from the utility company's generating plant to the user through wires, Figure 23–10. The wires resist the flow of electricity. To overcome this resistance, the electricity is transmitted at high voltage — sometimes as much as

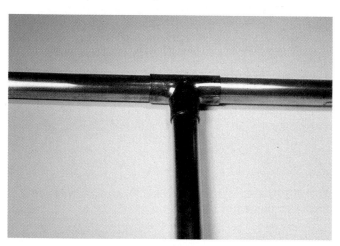

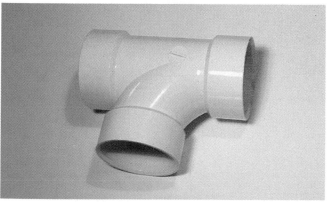

FIGURE 23–8 Tees (top) and wyes (bottom) are used to join three pieces of pipe.

FIGURE 23–9 Valves control the flow of water in pipes. Faucets are another example of valves.

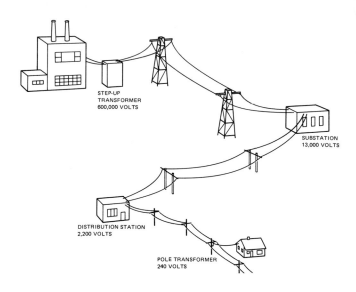

FIGURE 23–10 Electricity is transmitted at a high voltage, then stepped down for use.

60,000 volts. The voltage is reduced to 13,000 volts at substations. The substations supply distribution stations, where the voltage is further stepped down to 2,200 volts. From the distribution station it is transmitted only a few miles to homes and businesses. There it is stepped down to a more manageable voltage. Some industries use 480-volt electricity. Homes use 240 and 120 volts. From the last transformer, a cable, called a service, carries the current to the utility company's meter. This is where the utility company's responsibility ends and the customer's starts.

Parts of an Electrical System

The main parts of an electrical system in a house include the service panel, circuit wiring, outlets, and electrical boxes. The **service panel** is a box in the house which is connected to the service line outside. It contains a main disconnect and overcurrent protection devices, Figure 23–11.

The main disconnect is a large switch. It allows all of the electricity to the building to be disconnected. Usually the overcurrent protection devices are circuit breakers. A number of separate circuits run to the various parts of a house out of the service panel. Each circuit has its own wiring and circuit breaker. Each circuit breaker is used to disconnect the electricity to its circuit. Circuit breakers switch off automatically if an excess of current tries to flow.

An outlet is a point in the system where appliances can be plugged in or permanently wired into the system. The most common outlets in a house are light fixtures, switches, and wall outlets.

All connections in the wiring are made inside an electrical box. This includes connections with outlets, switches, light fixtures, appliances, and other wires. Electrical boxes are made of steel or high-impact plastic. They protect the structure from fire in the event that an electrical spark occurs at the connection. Electrical boxes are made in many shapes and sizes, Figure 23–12.

Grounding

All users of electricity must be **grounded**. Grounding involves connecting electrical wiring to a metal rod that is driven into the earth. In this manner, the earth provides a path for current to return to the source. If a live conductor accidentally comes in contact with the frame of the device, the current is directed through the ground rather than through the user. Some devices are grounded by attaching a ground wire to a metal water pipe. Even large electrical power generating plants are grounded.

Types of Wiring

The two main types of wiring commonly used in electrical systems are nonmetallic sheathed cable and armored cable, Figure 23–13. Nonmetallic sheathed cable, commonly called romex, is made of copper or aluminum conductors covered with plastic insulation. Armored cable, commonly called BX, is made of separately insulated conductors encased in a spiral-wound steel covering.

Often, nonmetallic wires are placed inside conduit for protection. **Conduit** is metal or plastic tubing with wires running inside of it. Conduit is primarily used in nonresidential construction.

Designing an Electrical System

Architects and electricians consider many factors in designing electrical systems. The system

FIGURE 23–11 This electrician is installing circuit breakers in a service panel. *(Photo by Richard T. Kreh, Sr.)*

FIGURE 23–12 Electrical boxes are made of steel or plastic and come in a variety of shapes and sizes.

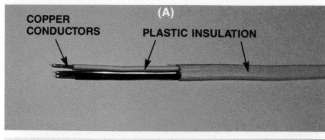

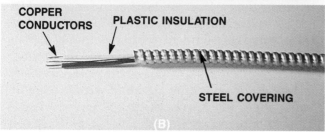

FIGURE 23–13 Two common types of wiring are (A) nonmetallic sheathed cable and (B) armored cable.

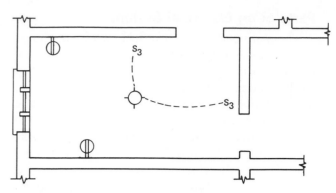

FIGURE 23–14 Electrical devices are shown on an electrical floor plan with the use of symbols.

must provide power for all the electrical devices in a structure. It must also be safe. The National Fire Protection Association publishes the **National Electrical Code**. It specifies the design of safe electrical systems. Electricians must know this code and any state or local codes that apply.

Electricans must also be able to read working drawings. The electrical plans for many homes are included on the floor plans, Figure 23–14. Special symbols are used to identify the various parts of the electrical system, Figure 23–15.

Roughing In Wiring

Wiring is roughed in after the framing is complete. This is done at about the same time as rough plumbing.

The service panel is installed first. Then the locations of all devices are marked on the studs and joists. At each point where an outlet or switch is to be installed, a box is fastened to the building frame.

When all of the boxes are installed, the electricians drill holes through the framing members and install the wires, Figure 23–16. Wires are connected to one another with wire nuts. These are threaded plastic fittings that are screwed on to the bare ends of the conductors to make a connection.

Before any other work is done on the electrical system, it is inspected. Electrical inspectors check the roughed-in wiring to make sure it meets the electrical codes.

Climate Control Systems

Buildings of all types use climate control systems. In nearly all parts of the country some means of heating is needed in the winter. In most parts, cooling is necessary in the summer. Heating, ventilating, and air conditioning (**HVAC**) is the common name given to these climate control systems.

Heat Transfer

Controlling air temperature is a process of transferring heat into or out of a building space. This heat transfer can be done by convection, radiation, evaporation, or gravity flow.

- **Convection:** Heat flows from a warm surface to a cold surface. For example, heat flows from warm air to a cold wall.
- **Radiation:** The movement of heat by heat rays. This does not require air or other medium. The sun's heat travels through space by radiation.
- **Evaporation:** As moisture evaporates it uses heat, thereby cooling the surface from which it evaporated. This his how perspiration cools the body.
- **Gravity flow:** Cool air is denser than warm air. Therefore, warm air rises and cool air settles. Due to gravity flow, the air near a ceiling is warmer than the air at the floor.

The Air Cycle

One of the most common systems for climate control circulates air from the living spaces past heating or cooling devices, Figure 23–17. A fan

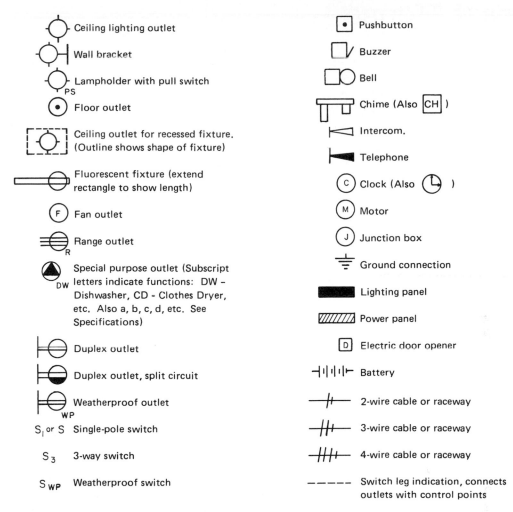

Ceiling lighting outlet

Wall bracket

Lampholder with pull switch
PS

Floor outlet

Ceiling outlet for recessed fixture.
(Outline shows shape of fixture)

Fluorescent fixture (extend
rectangle to show length)

F Fan outlet

Range outlet
R

Special purpose outlet (Subscript
letters indicate functions: DW –
Dishwasher, CD - Clothes Dryer,
etc. Also a, b, c, d, etc. See
Specifications)
DW

Duplex outlet

Duplex outlet, split circuit

Weatherproof outlet
WP

S_1 or S Single-pole switch

S_3 3-way switch

S_{WP} Weatherproof switch

Pushbutton

Buzzer

Bell

Chime (Also CH)

Intercom.

Telephone

C Clock (Also ⌚)

M Motor

J Junction box

Ground connection

Lighting panel

Power panel

D Electric door opener

Battery

2-wire cable or raceway

3-wire cable or raceway

4-wire cable or raceway

Switch leg indication, connects
outlets with control points

FIGURE 23–15 These are some of the common electrical symbols.

**FIGURE 23–16 The first steps in installing electricity
are to install the electrical boxes and run the wiring.**
(Photo by Larry Jeffus)

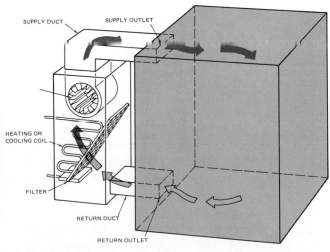

SUPPLY DUCT SUPPLY OUTLET

HEATING OR
COOLING COIL

FILTER

RETURN DUCT

RETURN OUTLET

**FIGURE 23–17 The air cycle is used to heat or cool air
in a home.**

Smart House

The National Association of Home Builders Research Foundation is conducting a project that will change electrical systems in homes. The project is SMART HOUSE.

The SMART HOUSE systems will use one set of wiring for all electrical functions. Each appliance or outlet will tell a central computer control system how much power is needed and if everything is working correctly. The control system will be able to deliver or cut off power exactly when needed. Some of the features of SMART HOUSE include (refer to Figure 1):

1. SMART HOUSE provides important safety features. The control computer identifies what is being plugged into any outlet. Only "authorized" devices receive current. A baby sticking a finger into an outlet would not be harmed because its finger is not an "authorized" device.

2. A person trying to play an electric guitar in the shower would also be protected. The power will shut off automatically. Likewise, a lamp with a frayed cord would not operate, and anyone sticking a knife into a toaster would not be harmed.

3. Speakers can be hooked up to a stereo by plugging into any electrical outlet in the home. As shown, the stereo in the living room can provide music to speakers in the bedroom and the basement at the same time.

4. Any SMART HOUSE device can operate in any outlet. A person in the basement can unplug the stereo speaker and plug a telephone into the same outlet to make a call.

5. SMART HOUSE can turn off lights automatically when a person leaves a room, and turn them on when the person reenters.

6. Monitors or simple alarms can alert people when the refrigerator door

FIGURE 1. Key features of SMART HOUSE.

has been left open, when oven or stove burners have been left on, when the front door is unlocked, and so on.

7. Other alarms can sense unwanted intruders. The system is so sensitive that a household member going to the kitchen for a snack would not trip the alarm.

8. Sensors can also detect smoke or fire in any part of the house.

9. The entire SMART HOUSE system can be controlled from any location by telephone. You can call from anywhere and check the system and turn on or off any device.

10. All SMART HOUSE devices can be controlled from anywhere in the house. A control panel next to a bed can be used to start a coffee-maker before you get out of bed.

Does this sound like science fiction? It is all possible with today's technology. In fact, SMART HOUSE technology has been around since 1986! Someday your house may be a SMART HOUSE.

forces the air into large sheet metal or plastic pipes called supply **ducts**. These ducts supply cool or warm air to the rooms, as needed. Air then flows from the room into the return duct. The return duct directs the air from the room over a heating or cooling device and the air cycle repeats.

Heating and Cooling Systems

If the air cycle is used for heating the air, the heat is generated in a furnace. Furnaces produce heat by burning fuel oil or natural gas inside a combustion chamber. The air to be heated does not enter the combustion chamber, but absorbs heat from the chamber's outer surface. The gases given off by the combustion are vented through a chimney.

If the air from the room is to be cooled, it is passed over a cooling coil. Cooling systems are called air conditioners. Air conditioning systems work by expanding a liquid to make it cool. This cooled liquid is passed through a coil where it picks up heat from the warm air in the house.

A **thermostat** is used to sense the temperature in the house. The thermostat is an automatic switch. For heat, the thermostat starts the furnace when the temperature drops below a preset level. Then, it turns the furnace off when the preset temperature is reached. The thermostat works in the same manner for the cooling system.

Hot Water Boiler System

Many buildings are heated by hot water systems. In these systems, an oil- or gas-fired boiler is used, Figure 23-18. Water is circulated around a combustion chamber inside the boiler to absorb heat. The hot water is then pumped to various parts of the house through a plumbing system hooked up to the boiler. Radiators or convectors transfer the heat from the water to the air in the room.

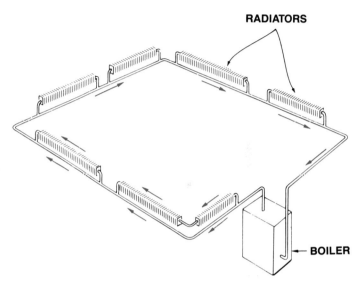

FIGURE 23-18 The main parts of a hot water boiler system are the boiler and the radiators.

Electrical Resistance Heat

There are a number of heating system designs that rely on electric heating elements located in each room. Some of these systems have electric heating elements embedded in the floor or ceiling. Others have electric heating elements that look like hot water radiators.

Solar Heat

The sun is a large, free source of heat. However, it is rarely used for home heating. Systems have been designed and used in some cases. The simplest form of solar heating is called the **greenhouse effect**. Glass allows the heat of the sun to pass through easily, but does not allow reflected heat to pass. This is the principle that allows the sun to warm a greenhouse. Sunlight enters the greenhouse through the glass and is trapped, heating the inside air.

This principle can be used to heat homes. By planning the size and location of windows, this heat can be used to warm a building. In winter the sun is closer to the horizon for most of the day. Large windows placed on the south side of the building allow the sun to enter and warm the house, Figure 23–19A.

The sun is higher in the sky during the summer. A large overhang built over the south-facing windows will block the sun, Figure 23–19B. This will keep it from getting too hot inside in the summer.

A

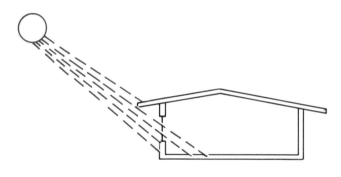

B

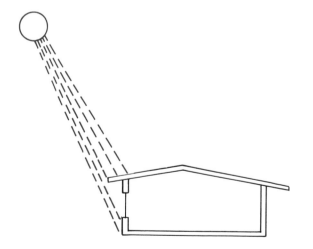

FIGURE 23–19 (A) In the winter, the sun is close to the horizon and can shine in a window to heat a home. (B) In the summer, the long overhang on the roof can block the high sun and prevent excess heat buildup in the home.

Solar Collectors

Solar energy can also be used as a source of heat for a hot water system. The sun is used to heat pipes carrying water to the heating system. Flat-plate solar collectors direct the sun's energy onto the heating system pipes.

A flat-plate solar collector is a wood or metal box containing several pipes, Figure 23–20. A glass or clear plastic top covers the pipes. The glass takes advantage of the greenhouse effect to help heat water flowing in the pipes. A layer of dull black sheet metal is placed between the glass and the pipes. The dark surface on the sheet metal absorbs a large amount of heat from the sun. This heat is transferred to the water in the pipes. The hot water is stored in tanks for general hot water needs or a hot water heating system.

Thermal Insulation

It is important to prevent heat from entering a cooled building in the summer and from leaving a heated building in the winter. All materials conduct heat. The purpose of thermal insulation is to resist the flow of heat. The resistance of a material to the flow of heat is its **R value**.

The most common insulating materials for buildings are fiberglass and foamed polystyrene (Styrofoam®). Fiberglass insulation is made in rolls and batts (pieces a few feet long). Foamed-polystyrene insulation is made in sheets from 1 to 4 feet wide, and 4 to 8 feet long.

FIGURE 23–20 Flat-plate solar collectors are often placed on a roof so they will face the sun. (*Courtesy of Advanced Energy Technologies, Clifton Park, NY*)

Applying Mathematics: Calculating Thermal Resistance

It is impossible to completely stop the flow of heat through a building section, but insulation can greatly reduce it. Insulating and building materials vary in their ability to restrict the flow of heat. Even air has some resistance to the flow of heat.

Figure 1 compares the thermal resistance of two building sections. The one on the left does not use fiberglass or foamed-polystyrene insulation; the one on the right does.

The R values show how the total thermal resistance would be calculated for each. The R values of building components are simply added together.

The section on the left (without insulation) has a total R value of 4.59. The insulated section has 3½ inches of fiberglass between the wall studs. This raises the total R value to 16.95. The higher the R value, the greater the resistance to the flow of heat.

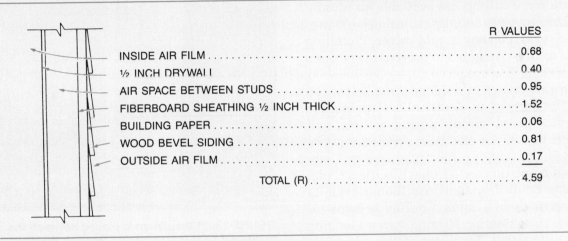

R VALUES

INSIDE AIR FILM	0.68
½ INCH DRYWALL	0.40
AIR SPACE BETWEEN STUDS	0.95
FIBERBOARD SHEATHING ½ INCH THICK	1.52
BUILDING PAPER	0.06
WOOD BEVEL SIDING	0.81
OUTSIDE AIR FILM	0.17
TOTAL (R)	4.59

R VALUES

13.31	3½" INSULATION
0.68	INSIDE AIR FILM
0.40	½ INCH DRYWALL
1.52	FIBERBOARD SHEATHING ½ INCH THICK
0.06	BUILDING PAPER
0.81	WOOD BEVEL SIDING
0.17	OUTSIDE AIR FILM
16.95	TOTAL (R)

FIGURE 1. Adding insulation improved the thermal insulating qualities of this wall section.

Insulating materials are placed between the framing members. Care must be taken so no openings are left through which heat can easily pass, Figure 23–21.

Summary

People require utilities in their homes and work places. This chapter explained three common utilities: plumbing, electrical, and climate control.

The two basic sections of the plumbing system are the supply plumbing and sewage plumbing. Supply plumbing provides fresh water. Sewage plumbing removes wastes. Plumbing parts are made from copper, plastic, or cast iron. A number of different fittings are available for plumbing. These include couplings, unions, elbows, tees and wyes, cleanouts, and valves. Plumbing systems must be designed to prevent sewer gases from entering the home.

Electrical systems in residences use 240 or 120 volts. The parts of an electrical system include the service panel, circuit wiring, outlets, and electrical boxes. All electrical systems must be grounded for safety. Electrical systems are designed based on the National Electrical Code.

Climate control systems use the air cycle to heat or cool air in a home. Cooling systems are called air conditioning. Heating systems include the use of furnaces, hot water boilers, electrical resistance heaters, and solar heat. Thermal insulation is very important to resist the flow of heat into and out of a residence.

FIGURE 23–21 Insulation is placed between the studs to prevent heat from entering through the walls in the summer and leaving in the winter. *(Photo by Larry Jeffus)*

DISCUSSION QUESTIONS

1. What are some of the utilities available in your home?
2. What are the differences between rough and finish plumbing and supply and sewage plumbing?
3. How do plumbing systems prevent sewer gases from entering a home?
4. What are some of the common fittings used in plumbing systems?
5. How is electricity transmitted from the generating plant to your home? What voltages and substations are used?
6. Can you explain the various parts of an electrical system from the outside service line to the various points of use in a home?
7. What are some of the features of SMART HOUSE?
8. What four heating systems were described in this chapter?
9. How is R value important to thermal insulation?

CHAPTER ACTIVITIES

 Flat-Plate Solar Collector

OBJECTIVE

Flate-plate solar collectors can use the energy of the sun to heat water. In this activity, you will produce a flat-plate solar collector, Figure 1. Then you will test its ability to heat water.

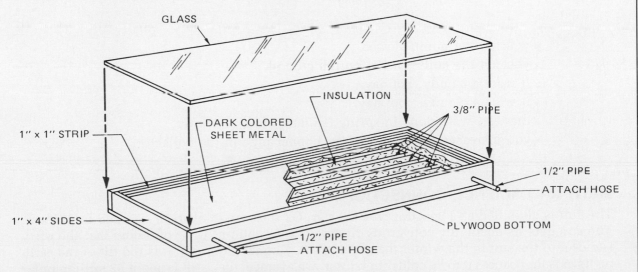

FIGURE 1. Plans for a flat-plate solar collector.

EQUIPMENT AND SUPPLIES

8 linear feet of 1″ x 4″ lumber
8 linear feet of 1″ x 1″ lumber
1′ x 3′ plywood or hardboard (any thickness)
1′ x 3′ glass or clear plastic
1′ x 3′ sheet metal
11 feet of ⅜-inch copper pipe
6 feet of ½-inch copper pipe
6 copper tees, ½″ x ⅜″
Small water pump (powered by an electric drill)
4 feet of ½-inch hose
Saw

Supply of 4d common nails
Drill and ¾″ bit
Solder, flux, and torch
3 square feet of fiberglass insulation
Safety glasses
3 hose clamps
Hammer
Tape measure
Thermometer
5-gallon bucket

PROCEDURE

1. Construct the flat-plate solar collector as shown in Figure 1. Be sure to follow all safety rules.
2. Position the collector where the glass surface gets direct sunlight. The collector should be tilted to face directly into the sun.
3. Connect the pump with a piece of the hose so that it pumps water into the collector.
4. Fill the bucket with room-temperature water. Record the temperature with the thermometer.
5. Attach a short piece of hose to the pump inlet and another to the collector outlet. These hoses are to be kept in the bucket of water.
6. Run the pump for 15 minutes and record the water temperature.
7. Conduct experiments to find out what would happen if

  ▪ you ran the pump slower.
  ▪ you put the collector flat on the ground.
  ▪ it were a cloudy, hot day.
  ▪ it were a cloudy, cool day.
  ▪ the tests were done in spring, summer, fall, or winter.
  ▪ you increased the length of the copper pipe in the collector.

MATH AND SCIENCE CONNECTIONS

The sun is classified as a spectral type G2 star. G2 stars are of average size and brightness. The sun gets its energy by converting hydrogen into helium. Did you know that the earth is closer to the sun in the winter than it is in the summer? It's true! It is the tilt of the earth on its axis that causes it to be colder in winter. On average, the earth is about 92 million miles away from the sun.

RELATED QUESTIONS

1. Why is it important to use thermal insulation?
2. What are some common insulation materials?
3. How would you explain R value to someone who had never heard of it before?
4. Does your home have insulation?

 ## Calculating R Values

OBJECTIVE

R values are a measure of the resistance of a material to the flow of heat. You can calculate the R value for walls, roofs, or floors by adding together the R values of the materials used to make them.

EQUIPMENT AND SUPPLIES

R Values Worksheet (from instructor)
R Values for Construction Materials Information Sheet (from instructor)
R Values Map (from instructor)
Calculator

PROCEDURE

1. Obtain the worksheet and information sheets from your instructor.
2. For the wall section in the worksheet, calculate the total R value.
3. Find the R value of each part of the wall on the information sheet. Write each R value in the space provided.
4. When you have listed all of the R values, add them together to find the total R value for the wall.
5. Compare the total R value number you obtained against the recommended R value for your area of the country on the R Value Map.

MATH AND SCIENCE CONNECTIONS

Heat is measured by Btu (British thermal units). One Btu is the amount of heat needed to raise the temperature of one pound of water one degree Fahrenheit. R value numbers give the resistance to the flow of heat. Another number, the K factor, gives the rate at which a material conducts heat.

RELATED QUESTIONS

1. Why is it important to use thermal insulation?
2. What are some common insulation materials?
3. How would you explain R value to someone who had never heard of it before?
4. Does your home have insulation?

CHAPTER 24

Completing the Structure: Finishing and Landscaping

OBJECTIVES

After completing this chapter, you will be able to:

■ Identify the various materials used to finish ceilings, walls, and floors.

■ Describe the use of molding in finishing a structure.

■ Describe the paints and finishes used in construction.

■ Describe the basic processes involved in landscaping.

■ Explain the importance of maintenance and remodeling.

KEY TERMS

Acrylic latex   Landscape plan   Primer
Alkyd-resin   Maintenance   Remodeling
Casing   Molding   Shellac
Grout   Pigments   Sod
Joint compound   Plastic laminate   Vehicle

When all of the framing is complete and the utilities have been roughed in, the final steps can be taken to complete the structure. This includes finishing and installing the following:

■ Ceilings
■ Wall Coverings
■ Molding
■ Cabinets and Countertops
■ Paints and Clear Finishes
■ Landscaping

In general, workers who specialize in these jobs must work to more precise dimensions and use more caution than workers who do framing and rough in utilities.

Ceilings

The ceilings are usually the first interior surface to be covered. Two common ceiling coverings are gypsum wallboard and suspended ceilings.

Gypsum wallboard is fastened to the ceiling joists with special nails or screws, Figure 24–1. The heads of the nails or screws are driven slightly below the surface of the wallboard. This leaves a shallow dent which is filled with **joint compound** to produce a smooth surface.

Gypsum wallboard is made of a plaster core of ⅜-, ½-, or ⅝-inch thick sheets with a strong paper covering. It gives a sound surface for painting. It also resists the spread of fire. Gypsum wallboard sheets are made in 4-foot widths, and are 8 to 16 feet long. The long edges are tapered to permit covering the joints.

The joints are covered with paper tape and joint compound. Several thin layers of compound are applied with a joint knife or trowel to produce an even finish, Figure 24–2. The final coat is sanded to produce a smooth surface for painting. Paint does not hide rough spots.

Suspended ceilings are made up of a metal framework hung on wires from the ceiling joists. Fiber panels fit into the framework, Figure 24–3. When installing the framework for a suspended ceiling, it is important to make sure it is level. Until recently this leveling was done by hand

FIGURE 24–1 (A) Gypsum wallboard being attached to the ceiling. (B) Special nails or screws are used.

FIGURE 24–3 In a suspended ceiling, panels rest on a metal framework hung from the ceiling. *(Courtesy of Celotex)*

with a spirit level and chalkline. Today lasers are used to do the job more quickly and accurately.

Walls

Gypsum wallboard is also a common wall covering. The technique used to install and finish wallboard walls is the same as that used for ceilings. Wallpaper may also be used to cover walls. As with painting, it is important that the wall surface be even and sanded smooth. Wallpaper will not hide rough spots.

Another common wall covering is plywood or hardboard paneling. These materials are made up of sheets, usually 4 feet by 8 feet, of plywood or hardboard, with a decorative face. The decorative face may be hardwood veneer, wood grain printed on vinyl, or any attractive pattern on plastics.

Wall paneling can be installed either directly to the wall studs or over gypsum wallboard. Special colored nails are available for nailing wood-grained paneling and its trim. Plastic-faced paneling is usually cemented in place with special adhesives.

Floors

Floors are usually covered with hardwood flooring, carpet, vinyl flooring, or ceramic tile.

Hardwood flooring is usually made of strips of oak or maple. These strips have tongue-and-groove joints. As each piece is installed over the subfloor, it is driven up tight against the preceding one, Figure 24–4. It is nailed through the edge at the base of the tongue. The completed floor is sanded with a floor sander to prepare it for varnishing. Some hardwood flooring is prefinished.

When carpeting is installed, a layer of plywood underlayment is applied first over the subfloor. Next, a pad is placed over the underlayment, and finally the carpet is installed. The carpet and pad are sold by the square yard. Installers stretch the carpet to the edge of the room. It is held in place by special carpet grippers which have been fastened to the floor around the walls of the room. Some carpeting is glued directly to the floor. Being a carpet installer requires accurate measurement skills, Figure 24–5.

Seamless vinyl and vinyl tiles are often used where floors may get wet. Seamless vinyl and vinyl tiles are cemented to the plywood underlayment with a special adhesive.

Ceramic tile is also popular for use on bathroom floors. Ceramic tiles are first cemented to the floor. Then the joints between the tiles are filled with a mortarlike material called **grout**.

FIGURE 24–5 Carpet installers must do very accurate work. *(Photo by Larry Jeffus)*

FIGURE 24–4 Hardwood flooring being installed over the subfloor. *(Photo by Paul E. Meyers)*

Molding

Molding is used to create special effects on paneling, to cover joints between building parts, and to protect areas. When the interior walls are covered with paneling, special colored molding is used to match the paneling. To protect the walls from floor-cleaning equipment and furniture legs, base molding is installed near the base of the wall.

Window and door frames are trimmed with molding called **casing**, Figure 24–6. The casing is mitered (cut to a 45-degree angle) at the corners and nailed into the wall. Finishing nails are used. They are set below the surface and covered with wood putty.

FIGURE 24-6 Door molding trim is called casing.

FIGURE 24-7 Base cabinets rest on the floor, wall cabinets are hung on the wall.

Cabinets and Countertops

The working drawings and specs indicate the type of cabinets and where they are to be installed. Cabinets can be included in several rooms, but most are used in kitchens. The kitchen cabinet layout is carefully planned. Usually the kitchen includes base cabinets, which rest on the floor. A countertop is fastened to the top of the base cabinets. Wall cabinets are mounted above the base cabinets, Figure 24-7.

Good quality cabinets have strong, glued joints. Although most cabinets are factory-made, good quality cabinets can be constructed using carpentry tools.

The backs of cabinets are often made of thin hardboard with one solid wood crosspiece. The crosspiece is used to screw the cabinet to the wall studs.

Countertops are made of particle board or plywood covered with **plastic laminate**. Plastic laminate is a thin sheet of plastic that protects the wood used in making the cabinet. Factory-made countertops are available with the plastic

laminate molded over the edge, Figure 24-8. This type of countertop is simply cut to length and attached to the cabinets with screws.

When countertops are made by the carpenter at the site, the laminate is cut slightly larger than the countertop. Contact cement is applied to the back of the laminate and the countertop. After the cement is dry, the laminate is pressed to the countertop. The edge of the countertop is covered with laminate in the same manner. After the laminate is installed, the excess is trimmed off with an electric router. Carbide-tipped bits are used.

FIGURE 24-8 These countertops have a plastic laminate covering.

Paints

Surfaces are painted for several reasons. Decoration is the most obvious reason. Steel and iron are painted to prevent rust. Wood will warp, crack, and decay if it is not painted. Painted surfaces are also easier to clean than unpainted surfaces. Paint gives protection from the sun, wind, and rain, Figure 24–9.

The ingredients in paint include pigments, a vehicle, driers, and thinners. **Pigments** are the coloring materials used in the paint. The **vehicle** is the liquid in which the pigment and other ingredients are mixed. Driers are added to paint to speed it drying. Without driers, paint would dry very slowly, if at all. Thinners are chemicals added to make the paint more liquid and easier to apply. There are many different types of paints. They can be grouped as water-based or oil-based paints and interior or exterior paints.

Water-Based Paints

Water-based paints have a water-soluble vehicle. The thinner for water-based paints is water. This paint dries quickly (usually within 30 minutes) and covers well. Water-based paints are commonly called latex or **acrylic latex** paints.

Oil-Based Paints

Oil-based paints is a common name for paints that must be thinned with mineral spirits or turpentine. Another name for oil-based paints is **alkyd-resin** paints. Alkyd-resin paints have a vehicle of soya and alkyd resin (a type of plastic). Oil-based paints produce an extremely hard surface, but they take longer to dry than water-based paints.

Both water-based and oil-based paints can be used as interior or exterior paints. The specific applications are set by the manufacturer of the paint. Check with your paint dealer for the best paint to use.

Interior Paints

Paint that is used on interior surfaces must be able to be easily cleaned. It must produce a smooth, uniform surface. Interior paints are available in gloss (highly reflective), semi-gloss, and flat finishes (nonreflective). Most interior walls and ceilings are painted with flat paint. A flat paint is one that has no gloss when dry. Kitchen and bathroom walls are sometimes painted with semi-gloss paint. Woodwork is usually painted with gloss or semi-gloss paint.

Exterior Paint

Exterior paint must have different properties than interior paint. Exterior paints must protect the structure from sun, snow, sleet, and rain. Some white exterior house paints have a self-cleaning property. This paint is chalky so that the surface cleans itself when it rains.

Painting

To get good results from any paint job the surface should be clean, dry, and free of all loose paint, Figure 24–10. Wood surfaces should be sanded where necessary to smooth rough spots. Sanding is especially important on interior woodwork. Before painting begins, drop cloths are laid to protect electrical fixtures, finished floors, and unpainted work.

Primer is applied to newly painted surfaces. **Primer** is a special paint that sticks to the surface better than regular paint. The primer should be the one recommended by the paint manufacturer.

Even though the paint may have been mixed when it was purchased, it should be stirred before it is used. Paint can be applied with a brush, roller, or sprayer.

To brush paint on, dip the brush about one-third the length of the bristles into the paint.

FIGURE 24–9 Paint provides protection and improves appearance.

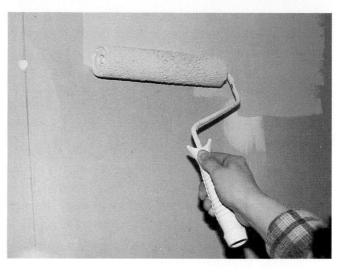

FIGURE 24–12 Large surfaces can be painted quickly with a roller.

FIGURE 24–10 To get good results from any paint job, the surface must be clean, dry, and free of old, loose paint. (Photo by Paul E. Meyers)

Remove the excess paint by tapping the bristles against the inside of the can, Figure 24–11. Flow the paint on the surface with long, full strokes.

To use a roller, pour a little paint into a roller pan. Work the roller back and forth in the pan until the roller cover is evenly covered with paint. Roll the paint in several directions on the surface. Finish by rolling in one direction, Figure 24–12.

Spraying is the fastest method of applying paint. Spray equipment is also the most expen-

sive and takes the most practice to learn. Spray equipment holds a supply of paint and uses air pressure to spray the paint evenly over the surfaces. The sprayer must be kept moving to prevent runs in the paint.

When finished, the painters clean their equipment. The correct thinner (water, mineral spirits, or turpentine) is used first, Figure 24–13. Then

FIGURE 24–11 Tap the bristles of the brush on the inside of the can to remove any excess paint.

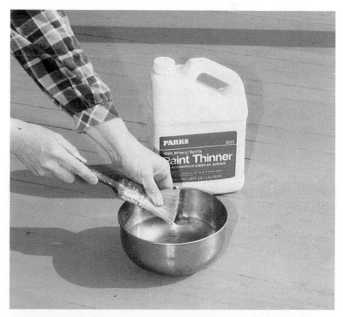

FIGURE 24–13 Clean painting equipment with the proper thinner for the paint used.

the equipment is washed with soap and water. Paint brushes are wrapped in paper for storage.

CAUTION: It is dangerous to use gasoline or kerosene to clean painting equipment. These fuels present a fire hazard and may cause severe skin irritation. Also, when using solvents, provide adequate ventilation.

Clear Finishes

There are many clear finishing materials available. Three common ones are shellac, lacquer, and varnish.

Shellac. Shellac produces a very fine finish on wood. However, it does not withstand heat, direct sunlight, or water spills well. An important use of shellac is for sealing knots in pine before painting, Figure 24–14. The shellac prevents the resin in the knot from discoloring the paint.

Lacquer. It is common to refer to any finishing material that dries very quickly through evaporation as lacquer. Many of the modern coatings in this category produce very tough, water-resistant surfaces. They are excellent for spray application because of their very short drying time.

Varnish. There are several kinds of varnish. Varnish produces an extremely tough, clear finish. A disadvantage of varnish is its slow drying time and tackiness (sticky quality). Varnish is difficult to apply without getting dust bubbles in its surface. The tough, durable surface produced by varnish makes it a popular finishing material.

Applying Clear Finishes

Clear finishes are applied in much the same manner as paints. The surface must be dry, clean, and sanded smooth, Figure 24–15. It is very important to clean up all the dust created by sanding since dust is easily seen in clear finishes. It is also important not to shake or stir varnish. This will create bubbles in the varnish in the can and on the surface being covered. Clear finishes are usually applied with brushes or spray equipment, Figure 24–16. Again, rollers create too many bubbles and should not be used.

FIGURE 24–14 Shellac is a good knot sealer.

FIGURE 24–15 Woodwork must be sanded smooth before applying varnish.

FIGURE 24–17 Landscaping includes planting trees, grass, and shrubs and constructing features like sidewalks and patios. *(Courtesy of the National Landscape Association)*

FIGURE 24–16 Flow the finish on with long strokes in the direction of the grain.

Landscaping

One of the final steps in completing any construction project is landscaping. This may be the simple grading and planting of grass and gardens. It may also involve constructed features like sidewalks or patios, Figure 24–17. Landscaping improves appearance, holds the soil in place, and provides access to structures.

All of the features of the landscape design are included on a working drawing called a **landscape plan**, Figure 24–18. This plan shows the location of buildings, driveways, parking lots, patios, and other aspects of the landscape.

Constructed Features

The first step in landscaping the site is to complete constructed features. These include such things as driveways, parking lots, and patios. Most of these features are constructed of concrete, stone, wood, or brick. The details of these features are shown on drawings like those used for the main structure.

Grading. Grading is the shaping of the contour of the site. Grading affects the appearance and

Landscape Architect

The landscape design begins early in the planning of the project. On small residential jobs the architect for the building designs the landscaping. On larger projects the planners rely on a landscape architect for this part of the design. Four or five years of college preparation are needed to become a landscape architect. The landscape architect considers the needs of the people served by the structure, the environmental surroundings, and advice from other professionals and experts in designing the landscape.

Environmental Design For:
SCHOHARIE BOCES CENTER
Schoharie, New York

FIGURE 24-18 Completed landscape plan.

controls the runoff of water. Rain and melting snow run downhill. On a properly graded site, water runoff is gradual until it reaches storm drains or streams.

Earthmoving equipment is used for rough grading where large amounts of earth must be moved, Figure 24-19. Rough grading does not prepare the site for planting. It simply contours the site. Finished grading often involves a great amount of handwork. Removing small stones

and smoothing the topsoil for planting requires the use of hand rakes, Figure 24-20. The raked surface must be smooth and completely free of unwanted stones. After the grading is complete, planting can begin.

Planting

Trees, shrubs, and grass are usually planted during landscaping. Each of these is an important part of the landscape design.

Trees provide shade, act as wind breaks, and create a natural, attractive appearance. Their roots also aid in controlling soil erosion. Trees growing naturally on the site are often included in the landscape design. Many trees are grown in nurseries and transplanted on the site, Figure 24-21.

Shrubs are smaller than trees. They are often used to line sidewalks or patios. Shrubs are also used to improve the appearance of gardens. Like trees, shrubs are grown in nurseries and transplanted on the site.

Grass is used to cover large yard areas of the site. Grass is easy to maintain and prevents soil

FIGURE 24-20 Finish grading requires a great amount of handwork.

FIGURE 24-19 Rough grading is done with heavy equipment. *(Courtesy of Deere and Co.)*

FIGURE 24–21 Many trees are grown in nurseries and then transplanted on the site.

erosion. Before grass is planted, the topsoil is prepared and fertilized.

Two common methods of planting grass are sodding and seeding. **Sod** is a blanket of existing grass that is grown on a sod farm. Rolled up strips of sod are transplanted on the site, Figure 24–22. Most grass is planted by seeding. Grass seed is spread evenly over the topsoil, then the seeds are forced into the soil with a roller.

Freshly planted grass must be treated with care until it is well established. A mulch of clean

FIGURE 24–22 Sod is transplanted in rolled-up strips.

straw or hay is often spread over the surface. Mulching protects the new grass from direct sunlight and helps the soil hold its moisture.

Landscape Maintenance

Like any living thing, the landscape requires care. Landscape gardeners mow lawns, prune trees and shrubs, and care for gardens. From time to time, these plants need fertilizing and other special care. Landscape gardeners analyze the needs of this expensive part of the owner's investment and provide the necessary care.

Maintaining the Structure

After the structure and landscaping are finished it is important to perform maintenance. **Maintenance**, also called servicing, means keeping the structure in good working condition. It also involves keeping the structure attractive and clean. Examples of maintenance include cleaning, repainting, and replacing worn parts. Both the inside and outside of a structure get dirty and need to be cleaned. Paint gets dirty, wears out, and fades and must be repainted. Windows, light switches, and other parts can wear out and need replacement.

Another important part of maintenance is checking the utilities. It is important to service climate control equipment by replacing filters, checking motors and belts, and checking its operation on a regular basis. Maintenance helps keep equipment operating efficiently and safely.

Remodeling

Remodeling involves changing the structure to improve its usefulness or appearance. As an example, a family living in a small house may add additional rooms as they have more children, Figure 24-23. Constructing the addition makes the home more useful for the growing family. Remodeling can be done on the inside or outside of a structure.

Installing new windows, doors, siding, or roofing is also a part of remodeling. These changes can be made to improve the appearance or the usefulness of the structure. Newer windows, for example, can add to the appearance. They may also be easier to open and close and more energy efficient. Other remodeling jobs done to improve

FIGURE 24–23 This structure is being remodeled to improve its appearance and to make it more energy efficient.

the usefulness of a structure include building a patio, installing a pool, adding a new bathroom, or installing insulation.

Summary

The final steps in completing a structure include finishing and installing ceilings, wall coverings, molding, cabinets and countertops, paints and clear finishes, and landscaping. Workers who do these jobs must do more precise work than those who frame the structure and install utilities.

When the structure and landscape are finished, it is important to perform maintenance. Remodeling is done to improve the appearance or usefulness of the structure.

Remodeling for Handicapped Access

A growing field in remodeling involves making structures more useful for handicapped people, Figure 1. The design of most structures does not consider the access problems for people in wheelchairs. Special designs are needed for hallways, closets, and kitchen and bathroom countertops and cabinets, among other parts. Most public structures have been remodeled to meet these needs. Often, however, homes and workplaces must be remodeled to provide unlimited access for handicapped people.

FIGURE 1. This structure has been remodeled to make it accessible to people in wheelchairs.

DISCUSSION QUESTIONS

1. What are two common methods of covering ceilings and walls?
2. What four materials are often used to cover floors?
3. What are the differences between water-based and oil-based paint and interior and exterior paint?
4. What three clear finishes were described in this chapter?
5. What tools are used to apply paints and clear finishes?
6. What are the various processes involved in landscaping?

CHAPTER ACTIVITIES

 ## Building Countertops

OBJECTIVE

Countertops are often finished with plastic laminate. The laminate protects the plywood or particleboard surfaces of the countertop from water and other substances. In this activity, you will build a countertop.

EQUIPMENT AND SUPPLIES

Plywood or particleboard, 2 feet square
1″ x 1″ x 24″ piece of lumber
Three 1¼ x 8 flathead steel screws
Electric or hand drill and selection of bits
Screwdriver

Contact cement
Plastic laminate to cover surface
Plastic laminate edge banding
Electric router and laminate trimmer bit
Waxed paper

PROCEDURE

1. To create an edge with the appearance of thicker material, attach a piece of solid wood to the underside of the surface at the edge, Figure 1.

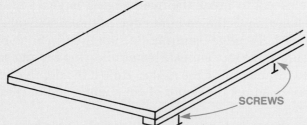

FIGURE 1. Attach a piece of solid wood to build up the edge of the counter.

2. Brush a uniform coat of contact cement on the back of the laminate and on the surface to be covered.

CAUTION: Some contact cement is highly flammable. Do not use contact cement near an open flame. Use adequate ventilation.

3. When the contact cement is no longer sticky to the touch, cover the surface with two pieces of waxed paper. Overlap the pieces of paper near the center. The waxed paper prevents the laminate from sticking until it is in position.

4. Position the laminate on top of the waxed paper. Allow a slight overhang at the edges. Raise one side of the laminate enough to remove one piece of waxed paper. Raise the other side and remove the remaining piece of waxed paper, Figure 2.

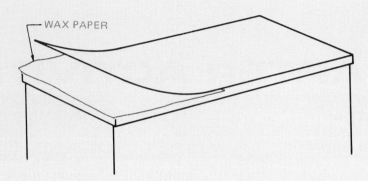

WAX PAPER

FIGURE 2. Raise the end of the laminate and pull out the second piece of waxed paper.

5. Apply pressure all over the surface to ensure good contact at all points. This can be done with a soft-rubber mallet or by rubbing the surface with the corner of a piece of soft pine. Be careful not to break the overhanging edges.

6. Cement the edge banding to the counter edge in the same way. It is not necessary to use waxed paper with the edge banding. Be sure the top edge of the edge banding is against the underside of the top laminate before allowing the cemented surfaces to touch.

7. Insert the laminate trimmer bit in the router. Adjust the depth of cut as shown in Figure 3. It is best, at first, to adjust the depth of cut slightly high. Then make a trial cut and readjust the depth.

8. Trim the overhang from the top laminate with the router base on the countertop and the trimmer-bit pilot against the edge banding.

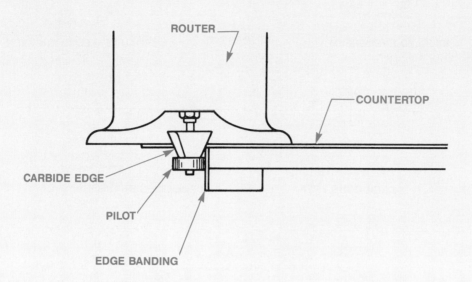

FIGURE 3. Adjust the router depth of cut so that the countertop will be flush with the edge banding.

MATH AND SCIENCE CONNECTIONS

The motor in a router can turn at over 20,000 revolutions per minute.

RELATED QUESTIONS

1. What does "laminate" mean?
2. Why do you think plastic laminates are trimmed with a router?
3. What are carbide-tipped router bits?
4. What is the purpose of the "pilot" in Figure 3?

 Designing Landscape

OBJECTIVE

Landscape architects arrange trees and shrubs with such constructed features as walkways, patios, terraces, and fences. They often supervise the necessary grading, construction, and planting. In order to do this, landscape architects must study construction techniques as well as horticulture and art.

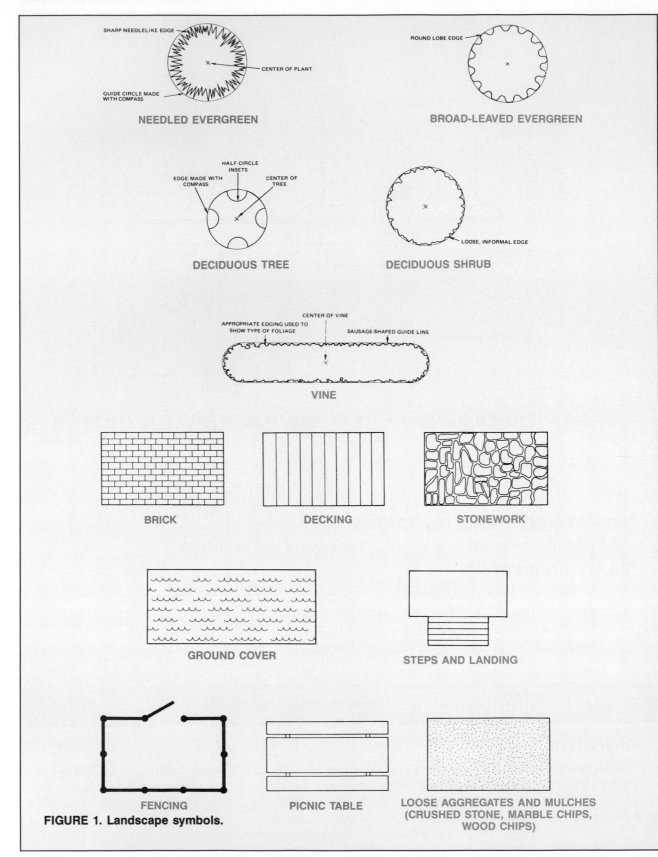

FIGURE 1. Landscape symbols.

EQUIPMENT AND SUPPLIES

Architect's scale   Straightedge
Pencils and paper   Compass

PROCEDURE

Using the symbols shown in Figure 1, draw a landscape design for your home. You may draw a plan of the existing landscape or completely redesign it. Figure 2 shows an example of a landscape plan. Your design should include the following:

1. General shape and location of buildings
2. Approximate size and shape of the area to be landscaped
3. Constructed features
4. Trees
5. Shrubs
6. Lawn

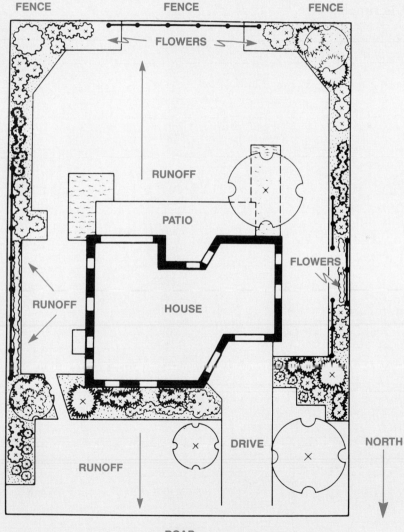

FIGURE 2. Sample landscape plan.

7. At least one ornamental garden
8. It is not necessary to include contour lines on this drawing. Indicate the direction of water runoff by arrows labeled "runoff."

MATH AND SCIENCE CONNECTIONS

Landscape architects must understand horticulture. Horticulture is the science and art of growing fruits, vegetables, trees, and flowers.

RELATED QUESTIONS

1. Why do you think a landscape architect should study art?
2. What was the most difficult part of this activity?
3. How could you get to see the results of your landscape design without making the actual full-size layout?
4. What is runoff?

CHAPTER 25

Heavy Construction: Civil Construction and Industrial Construction

OBJECTIVES

After completing this chapter, you will be able to:

- Point out the similarities and differences between light (residential) construction and heavy (civil and industrial) construction.
- Describe the various steps in constructing a highway.
- Identify the major parts of a bridge.
- List the major elements in a typical industrial construction project.

KEY TERMS

Abutment	Eminent domain	Industrial construction
Civil construction	Environmental impact statement	Light construction
Cut	Fill	Prestressed concrete beam
Embankment	Heavy construction	Pylon

Most of the construction technology described so far in this book is related to **light construction**. Light construction involves building houses, smaller apartment buildings, and smaller commercial and business buildings. Most of the work is done by people working with hand tools. This chapter deals with **heavy construction**. Heavy construction involves building large structures such as roads, highways, bridges, tunnels, dams, power plants, and manufacturing factories. Skyscrapers are also built with heavy construc-

tion. Much of the work done in heavy construction is done with heavy earthmoving equipment.

Most heavy construction can be divided into two types — civil and industrial. **Civil construction** is any field of construction that is closely related to earthwork, Figure 25-1. Highways, roads, dams, bridges, and tunnels are examples of civil construction. **Industrial construction** involves building facilities for processing or controlling materials, energy, or other resources. Power plants and manufacturing factories are examples.

FIGURE 25–1 Civil construction involves earthwork using heavy machinery. *(Courtesy of Texas Highways)*

FIGURE 25–2 Civil engineers design highways. *(Courtesy of Texas Highways)*

Civil Construction

Planning a Highway

Civil engineers design highways to satisfy needs, Figure 25-2. A highway meets the need of allowing traffic to travel safely and smoothly from one place to another.

Traffic surveys are used in the early planning stages. A traffic counter measures the number of vehicles passing a given point each hour or day. Traffic engineers look at the results of the traffic survey, the speed of the traffic, and the number of traffic lanes on each highway. These items are used to develop the basic design.

When the basic design is approved, surveyors measure the land, Figure 25-3. When all of the information is collected, drafters create plans for the project. These plans show where roadways and ramps will be built.

Highways are usually owned by the government. Any land where the road will be built that is not owned by the government must be bought. The government can buy needed land by exercising its right of **eminent domain**. Eminent domain is the right of the government to buy land for public uses and to pay a fair price for that land.

Civil Drawings

The final planning step is to create detail drawings and specs for the roadway, bridges, and other constructed features.

FIGURE 25–3 This surveyor is surveying the land where a new highway will be constructed. *(Courtesy of the Lietz Company, copyright © 1986)*

The drawings for a complete highway project may be well over one hundred pages long. Each part of the project requires different drawings. Section views are drawn to show more detail about the roadway and embankments. Complete

BART's Transbay Tube

The San Francisco Bay Area Rapid Transit (BART) is a 71.5 mile rapid-transit rail system, Figure 1. BART is made up of 19 subway and tunnel miles, 23 elevated railway miles, 25 surface miles, and nearly 4 miles in the transbay tube. The tube, which carries BART trains under the San Francisco Bay, has been recognized the world over as one of history's most outstanding civil engineering achievements.

BART's civil engineers planned to build the tube in sections. There were to be 57 tube sections, each averaging 330 feet in length, longer than a football field. These were made on dry-land shipyards, at the Bethlehem Shipyards in South San Francisco. From there they were launched, towed into the bay, and sunk in their proper position, Figure 2.

The tube sections look like huge binoculars in cross section. They are 24 feet high

FIGURE 2.

and 48 feet wide. Trackways in each tube carry trains in each direction. They are separated by an enclosed central corridor for pedestrian access, ventilation, and utilities. The shell of each tube is made from ⅜-inch steel plate reinforced with steel I-beams set six feet apart. The inside of the tubes is concrete reinforced with steel bars. Before the sections were sunk into position, watertight bulkheads were placed at each end. To lower the tubes to the bottom of the bay, 500 tons of gravel was placed on top of each tube section. The final weight of each section is about 10,000 tons.

Once in place, each new section was snugged tightly against the previous one. Four 50-ton hydraulic couplers were used to pull the new section into position. The joints between sections were sealed by neoprene rubber gaskets. The watertight bulkheads were removed to allow passage from one end of the bay to the other. Construction of BART was completed in 1973. The full cost of the project was $180 million in 1970 dollars.

FIGURE 1.
(Courtesy of Bay Area Rapid Transit / Photos by Gordon Kloess)

elevations, sections, and detail drawings are done for bridges.

Civil drawings are usually dimensioned in feet and decimal parts of a foot. They are drawn and read with a civil engineer's scale, Figure 25–4. These scales are graduated in multiples of ten. A small number near the zero end of the scale indicates the ratio of the scale. The number 10 indicates a 1:1 ratio, Figure 25–5A. The number 20 indicates a 1:2 or ½ scale, Figure 25–5B; and so on. Figure 25–6 shows several of the scale options that can be obtained using a civil engineer's scale.

Because highways cover so much area, it is sometimes necessary to use scales with one inch representing several hundred feet. The size of the area is one of the greatest differences between heavy and light construction.

Preparing the Site

With the design work complete, it is time to begin actual work on the site. The first step in site preparation is staking the boundary of the site. Highways require large amounts of earthwork. The roadway must be surveyed often to check the location of stakes.

When a highway is being built through wooded land, the site must be cleared. Large trees are cut

		CIVIL ENGINEER'S SCALE			
Divisions	Ratio	Scale Used With This Division			
10	1:1	1″ = 1″	1″ = 1′	1″ = 10′	1″ = 100′
20	1:2	1″ = 2″	1″ = 2′	1″ = 20′	1″ = 200′
30	1:3	1″ = 3″	1″ = 3′	1″ = 30′	1″ = 300′
40	1:4	1″ = 4″	1″ = 4′	1″ = 40′	1″ = 400′
50	1:5	1″ = 5″	1″ = 5′	1″ = 50′	1″ = 500′
60	1:6	1″ = 6″	1″ = 6′	1″ = 60′	1″ = 600′

FIGURE 25–6 Scale options with a triangular civil engineer's scale.

down with chainsaws. Smaller trees, stumps, and brush are cleared with bulldozers, Figure 25–7. Large boulders have a tendency to work their way to the surface. This is a disaster under a highway, so they must also be removed.

FIGURE 25–7 Bulldozers are used to clear the site for a new road. *(Courtesy of Caterpillar Tractor)*

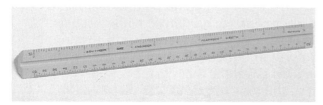

FIGURE 25–4 Civil engineer's scale. *(Courtesy of Koh-I-Noor Rapidograph, Inc.)*

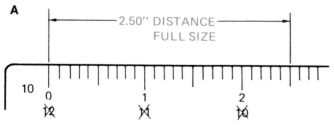

A

2.50″ DISTANCE
FULL SIZE

10

0   1   2

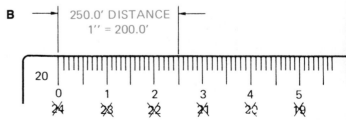

B

250.0′ DISTANCE
1″ = 200.0′

20

0   1   2   3   4   5

FIGURE 25–5 (A) Civil engineer's scale—1:1. (B) Civil engineer's scale—1:2.

Cut and Fill. In some cases, it may be necessary to remove or add earth to get the desired grade. Soil that is removed in order to create the desired grade is called **cut**. Engineers usually use the cut to **fill** low spots nearby, Figure 25–8. After trees and boulders have been removed, scrapers are used for most of the cut and fill work.

Embankments and Foundation Work

In highway construction, the foundation is made from soil. The superstructure is the pavement, Figure 25–9. When the roadway is elevated, such as at the approach to a bridge, an **embankment** is created. Embankments are built up in layers. Rollers are used to compact the soil as the embankment is built. A sheepsfoot roller, Figure 25–10, is often used for soil compaction.

Bridges

Bridges are needed to span streams or for one roadway to pass over another. Major highway construction projects often include bridges.

FIGURE 25–10 A sheepsfoot roller is used to compact soil. *(Courtesy of Texas Highways)*

The simplest type of bridge is made up of **abutments** and **pylons** with a deck supported on beams, Figure 25–11. In this type of bridge, the abutments are the foundation and the roadway is the superstructure. Bridges that span greater distances need either more complex superstructures or added foundations. Figure 25–12 shows three examples: the trussed superstructure, suspension superstructure, and multiple-pylon foundation.

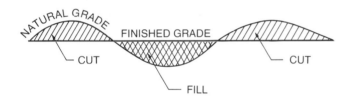

FIGURE 25–8 Cut is where earth must be removed. Fill is where earth must be added.

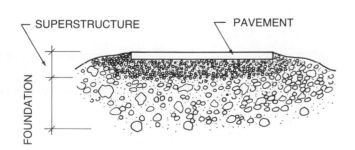

FIGURE 25–9 The foundation for a roadway is soil and stone. The superstructure is the pavement.

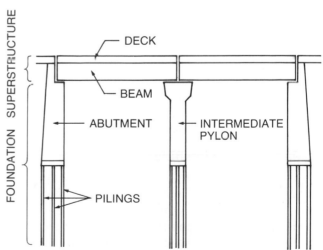

FIGURE 25–11 Major parts of a bridge.

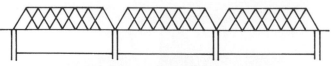

TRUSSED SUPERSTRUCTURE

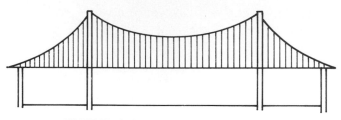

SUSPENSION SUPERSTRUCTURE

MULTIPLE-PYLON FOUNDATION

FIGURE 25–12 For long spans, these bridge designs are used.

FIGURE 25–13 This is a central pylon for a two-level bridge on a highway interchange. *(Courtesy of Texas Highways)*

The abutments and pylons for the bridge are built first, Figure 25–13. They are built of reinforced concrete on a heavy foundation. Next, the embankment is built on each end of the bridge. Then, the beams of the bridge are installed between the abutments and pylons, Figure 25–14. The beams are usually steel I-beams or prestressed concrete. Light steel decking is placed over the beams. It will be the form for the bottom surface of a concrete roadway.

Paving and Finishing

After the decking is in place, the roadway is ready for paving. Both portland-cement concrete and asphalt are widely used for pavement. Portland-cement concrete makes a very hard and wear-resistant pavement. However, it needs much more curing time and is less flexible than asphalt. That is why asphalt is used slightly more often than concrete.

FIGURE 25-14 Steel I-beams are added to the pylons. *(Courtesy of Texas Highways)*

For paving, asphalt is mixed, loaded into dump trucks, and delivered to the site. The asphalt is dumped into a paving machine, Figure

Prestressed Concrete Beams

Prestressed concrete beams are cast with steel cables inside them. The term "prestressed" is used because the cables are tightened (-stressed) before (pre-) the concrete cures. When the concrete cures, it bonds to the cable, holding it in stress. The stressed cable pulls against the inside of the concrete, helping to hold it together.

The effect of the cable pulling against the concrete beam is similar to the effect of holding a row of books together tightly enough to lift them as a single unit, Figure 1. The force exerted on the books holds them together. The stressed cable inside the concrete does the same thing. The force it exerts on the concrete helps make the beam more rigid.

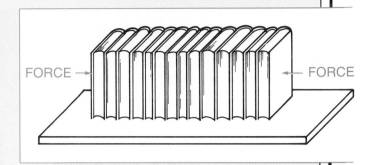

FIGURE 1. The principle of prestressed concrete.

25–15. The paving machine spreads the asphalt, screeds it, and compresses it. A smooth roller finishes the paving operation.

There are many other jobs required to finish the highway. Guard rails are installed, traffic lines are painted, signs are posted, and landscaping is done. All the work is inspected before the highway is ready for use.

Industrial Construction

It would be impossible in a few pages to describe all of the possible kinds of industrial construction. However, an electrical power generation plant includes most of the major elements. This unit describes the construction of one hydroelectric power generation plant on the Black River.

Environmental Impacts

Before any heavy industrial construction project can be started, engineers prepare an **environmental impact statement**. This is a report of all important aspects of the environment where construction will take place. It describes how construction will affect the environment.

FIGURE 25–15 A paving machine receiving asphalt from a dump truck. *(Courtesy of Texas Highways)*

Let's look at the environmental impact statement for the Black River.

Environmental Impacts on the Black River

Two major impacts were considered on the Black River project. First, the Black River is a fast-flowing, medium-sized river in a wilderness area. It is important to fishermen, white-water rafters, and sightseers, Figure 25–16. If the water from the river were used only for the power plant, the river would be destroyed for all of these recreational purposes. The design of the power plant had to maintain a constant flow of water in the river.

It was also determined that an endangered species might be affected by the project. The Indiana brown bat is an endangered species that lives in caves near the construction site. It was feared that blasting large amounts of rock would disturb the sleeping bats. A method had to be found to keep track of the bats' sleep patterns and to blast rock without disturbing them.

A Solution for Recreation. To protect the recreational value of the river, the water was diverted into a long canal running parallel to the river, Figure 25–17. Also, the power plant was designed in two parts; an upper powerhouse and main powerhouse. This design had little effect on the recreational uses of the river and still provided the water needed to generate electricity.

A Solution for the Bats. Little information was available on the sleep habits of the Indiana brown bats. Scientists studied the bats to learn when they were awake so blasting could be done. Small blasts were done at first. The size of the dynamite charges was increased slowly. During blasting, the bats were monitored with video and electronic equipment for changes in behavior. The

FIGURE 25–16 The Black River is important for recreation. *(Courtesy of James Besha Associates)*

scientists were very careful to make sure the normal patterns of the bats were not disturbed.

Excavating the Canal

It was a huge excavation job to create the 70-foot-deep by quarter-mile-long canal. About one-half million cubic yards of rock had to be excavated with blasting and heavy equipment, Figure 25–18. The blasting was used to break the rock into pieces that could be loaded onto trucks and removed.

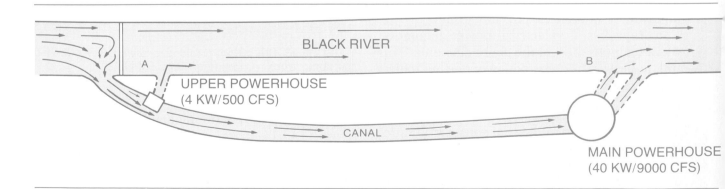

BLACK RIVER

A

UPPER POWERHOUSE
(4 KW/500 CFS)

B

CANAL

MAIN POWERHOUSE
(40 KW/9000 CFS)

FIGURE 25–17 Plan for the canal and two powerhouses on the Black River.

FIGURE 25–18 The canal was dug 70 feet deep and one-quarter mile long. *(Courtesy of James Besha Associates)*

Building the Powerhouses

The two powerhouses are buildings with much reinforced concrete and some structural steel. Both exist only to house a turbine-powered electrical generator and the necessary controls.

The powerhouses are built on a foundation of reinforced concrete. The walls are a reinforced concrete shell with a metal geodesic-dome roof, Figure 25–19. The geodesic dome was used because it is a very efficient use of building materials.

FIGURE 25–19 Geodesic-dome roof on the main powerhouse. *(Courtesy of James Besha Associates)*

The Powerplant Equipment

Figure 25–20 shows a cross-sectional view of the inside of the main powerhouse. Water enters from the canal, turns a huge turbine blade, and is returned to the river. The turbine is attached to a generator. The generator is used to create the needed electricity.

The turbines are located 66 feet below the surface of the water. Large concrete tunnels had to be built to let the water enter and exit under the powerhouse, Figure 25–21.

The electric generators are so large that it was impractical to build them in a factory and truck them to the site. Instead, they were built on site. It is common in industrial construction to see large machinery assembled at the construction site.

Electricity Transmission Towers

The final constructed features for the power plant are the towers used to transmit the electricity to nearby communities, Figure 25–22. Near the power plant, the towers are large, trussed structures made of steel. Closer to the local homes, single wooden poles with horizontal bars

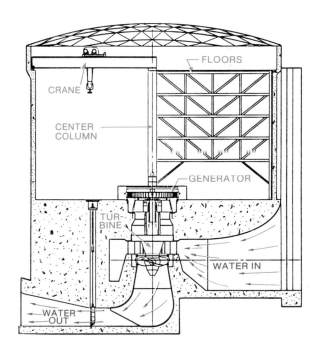

FIGURE 25–20 Cross section of the main powerhouse.

FIGURE 25–21 Discharge tunnels from the main powerhouse. *(Courtesy of James Besha Associates)*

FIGURE 25–22 Tower for the electrical power lines.

on top are used to support the lines. The lines near the plant carry higher voltages and must be heavier. This is why trussed steel towers are used. Near the homes, the voltage is smaller and the lines can be lighter weight.

Trussed steel transmission towers come in several different designs. Their main purpose is to support the heavy weight of the electrical lines. They must also be strong enough to support the lines when snow or ice builds up. The effect of winds must also be considered.

Like all constructed structures, the trussed towers require a strong foundation. Reinforced concrete foundations are used to anchor the base of the tower. The lines can exert a tremendous weight on the foundation. Engineers must also consider the side forces created when the lines make a turn. These side forces will try to pull over the tower. The structural steel used to make the tower is usually welded or bolted together.

Summary

Light construction involves building homes and apartments. Workers mostly use hand tools for light construction. Heavy construction involves building large projects like highways and dams. Much of the work in heavy construction is done with heavy earthmoving equipment.

The two types of heavy construction described in this chapter are civil construction and industrial construction. Civil construction involves making roads, highways, bridges, tunnels, and dams. It is closely related to earthwork. Industrial construction involves building power plants and manufacturing factories.

The size of a civil or industrial construction project is one major difference between heavy and light construction. Highway and industrial projects can cover miles of land.

DISCUSSION QUESTIONS

1. What profession is most directly involved in designing the construction details of a roadway?
2. What factors are considered in the design of a highway exchange?
3. What is the job of surveyors in a highway project?
4. What is the term for the power of the government to take over land for public use?
5. What are some of the scales used on a civil engineer's scale?
6. What environmental impacts were considered on the Black River power plant project?
7. Do you think the designers of the power plant should have been concerned for the recreational uses of the river? Should they have been concerned about the Indiana brown bats? Why or why not?
8. How did construction and manufacturing work together in the Black River power plant project?

CHAPTER ACTIVITIES

 Prestressed Beam

OBJECTIVE

Prestressed concrete beams are often used in bridge construction. A stressed cable pulls against the inside of the concrete, helping to make it stronger. In this activity, you will make a model which can be used to demonstrate the prestressed beam concept.

EQUIPMENT AND SUPPLIES

¾" plywood
Small diameter all thread
Nuts and washers for all thread
Table saw
Sandpaper
Drill press
Drill bit (diameter slightly larger than all thread)
Safety glasses

PROCEDURE

1. Cut the plywood into 2-inch squares. You will need about 20 to 30 squares. Sand the edges of the squares smooth.
2. Drill two holes in each square. See Figure 1 for the location of the holes. Your teacher may have you design a fixture to make sure the holes are properly located.
3. Place two pieces of all thread through the holes in the plywood squares, Figure 2. The all thread will represent the stretched reinforcing cables in prestressed concrete beams.

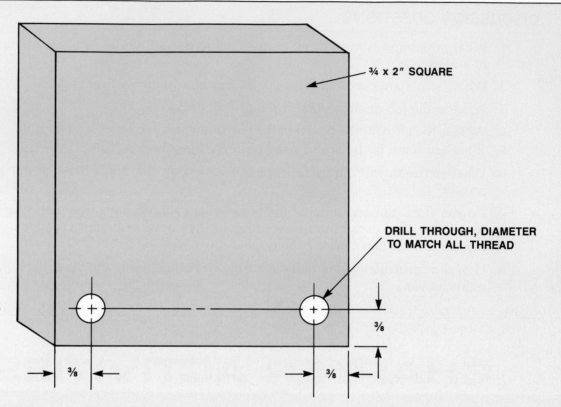

FIGURE 1. Drill two holes in each plywood square.

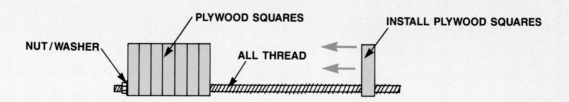

FIGURE 2. Place all thread through the two holes.

4. Secure the all thread on the ends of the beam with washers and nuts.
5. Your prestressed concrete beam model is ready for testing.

INVESTIGATION

1. Take about 10 of the plywood squares not fastened on all thread. Place them in a row like books on a bookshelf.
2. Press in on each end of the row and try to pick up the beam. Were you able to put enough force on the pieces to hold them in place as you lifted the beam?

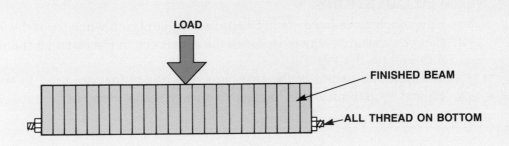

FIGURE 3. Test your simulated prestressed beam.

3. Place your prestressed beam between two tables or bench tops. Position the tables so about 2 inches of your beam rests on them on each end. Also, make sure the all thread is located on the bottom of the beam, Figure 3.
4. What do you think will happen if you apply a small downward force on the center of the beam? Apply the force to check your conclusion. Was your conclusion correct?
5. What do you think will happen if you turn the beam over so the all thread is near the top of the beam and apply the same force? Test your conclusion.

MATH AND SCIENCE CONNECTIONS

Your prestressed beam should have held a larger force when the reinforcing rod was near the bottom. When the reinforcment was near the top of the beam, spaces between the plywood should have been opened up as the force was applied. This can be explained by tension and compression forces. A beam experiences both tension and compression when a force is applied, Figure 4. The compression (squeezing) is near the top, while the tension (pulling apart) is near the bottom. The plywood squares in the beam can take compression and stay in place. However, tension causes the squares to move apart. When the rod is on the bottom of the beam, it resists tension and holds the squares together.

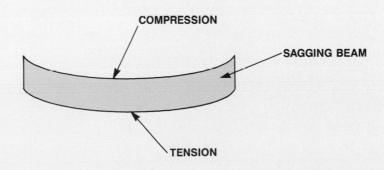

FIGURE 4. Tension and compression in a loaded beam.

RELATED QUESTIONS

1. How much more force could be applied to your beam when the rod was on the bottom?
2. Can you design a way to measure the difference in forces when the rod is on the bottom and on the top?
3. What are the names of the two parts of a bridge that are used to support beams?
4. What do you think would happen if concrete beams were built without reinforcing rods?

SECTION SIX

PRODUCTION IN THE FUTURE

(Courtesy of Lockheed Corporation/Photo by Dick Luria)

CHAPTER 26

Environmental Impacts: Scrap, Waste, and Pollution

OBJECTIVES

After completing this chapter, you will be able to:

- Describe various environmental impacts of scrap, waste, and pollution.
- Describe the differences between scrap, waste, and pollution.
- Explain how recycling programs can be used to help reduce waste and pollution.
- Identify various hazardous household wastes and their potential impacts.
- Describe the creation of acid rain and its environmental impacts.
- Describe the greenhouse effect and its environmental impacts.

KEY TERMS

Acid rain	Hazardous waste	Reclamation
Air pollution	Heap leach mining	Recycling programs
Biopesticides	Herbicides	Scrap
Digester	Pesticides	Transfer station
Greenhouse effect	Pollution	Waste

Some people say we live in a "throwaway" society. This is only partially true. We do throw *out* large amounts of garbage, but we really can't throw it *away.* It stays around us for a long time. Love Canal in Niagara Falls, New York, is a good reminder that there is no "away" for our garbage. A chemical company dumped 21,800 tons of toxic waste near Love Canal between 1947 and 1952. Large metal cans filled with toxic chemicals were buried in the landfill. Thirty years of rusting finally caused leaks in the metal containers.

Chemicals seeped into groundwater as well as the land surface. After heavy rains, the chemicals rose to the surface in areas where children played. Many suffered acid burns. The terrible smell of chemicals was in the air. Trees and grass suffered burns. The air itself over Love Canal contained eighty-two different compounds, eleven of which were shown to cause or promote cancer. Finally in August of 1978, the New York State Health Commission stepped in and bought 239 homes from residents of the area. The homes were boarded up, and families were forced to move from the area, probably forever. It might be safe to say that the chemical company could not forecast the impact of their dumping at the time. We now know, however, that nowhere is far enough "away" when it comes to scrap, waste, and pollution, Figure 26–1.

Scrap is material that cannot easily be reused by the company that makes it without first being reprocessed. **Waste** is any material that is thrown away and never reused or recycled. **Pollution** is any harmful changes in air, water, or land caused by scrap or waste. An aluminum beverage can that is sent to a recycling center is considered scrap. That is, it will be reused to make more aluminum cans or other aluminum products. If the same aluminum can is sent to a landfill, it is considered waste. That is, the material is never reused or recycled. And, if the can causes changes in the environment (land, air, or water) when placed in the landfill, it is considered pollution.

Scrap

Scrap may take many forms. The most common forms of scrap are materials like steel, aluminum, plastics, ceramics, wood, and so on. Although scrap cannot easily be used by the company that made it, it can become a material input for another company. For example, the scrap metal created when golf club heads are made can become a raw material for tableware manufacturers producing spoons, forks, and knives. Scrap in the form of thermoplastics can be reformed into different products. An old automobile that is sent to a scrap yard may be crushed and reformed into other steel products, Figure 26–2. Many times, scrap from manufacturing is not reused and ends up in landfills or dumps. This material is considered waste.

Waste

Waste is material or products that have already been consumed and discarded. The United States produces three billion tons of solid waste material every year. Many of the products that enter the waste stream can be reused or recycled. Steel, aluminum, plastic, glass, and paper are easily recycled but often go to landfills, never to be recycled, Figure 26–3.

Recycling programs aid consumers in the recycling process. Consumers are asked to separate materials at home and take the materials

FIGURE 26–1 The New York State Department of Environmental Conservation drills to test wells at the Love Canal site. They test water in the ground for chemicals. *(Courtesy New York State Department of Environmental Conservation)*

FIGURE 26–2 A car body is lifted to the shredder at a scrap yard. The metal in the car will be used to make new cars or other products. *(Courtesy of Commercial Metals Company)*

FIGURE 26–3 Landfills are running out of space for the millions of tons of solid waste we throw out each year. *(Photo by Helena Frost)*

FIGURE 26–4 Recycling programs are being used around the country to aid consumers in reusing waste materials. *(Photo by Sonya Stang and Brent Miller)*

to recycling centers, Figure 26–4. The recycling company pays the consumer for the materials if they are reusable. The recycled materials are then processed to create raw materials for the manufacture of other products. The recyling material may also be picked up with the normal garbage collection. In some communities, special garbage bags or containers are used to identify waste materials that can be recycled, such as aluminum cans, glass, paper, and plastics. In other communities, consumers do not separate the reusable materials. In these cases, separation is done at a

Biotechnology: Heap Leach Mining

The newest mining process that directly uses biotechnology is called **heap leach mining.** It is used for mining copper, silver, and gold. The heap leach mining process involves chemicals percolating through a pile, or heap, of ore. The percolation (per coe 'lay shun) of a chemical through the ore is similar to running water through coffee grounds. Sulfuric acid solutions are used for copper, and cyanide solutions are used for silver and gold. The percolating solution dissolves the metal in the ore heap and lets it run into a storage basin. The storage basin may be a very large pond. The solution, now full of the dissolved metal, is then pumped through a metal recovery system. In the metal recovery system, microbes are used to leach out the copper, silver, or gold.

After the metals are leached out, the remaining water is very acidic. It is acidic because it contains waste material such as iron sulfide and pyrite. The acidic waste water could pollute the surrounding environment. In the Appalachian coal mine regions of the United States, over 6,500 miles of streams have been polluted by acid waste from coal mines. The waste water is usually rusty in color, and is unfit to drink. It can kill or damage plants and animals. Because of its acidity, it is very corrosive. Metal bridges and boats can be damaged by acid water waste.

In order to prevent the pollution of the environment by the acid waste water, a process known as **reclamation** is used. The reclamation process also uses microbes to leach the waste water. This makes the waste water less acidic and safe for the environment.

transfer station, Figure 26–5. Inside the transfer station is a long conveyor belt. People stand on each side of the conveyor belt, open the bags of garbage, and sort out the materials that can be reused. Materials that can not be recycled are burned in a **digester**. The material left over after burning in a digester is called humus. Humus is an earthy brown product that can be used as a fertilizer.

Household Waste

Consumer dumping of waste is a serious problem that few people are aware of. When thrown away with the garbage, certain household wastes can pollute water supplies. Many of the household waste materials we throw away are considered **hazardous waste**. Hazardous waste includes materials that are explosive or highly reactive, flammable, corrosive, toxic, or poisonous. Common hazardous wastes that are often thrown out with the normal garbage include:

- paints
- cleaners
- wood preservatives
- motor oil
- antifreeze
- solvents
- pesticides

Paints contain solvents and metals that may enter water supplies if not disposed of properly. Serious pollution problems result when paints are poured down the drain, dumped on the ground, or thrown in the trash, Figure 26–6. If there is no outlet for unused paints, the proper action is to remove the lid of the paint can, place the can outside, and let the paint harden.

Cleaners with labels that say ''flammable'' or ''poison'' should not be poured down the drain. The best way to dispose of cleaners is to use them up for their intended purposes. Never mix cleaners with ammonia and chlorine bleach because the result will be toxic fumes.

Wood preservatives are products containing chemicals that are toxic to organisms that cause wood to decay. Wood preservatives with pentachlorophenol mixtures in their ingredients are the most dangerous and have no safe means of disposal. Even a small amount of this substance will pollute large amounts of water. The best approach is to use the wood preservative up by applying a second or third coat if necessary.

Motor oil contains a number of dangerous chemicals. Waste motor oil is often dumped on the ground. It then seeps into water supplies. Waste motor oil can be recycled by a used motor oil recycling location. Since 1979, stores that sell oil must provide a collection tank for used motor oil or post a sign that identifies the location of the nearest recycling center. Most service stations

FIGURE 26–5 This transfer station plant can separate up to 2000 tons of waste per day. *(Photo by Sonya Stang and Brent Miller)*

FIGURE 26–6 Unused paint is considered a hazardous waste. Paint should be allowed to dry in the can before disposal. *(Photo by Sonya Stang and Brent Miller)*

Biopesticides Moving into the Mainstream

One of the major problems a farmer must face is the destruction of valuable crops by pests and insects. A pest management system is needed by all farmers in order to overcome this problem.

In the past, the most common management technique was to spray fields and crops with chemical pesticides. A chemical **pesticide** is designed to kill the pests or make it difficult for them to find and eat the crop. Some of the chemicals sprayed to kill the pest resulted in contaminated food. Contaminated food is unfit to sell at the market. The soil and surrounding water supplies could become contaminated as well. Also, the pests would sometimes become immune to the chemical spray. This meant that every year, a new chemical pesticide would have to be used.

Through research in biotechnology, pesticide companies began to develop genetically engineered products **(biopesticides)** to help end the pest problem. The first biopesticide was called BT and was used in the 1960s. BT is a naturally occurring bacteria that kills most insects trying to eat a crop. BT is not poisonous to humans. It is still widely used today.

Another result of genetic engineering was the development of crop seeds that are resistant to certain herbicides. **(Herbicides** are chemicals that kill weeds, allowing the crop to grow.)

Today, research and development have shown that crops can be genetically altered to contain bacteria that, when eaten by an insect, will kill the insect.

It is hoped that the contamination of the crops, soil, and water by chemical pesticides will end due to biotechnology research and development. The future of biopesticides and crop seeds that will manage pests in an environmentally sound way looks bright.

Genetic engineering techniques are used by Sandoz Crop Protection scientists to help improve the company's line of biopesticides and in developing new types of seeds. *(Courtesy of Sandoz Crop Protection)*

will accept used motor oil because they sell it to recyclers.

Antifreeze is many times discarded on the ground during the flushing of an automobile radiator. This is often dangerous to small animals who are attracted to the sweet taste of antifreeze. The best way to discard antifreeze is to dump it in a sanitary sewer drain. Some service stations will take used antifreeze.

Solvents can be flammable and poisonous. They can also pose a serious health risk when inhaled or left in contact with the skin. The best way to discard leftover solvents is to use them up for their intended purposes. Never pour solvents down the drain.

Pesticides are chemicals designed to kill insects and rodents, Figure 26–7. Many have been found to cause long-range environmental damage. Pesticides should never be buried, mixed together, poured on the ground, or dumped down the drain. Many pesticides have been banned. There is no safe way to dispose of pesticides. The nearest pollution control authority can advise users on how to store unused pesticides.

Pollution

One of the most important types of pollution is **air pollution**. Air pollution is a very serious impact of manufacturing because clean air is one of the most important parts of life. Each day humans breathe about 23,000 times and inhale about 525 gallons of air. Air pollution occurs whenever fuel or other materials are burned, Figure 26–8. Burning fuels and other materials is common during manufacturing processes. Many manufacturing plants have large stacks or chimneys that put smoke and chemicals into the air. Manufacturing processes often require great amounts of electrical energy. This energy may also be generated by burning fuels. Two common types of air pollution effects include acid rain and the greenhouse effect.

Acid Rain

Acid rain is a type of air pollution that eventually turns into water pollution. Chemicals put in the air by industrial processes combine with rainwater to create acid rain. A pH scale indicates the acidity of acid rain (acid snow and acid fog

FIGURE 26–7 The use of pesticides in the production of food is a major source of pollution. *(Photo by Andrew Horton)*

FIGURE 26–8 Emissions from smokestacks can contribute to air pollution problems. *(Photo by Sonya Stang and Brent Miller)*

are also considered acid rain), Figure 26–9. Distilled water has a pH value of 7. Pure rainwater has a pH of 5.6. Rainwater with a pH lower than 5.6 is considered acid rain. When pH changes one value lower on the scale, it represents a tenfold increase in acidity. For example, rainwater with a pH of 5 is ten times more acidic than rainwater with a pH of 6.

Notice how the pH of water in the lakes of the Adirondack region of New York State changed from the 1930s to 1975. Scientists believe that acid rain is caused by power plants, certain industries, and transportation emissions. It is difficult to trace what is happening in the Adirondacks to one specific source. No one source is held responsible for cleaning up the problem. Acid rain has a negative impact on aquatic life. Plants and fish can die quickly in acidic water. Birds that eat the fish and plants can also die or leave the area because of a lack of food. The effects of acid rain can be very misleading. When you look at lakes or streams affected by acid rain, they can be clear blue and look perfectly clean. Under the surface, however, the water might have no life, Figure 26–10. Lakes, streams, and forests in the Adirondack Mountains, Canada, and other parts of the world continue to die while governments try to study what is happening.

The Greenhouse Effect

A parked car that has the windows rolled up shows the **greenhouse effect**, Figure 26–11. Sunlight passes through the glass and strikes the inside of the car. The sunlight is absorbed by the seats and other parts inside the car and turned

FIGURE 26–10 Streams and lakes affected by acid rain may look clean and healthy. *(Photo by Edith Raviola)*

FIGURE 26–11 Heat trapped inside the car warms the inside. This is a simple example of the greenhouse effect.

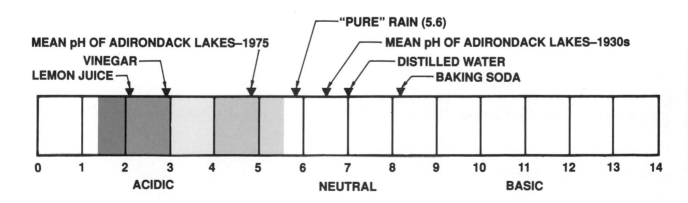

FIGURE 26–9 A pH scale indicates acid-base level of materials.

Environmental Checklist

Conserving materials conserves land, forests, and energy and reduces stress on living systems. Check your conservation awareness.

In my house, I
_____ recycle newspaper
_____ recycle plastic
_____ recycle glass
_____ recycle aluminum cans
_____ recycle aluminum foil
_____ reuse paper bags
_____ use hand-me-downs
_____ use recycled paper
_____ take my own bags to the grocery store
_____ repair and reuse things
_____ compost kitchen scraps

In my house, I conserve energy by
_____ having adequate insulation
_____ keeping the thermostat at or below 65 °F in the winter
_____ wearing warm enough clothing for the above
_____ using hot water sparingly
_____ avoiding air conditioning
_____ avoiding unnecessary electrical gadgets
_____ using high-efficiency electrical appliances
_____ using electricity sparingly in lights, range, tv, radios, and so on

To reduce pollution in my house, I
_____ do not dispose of toxic substances in the trash or down the drain

_____ use phosphate-free soap
_____ do not use pesticides
_____ clean windows with vinegar and water, not chemicals
_____ follow guidelines for disposing of waste products
_____ compost organic matter in an outdoor compost pile
_____ grow an organic garden using recycled compost as fertilizer

In my transportation, I
_____ have a high-miles-per-gallon car
_____ maintain efficient exhaust emissions control
_____ keep the car tuned up
_____ carpool when feasible
_____ walk or bike for short trips when feasible
_____ recycle used motor oil
_____ recycle used tires
_____ use public transportation

In my shopping, I
_____ decline excess packaging from merchants
_____ avoid styrofoam
_____ cut back on disposable products
_____ read labels and buy least toxic products
_____ avoid buying unnecessary products
_____ buy products that can be repaired easily
_____ buy durable products that may cost a little more but last much longer

into heat energy. In the summertime, the temperature inside the car can reach higher than 150 degrees. Glass lets the sunlight pass through. After it changes into heat, the glass blocks the energy from escaping.

Chemicals in the air around the earth perform the same function as the glass in the car. They let sunlight pass through to strike the earth. The sunlight causes the earth and air to heat up. Carbon dioxide in the atmosphere then prevents the heat from escaping. The more carbon dioxide in the air, the greater the warming of the earth's atmosphere. The burning of fuels, such as coal, oil, and gas, creates carbon dioxide.

Scientists fear that as we continue to add carbon dioxide to the atmosphere, the polar ice caps will melt from the excess heat. This would raise ocean water levels about eighty feet. Massive flooding of coastal and low regions might occur. Increased temperatures might also lead to less rainfall. The 1988 summer drought in the midwestern United States stirred debate about the impact of the greenhouse effect on rainfall, Figure 26–12.

FIGURE 26–12 During the drought of 1988, a farmer inspects a crop that is usually knee-high by the fourth of July. *(Courtesy U.S. Department of Agriculture)*

Summary

Some people say we live in a throwaway society. But we can not throw away our wastes. They stay with us in landfills and garbage dumps for decades. The three types of manufacturing outputs that cause environmental impacts are scrap, waste, and pollution. Scrap is material that cannot be easily reused by the company that makes it. Scrap can be used as a raw material input by another manufacturer. Waste is products or materials that are never reused or recycled. Recycling programs are helping to reduce waste. There are many household wastes, such as paints, solvents, and motor oil, that are considered hazardous. These must be disposed of properly in order to reduce possible negative impacts. Pollution is any harmful change in the water, air, or land caused by scrap or waste. Acid rain and the greenhouse effect are two examples of air pollution caused by industrial processes.

DISCUSSION QUESTIONS

1. What are some of the possible environmental impacts of scrap, waste, and pollution?
2. What is the difference between scrap, waste, and pollution?
3. How are recycling programs helping to reduce waste and pollution? Are there any recyling programs in your community?
4. What potentially hazardous waste materials are there in your home or school? What would you do to reduce their potential impact on the environment?
5. What causes acid rain? How does it affect the environment?
6. What is the greenhouse effect?

CHAPTER ACTIVITIES

 ## Use It — Don't Scrap It

OBJECTIVE
In this activity, you will study the different types of scrap that are made by your production classroom and laboratory. You will then decide what products could be made out of the scrap or make a plan for reusing or recycling the scrap.

EQUIPMENT AND SUPPLIES
1. Recycled paper and pencils
2. Scrap made by your production classroom and lab

PROCEDURE
1. Form groups of four or five and select a team leader.
2. Each group should have a sample of the scrap made by the materials processing that is done in the lab. You might have wood cutoffs, sawdust, metal scrap, and/or plastic scrap.
3. Hold a brainstorming session with your group. Decide what other uses there are for the scrap. Try to create a new product that can be made from the scrap. NOTE: Sending the scrap to a landfill is not acceptable. Find other solutions to the garbage problem.
4. If the scrap cannot be turned into a product, then each team must create a reuse plan or recycling plan.
5. Team leaders will present their product ideas and reuse or recycling plans to the class.

MATH AND SCIENCE CONNECTIONS
Today many companies use recycled materials. It is not uncommon to see paper in letters, envelopes, and packages stamped ''made from recycled materials.'' Look around your home, school, or community for evidence of products made from recycled materials.

RELATED QUESTIONS

1. Where does the scrap made in your school lab currently go? Trace where the various scrap products end up. Are these scraps polluting the local environment?
2. Try this mini research project. Visit a scrap yard and decide what materials could be reused. Write a list of all the ways that reusing these materials could benefit the local community.

 ## Acid Rain

OBJECTIVE

In this activity, you will learn more about acid rain. You will measure the pH level of local rainfall to determine if acid rain is a problem in your community. You will also produce simulated acid rain and observe its effects on seedlings and materials.

EQUIPMENT AND SUPPLIES

1. Safety glasses
2. Package of radish seeds
3. Package of plastic cups
4. Potting soil
5. Vinegar
6. Distilled water
7. Four test tube racks and 16 test tubes
8. pH meter, or test strip kit
9. Strips of copper, aluminum, steel, galvanized sheet metal
10. Magnifying glass

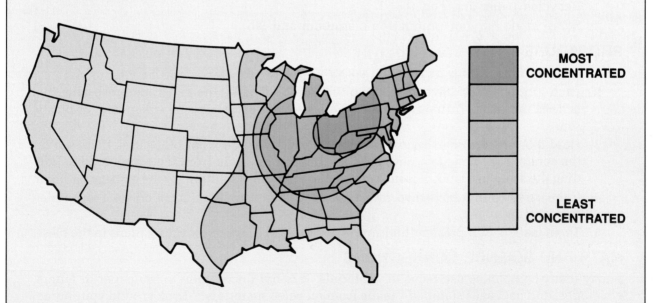

FIGURE 1. This map shows the acid rain concentrations throughout the United States.

PROCEDURE

Part I: Growing Seedlings

1. Fill ten plastic cups with potting soil. Plant three radish seeds in each cup.
2. Label the plants so that you can keep accurate records.
3. Provide light and proper moisture using distilled water for two to three weeks.
4. Record the height of each plant.

Part II: Monitoring Rainfall

1. Follow local weather reports to find the next two rainy days.
2. Clean three plastic cups by washing in distilled water. Store in a clean plastic bag.
3. On the next rainy day, place the cups outside and away from buildings and automobile traffic.
4. Bring the samples inside. Test the pH of the rainwater by using a pH meter or pH test strips. NOTE: Follow the directions provided with the meter or test strips to get accurate results.
5. Record your data. Repeat on the next rainy day. Compare results. Determine if your local rainfall is acid rain.

Part III: Seedlings and Acid Rain

1. Complete this part of the activity two to three weeks after planting the radish seeds. NOTE: Be sure to wear safety glasses during this part of the activity.
2. Use household vinegar and distilled water to make solutions with pHs of 3, 4, 5, and 6. Store the solutions in clean, labeled containers.
3. Treat two plants with each pH solution for ten days. Provide only distilled water to two plants as a control.
4. Keep a log that includes the heights and general appearance of the plants each day.
5. After about a week, write a summary of the results in your log. Share this information with your class.

Part IV: Effects of Acid Rain on Materials

1. Use distilled water and vinegar to prepare solutions with pHs of 3, 4, 5, and 6. NOTE: Wear safety glasses during this activity.
2. Cut ¼ ″ x 1½ ″ strips of copper, aluminum, steel, and galvanized sheet metal.
3. Use a magnifying glass to look at each metal. Look closely at the surface of the metal. Write down your observations.
4. Set up and label four racks of test tubes. One sample of each metal should be placed in a test tube with each pH solution. Record your observations each day.
5. After a week, remove the samples one at a time. Wash with water and use a magnifying glass to examine each sample. Compare the appearance of the metals after being exposed to acid rain (water) with their original appearance.

MATH AND SCIENCE CONNECTIONS

The pH scale ranges from 0 to 14. A pH of 1 is very acidic, while a pH of 14 is very alkaline. The pH of several common items is shown in Figure 2.

RELATED QUESTIONS

1. Describe two ways that power plants can reduce the emissions that cause acid rain.

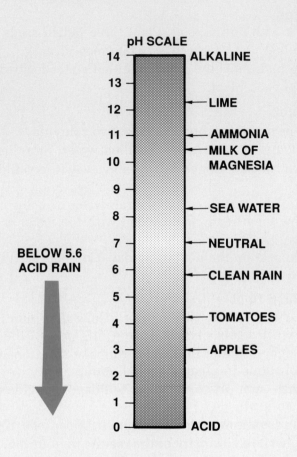

FIGURE 2.

2. Why do you think power companies resist using low-sulfur coal as a fuel?
3. Who should be held responsible for the damage done to lakes and streams from acid rain? Why?
4. Is acid rain a problem in your community? If not, should you be concerned about acid rain? Why or why not?
5. What do you think can be done to reduce the damage acid rain causes?

CHAPTER 27

Production in Space

OBJECTIVES

After completing this chapter, you will be able to:

■ Describe the three phases required for space industrialization.

■ Identify the new materials and products that may be manufactured in space.

■ Describe the construction techniques that may be used to build structures in space.

KEY TERMS

Automated beam builder
Deployable
Levitator

Microgravity
Microspheres
Neutral buoyancy tank

Photovoltaics
Superalloy

One important development that may shape the future of production is space exploration. Someday, people might work in space stations, manufacturing products and constructing structures. Let's look at how manufacturing and construction might be used in space.

Manufacturing in Space

For several years now, NASA (National Aeronautics and Space Administration) has been predicting that, someday, products may be manufactured in space. As you probably know, a space station would be built to orbit the Earth. One of the main uses of the space station will be to conduct manufacturing experiments.

The biggest problem with the space station is providing the basic life-support systems humans need. The space environment is very different from that found on Earth. In the space station, a human-made environment must provide all the

oxygen, food, water, and energy needed. A number of plans have been proposed, including transporting these necessities to the space station in the space shuttle. Long-range plans call for oxygen to be produced in space from other substances. One possibility is making oxygen from a material found on the moon called ilmenite. Miners would extract this mineral from the surface of the moon and process it to produce oxygen. Without the gravity of Earth, mining will be much easier on the moon, Figure 27–1.

Energy is the easiest problem to solve. NASA has spent years developing solar photovoltaic panels. **Photovoltaic** panels convert the light energy of the sun into electrical current. Huge panels would be needed to meet all the electrical needs of the space station, Figure 27–2.

The space station would be made up of pressurized modules about the size of a large bus (44.5′ x 13.8′), Figure 27–3. Astronauts would have different modules for work, sleep, and recreation.

FIGURE 27–1 One possibilty for producing oxygen for people who work in space is to produce oxygen from ilmenite mined from the moon. (Courtesy of NASA)

FIGURE 27–3 This mock-up of one module for the space station is about the size of a bus. Modules will be built for sleeping, working, and recreation facilities. (Courtesy of NASA)

FIGURE 27–4 This industrial spacecraft, which will be smaller than the space station, would house automatic and robotic equipment, not people. (Courtesy of NASA)

FIGURE 27–2 The space station will provide astronauts with a home in space. The eight blue rectangles on the ends of the station are solar photovoltaic panels that will produce all the electricity the astronauts need. (Courtesy of NASA)

Another plan, separate from the space station calls for an industrial spacecraft to be built, Figure 27–4. Under normal conditions, no life support systems would be provided on this craft. Experiments and processes would be performed automatically by robots and computers. Every few months, astronauts would fly to the industrial spacecraft in the shuttle and provide new raw materials, check out production, and make any necessary adjustments. When hooked to the shuttle, the industrial spacecraft would provide full life support for the astronauts.

Putting a space station or industrial spacecraft in orbit around Earth is just the first step to a long process of manufacturing in space. Engineers have identified three phases in their plans for space industrialization.

Phase 1. Basic Scientific Research

Scientists cannot accurately predict exactly how things will work in the space environment. In space, the effects of gravity are almost nonexistent. There is a very small amount of gravity in space. Astronauts and aerospace engineers call it **microgravity**. As an example, something that weighs 200 pounds on Earth would not be totally weightless in space, but it would weigh less than one-half ounce. We don't know all the effects that microgravity will have on people working in space or on manufacturing processes. Microgravity has been simulated on Earth using airplanes flying in parabolic curves, Figure 27–5. During descents, microgravity has been achieved for twenty to sixty seconds. Astronauts train in these airplanes to learn to function in microgravity, Figure 27–6. Most of the early experiments in space will help scientists and engineers better understand the effects of microgravity.

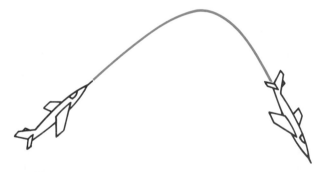

FIGURE 27–5 Planes flying in parabolic curves in the earth's atmosphere can be used to simulate microgravity. As the plane descends, microgravity is artificially produced for twenty to sixty seconds.

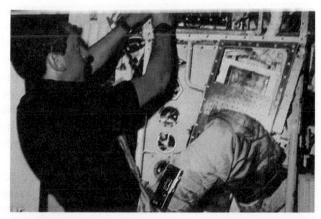

FIGURE 27–6 This astronaut must adapt to working in a weightless atmosphere. (*Courtesy of NASA*)

Phase 2. New Technology

Once we better understand how things work in space, engineers will design and create new technology that will use microgravity. From tests that have already been conducted, many new tools and equipment have been developed. An obvious example is portable power hand tools. In space, astronauts can't just string out an extension cord for electrical power. The popular cordless, portable, battery-powered screwdrivers and drills are examples of spinoffs from space research.

Another example are the furnaces that will be used to melt and process materials in space. Melting materials like glass or metal will be very different in space. For one thing, in microgravity, molten materials will not pour or spill out of a container. So, containerless material processing can be done. In fact, some **levitators** have already been developed. A levitator is a system for processing materials without containers. On earth, containers, like crucibles, can contaminate melted materials. No matter how clean, small particles of the crucible can mix with the molten material and cause defects, Figure 27–7. In space, levitators will use acoustical energy (sound waves) to levitate or suspend molten glass or metal in mid-space while it is being processed. Without containers, there will be no contamination. Other types of levitators would use electricity or magnetism to suspend materials.

Many other types of tools and machines will have to be created. Look around at the tools,

FIGURE 27–7 On earth, molten materials always pick up contaminants during processing. (*Courtesy of Allegheny Ludlum Corporation*)

machines, and equipment in your production lab. They may not work the same way in space.

Phase 3. Product Manufacturing

Once the new tools have been created, manufacturing industries will start to make new products. Based on past space experiments, NASA expects the first developments to be in new and improved materials. Research in metal, glass, and crystal manufacturing has shown much promise.

Manufacturing Metals in Space. By combining two or more different metals, a new metal alloy with improved qualities can be produced. On Earth, some alloys cannot be made. A good example is lead and aluminum. If these two metals were combined, a long-lasting, self-lubricating alloy would be produced. However, because of the difference in the weights of lead and aluminum, the lead/aluminum alloy cannot be made. Lead is four times heavier than aluminum. When these two metals are heated together for melting, the lead sinks to the bottom of the crucible while the lighter aluminum rises to the top, Figure 27–8. Mixing the molten metals in the crucible does not help. In space, however, because of

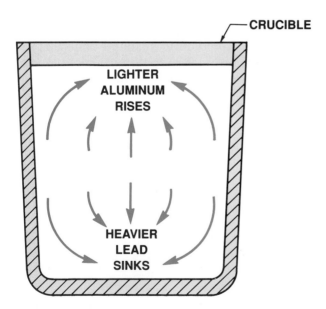

FIGURE 27–8 Because of the gravity on earth, certain metals cannot be alloyed. Lead and aluminum are examples. The heavier lead sinks, and the lighter aluminum rises; mixing doesn't help. In the microgravity of space, this alloy and hundreds of others will be possible.

microgravity, the difference in weights would not exist. In microgravity, lead and aluminum should mix to create a new **superalloy**. Engineers and scientists expect to create up to 400 new superalloys by combining two or more metals that cannot be combined on Earth. These superalloys could greatly improve the strength, cost, weight, and wear characteristics in many of our manufactured products.

Manufacturing Glass in Space. Any glass made on Earth, no matter how perfect, has impurities and imperfections. As mentioned before, during processing, the molten glass picks up impurities from the crucible. Also, because of gravity, molten glass will resolidify (harden) in uneven densities, causing imperfections. In space, the combined benefits of containerless processing using levitators and microgravity will permit the production of super-pure glass. This super-pure glass could be made into fibers for improved fiber-optic communication technology. In fact, futurists believe that the pure glass fibers may some day replace metal circuitry in computers. This would lead to the development of supercomputers that use light instead of electric current, Figure 27–9.

Manufacturing Computer Chips in Space. Computer chips are made from silicon crystals. Cylinder-shaped crystals are grown by melting a batch of silicon in a crucible, putting a cool "seed" crystal in the batch, and letting the silicon grow off the seed, Figure 27–10. As with glass and metal, no matter how clean the workplace, there are always some impurities in the crystals. Another problem is gravity. During the melting cycle, the hot silicon rises, while the cool silicon sinks. This heating and cooling cycle creates tensions in the crystals that can cause flaws and cracks. Usually, only 21 percent of these crystals are usable for computer chips. This means almost 80 percent of these crystals must be thrown out by the manufacturers. With such a high amount of waste, the cost of producing computer chips is high. Once again, microgravity and containerless processing will mean perfect crystals for computer chips. Astronauts have already demonstrated this by growing perfect crystals in space on several space shuttle missions.

Manufacturing Plastics in Space. The first products actually manufactured in space were tiny plastic beads called **microspheres**. These beads are 1/2500th of an inch in diameter (about 1/40th the diameter of an average human hair). Fifteen million microspheres can be placed in a tube about the size of your index finger. Astronauts made millions of microspheres in space during several space shuttle missions dating back to 1982. These tiny beads have multiple uses as measurement instruments for processes such as calibrating microscopes and other precision instruments and measuring paint pigment and gunpowder particles. As with other space-manufactured materials, each microsphere produced on the shuttle was perfect in size and shape. On Earth, making perfect microspheres is impossible because of the effects of gravity, Figure 27–11.

Manufacturing in space seems to hold great promise for the future. Obviously, many improved and new products will be produced. But putting a space station in orbit and conducting the necessary research will cost billions of tax dollars and take many years.

FIGURE 27–9 Research into glass manufacturing in space may someday lead to fiber-optic cables and light replacing wires and electricity in computers. *(Courtesy of US Sprint Communications Company)*

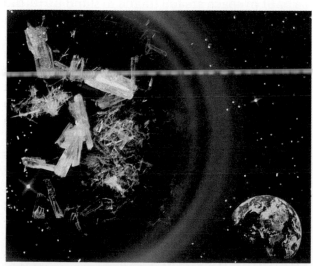

FIGURE 27–10 Perfect crystals have already been grown in space. These crystals are used to manufacture computer chips. *(Courtesy of NASA)*

FIGURE 27–11 Microspheres made in space are perfect in size and shape. These microspheres were the first product manufactured in space. *(Courtesy of NASA)*

Constructing the Space Station

Many designs have been proposed and rejected for the space station. Engineers studied many different designs and styles. The first space station designs were completed in March 1986. NASA selected a "dual-keel" design, Figure 27–12. The dual-keel space station is rectangular in shape. It has two parallel 297-foot tall vertical keels. Between the keels is a single horizontal beam which supports the solar-powered energy system. The pressurized modules would provide comfortable living quarters, laboratories, and work spaces for the crew. To provide easy access and maximum privacy, the modules are arranged in a raft pattern. They would be connected by external nodes which will house the command and control systems. External airlocks free up space inside for equipment and work space. The atmosphere inside the modules will be nearly identical to Earth's. An environment control and life support system (ECLSS) will provide the crew with a breathable atmosphere; furnish water for drinking, bathing, and food preparation; remove air pollutants; and process wastes. The ECLSS will be a closed system. This will permit oxygen to be recovered from the carbon dioxide expelled by the crew. Only food and nitrogen will have to be resupplied from time to time.

NASA's plans call for the space station to be launched in segments in the space shuttle. The towers of the station will be assembled in orbit, Figure 27–13. Construction is planned to begin in 1993. The space station should be fully staffed by 1994.

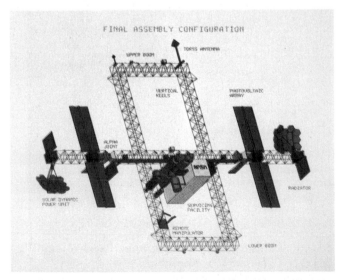

FIGURE 27–12 Current plans call for this dual-keel design. *(Courtesy of NASA)*

FIGURE 27–13 This tower was assembled by space shuttle crew members to demonstrate space construction. *(Courtesy of NASA)*

Building Large Antennas

The space shuttle enables us to move large and bulky cargo into space on a regular basis. With this capability, researchers are planning to build very large antennas in space. The shuttle's closed cargo bay carries up to 32 tons on each trip into space, Figure 27–14. The large antennas that will be built in space are too fragile to stand up under their own weight on Earth. They will be folded up in the cargo bay of the shuttle and transported into space. There they will be built into their final shapes in the microgravity of space. The first large space antennas will be **deployables**. This means that the entire antenna will fold into a compact container on Earth. In space, the antenna will deploy automatically. One type of antenna, the hoop-column, or maypole, antenna would open up in orbit much as an umbrella does, Figure 27–15. A cylinder no bigger than a school bus could be transformed within an hour into a gigantic antenna dish of two acres.

Building in Orbit

Deployable antennas will, in a sense, build themselves. They will unfold with the push of a button. The second type of structures planned for space are erectables, Figure 27–16. They will not build themselves; someone or something will have to put the separate pieces together. Some

FIGURE 27–14 The shuttle's cargo bay. *(Photo by Dick Luria; courtesy of Lockheed Corporation)*

FIGURE 27–16 Segments of a modular structure are removed from the cargo bay of the shuttle by remote manipulator arms and are unfolded before being attached to the main structure. *(Courtesy of NASA)*

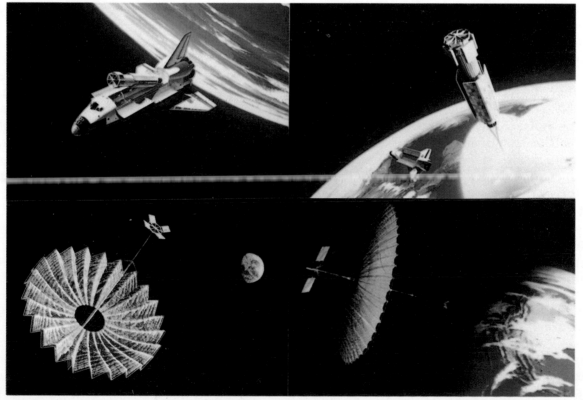

FIGURE 27–15 Deployment sequence for a 100-meter diameter hoop-column antenna. *(Courtesy of NASA)*

of the assembly projects will require several shuttle trips. This will mean having the first construction sites in space. And so, a new type of work for the human race will be devised. At various NASA centers, researchers are now determining the most efficient ways to build erectables. The testing is done under water in a huge **neutral buoyancy tank**, Figure 27–17. Working in the tank simulates the microgravity of space. Technicians practice the mechanical tasks necessary for the construction of these huge space structures. Many factors are taken into account — safety and fatigue of the astronauts, speed in moving from one place to another, and the requirement for simple tools and the need to restrain them so they do not float away. In one method of assembly, astronauts tethered to the shuttle would simply move from beam to column to module, manually snapping, locking, or latching everything together. Their travel time could be shortened by wearing jetpacks called manned maneuvering units (MMUs). Models of MMUs are used in the neutral buoyancy tank simulations. But it is not yet certain how manpower, machine operations, deployment, or assembly jobs to build space structures will be combined.

For some projects it might be more efficient to move astronauts around on a scaffold in a mobile work station instead of having them use the MMUs. The scaffold rests on a frame in the cargo bay and moves either up and down or right and left. As sections of the structure are finished, they are moved away from the station so that the next part to be built is always in reach. Astronauts might also stand in open cherry-pickers attached to the shuttle's 15-meter remote manipulator arm. They would be moved from beam joint to beam joint like telephone linemen working on high wires. Even more complex would be the closed cherry-pickers, where workers inside a comfortable chamber would work with remote control arms. For dangerous tasks, programmed robots called unmanned free-flying teleoperators could do work with their mechanical arms. There might also be assembler devices to form three-dimensional structures from struts by following simple repeatable steps. Maneuverable television units would transmit pictures to technicians in the shuttle control room so that they could direct work by remote control.

FIGURE 27–17 The neutral buoyancy tank allows divers to work in space suits under water to simulate the microgravity of space. *(Courtesy of NASA)*

These devices will most likely be used later in the shuttle era. In the meantime, astronauts will have to learn to erect structures the size of large stadiums in the peculiar world of microgravity. Seemingly easy tasks will become complicated — workers trying to turn ordinary bolts will be as likely to turn themselves as the bolts, due to the lack of leverage that comes with microgravity. These are precisely the problems studied during simulated assembly jobs in the neutral buoyancy tank. Those problems, in turn, influence the choice of technology, like using latches that snap firmly together with one quick motion instead of a series of twists and turns. The goal is to standardize hardware and assembly methods in order to get the job done as quickly and correctly as possible.

Automated Beam Builder

After the deployables and after the erectables, the next logical step is to build large structures completely from scratch. Individual building components will be assembled in space. A machine for that very purpose has already been designed. Called the **automated beam builder**, it will sit at one end of the shuttle's cargo bay. Spools of ultra-light material would be loaded into the machine on Earth and carried into orbit. Once at the space construction site, the beam builder would heat, shape, and weld the material into 3-foot-wide triangular beams that could then be cut to any length. The beams would then be latched together to build large structures, Figure 27–18. By loading the cargo bay with extra spools, enough material could be carried up in one trip to build thousands of yards of beams.

Now, with the beam builder, we will advance from the dreams of science fiction to practical blueprints for large structures that will dwarf the space shuttle flying around them. As the platforms grow in size, they will carry more science instruments and will grow even more gangly with their cross arms, dishes, and wing-like solar panels, Figure 27–19.

New Materials

Whatever their shape, large space structures will put great demands on the materials from which they are made. Even though they will be free from the weight stresses of Earth's gravity, there will be other strains from tight packaging, solar radiation, and the frigid temperatures of space. Engineers will need to build with new materials

FIGURE 27–18 An automated beam builder fabricating triangular-shaped trussed beams from the cargo bay of the shuttle. *(Courtesy of Grumman Corporation)*

for a new age. Materials will need to be light, super-strong, flexible or rigid (depending on the use), and thermally stable. For example, telescoping masts must be light, yet stay very stiff.

And antenna ribs need to be strong, but should be flexible enough to wrap around their hub. Everything needs to remain fixed in position in both the hot sun and cold shadow. If a structure were to expand with heat, for example, it would ruin the precise shape of an antenna.

One substance that meets these demands quite well is graphite-epoxy composite. This material is now being used in lightweight tennis rackets, golf clubs, airplane parts, and in the space shuttle itself, Figure 27–20. A three-meter-long hollow tube of this material can be lifted with one finger. Still, it is ten times stronger than steel!

There are other materials that will suit specific jobs. The hundreds of threads that pull and stretch a hoop-column antenna into shape might be made of a quartz filament, because quartz is very stable. The dishes themselves should be made of fabrics that fold like cloth before they are deployed. These would be metal meshes woven like nylon stockings or soft porch screening and coated with gold for reflectivity. A finer mesh will be used for dishes that deal in smaller wavelengths. For very small wavelengths, there are the ultra-thin membranes made of transparent film coated with metals. These look and feel like sheets of Christmas tinsel.

FIGURE 27–20 Engineers are experimenting with assembling possible structures for space use out of lightweight graphite-epoxy composite cones. (Courtesy of NASA)

Summary

Someday production workers might manufacture products and construct structures in space. Researchers suggest three phases for space industrialization — basic scientific research, new technology, and product manufacturing.

Basic scientific research involves studying the effects of microgravity on people, tools, and processes. New technology, such as new tools like the levitator, would be made to take advantage of microgravity. Some of the first products that would be manufactured in space include super-alloy metals, super-pure glass, computer chips, and plastics, such as microspheres.

The first types of structures that will be built in space are deployables. These would be umbrella-like structures that would practically build themselves. The next structures would be erectables that would be built piece by piece by astronauts or robots. NASA simulates building structures in space by using a neutral buoyancy tank, which is an underwater building tank. Eventually, an automated beam builder would be used to construct large space structures. New materials, such as graphite-epoxy composites, will be needed to build structures in space.

DISCUSSION QUESTIONS

1. What are the three phases required for space industrialization?
2. What effects might microgravity have on production workers, tools, and processes?
3. What new materials and products might be manufactured in space?
4. How do NASA researchers simulate the effects of microgravity on Earth?
5. What construction techniques might be used to build structures in space?
6. What is the difference between a deployable and an erectable structure?

CHAPTER ACTIVITIES

 Model Space Station

OBJECTIVE

The proposed space station would be made of lightweight beams. In space, automated beam builders will be used to construct lightweight beams. The beams would eventually be used to construct space stations and other large space structures.

In this activity, you will design and construct a model of the space station, Figure 1.

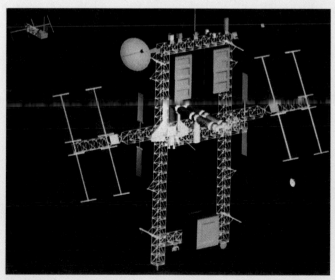

FIGURE 1. Proposed space station design. *(Courtesy of NASA)*

EQUIPMENT AND SUPPLIES

1. Building materials: drinking straws, dowel rods, balsa wood, wire, string, cardboard, styrofoam, clean used beverage cans, aluminum foil, and so on
2. Fasteners: hot glue or other adhesives, brads, pins, and so on
3. Measuring tools
4. Assorted production tools and machines
5. Safety glasses

PROCEDURE

1. Work in small groups.
2. Review the proposed design of the space station. Look at how small pieces of material are connected to make the large beams.
3. Decide how you can make a model of the space station. Consider these ideas:
   - beams can be made from dowel rods, balsa, straws, and so on.
   - solar panels can be wood or cardboard covered with aluminum foil.
   - modules can be made from blocks of wood or empty cans.
4. Decide how big your model will be. Consider making your model to scale. The dual keels of the proposed station are 297 feet in length.
5. Gather the necessary materials. Divide the work among group members. Each member can make a certain part of the station.
6. Construct your model. Consider making a stand for the space station or hanging it from the ceiling.
7. Display your space station model.

MATH AND SCIENCE CONNECTIONS

The dual keels of the proposed space station are 297 feet in length. Assume you added three feet and said they were 300 feet long. If you made your model space station keels three feet long, your model would be built at a 1/100 scale. Objects on your model would be 100 times smaller than the actual space station.

RELATED QUESTIONS

1. What problems will engineers have to overcome in order to build a space station?
2. What types of materials will be used to build the space station?
3. Do you think the space station will be completed before you graduate from high school? Why or why not?
4. What uses would be made of the space station?

CHAPTER 28

Future of Production

OBJECTIVES

After completing this chapter, you will be able to:

- Describe six trends that may predict the future of production.
- Explain the recent exponential rate of growth in technology.
- Describe the process of manufactured construction.
- Identify several methods that can be used to forecast the future.
- Describe the possibilities of our society becoming a technocracy.
- Differentiate between hard and soft technologies.

KEY TERMS

Appropriate technology	Manufactured construction	Technocrats
Exponential rate of growth	Mapping	Technopeasants
Forecasting	Scenario	Trend
Futurist	Simulation	Trend analysis
Global competition	Soft technology	
Hard technology	Technocracy	

What will the future hold? No one can know for sure, but we can find some clues by looking at the past and extending recent trends to the future. A **trend** is a pattern in recent events in society. Throughout history, one important trend is that production has been an important part of the development of societies. In the future, production will continue to help shape the standard of living and financial health of our society. The only thing we can say for sure is that the future will be different from our world today. Change is the only constant. It is sure to occur in every area of society. Look, for example, at the trend in shortening the length of the average work week, Figure 28–1. In the early 1900s, people worked sixty to seventy hours per week. Today, the average work week lasts about thirty-five to forty hours. If we extend this trend to the future, people may work only fifteen to twenty hours a week in the year 2070. We may not be able to predict the exact number of hours, but there is a pretty good chance that the trend toward fewer hours will continue. Today's trends can help us predict the future.

TREND—DECLINE IN AVERAGE HOURS IN A WORK WEEK

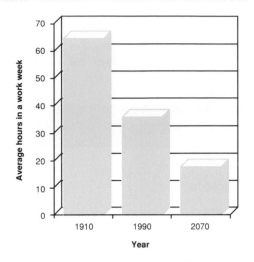

FIGURE 28-1 If present trends continue, people may only work 15 to 20 hours per week in the future.

Exponential Rate of Growth

Futurists are people who study the future. They look at trends and try to predict future events. An important factor they must consider is the increasing rate of change in our technological society. The time it takes for new technologies to be accepted and used is rapidly decreasing. As an example, the Wright brothers flew the first powered aircraft in 1903. By 1969, a span of sixty-six years, Neil Armstrong flew to the moon and back. In 1946, ENIAC, the first electronic computer, was invented. Today, less than fifty years later, having a personal computer at home is common. Finally, the first laser was invented in 1960. Today, less than thirty-five years later, lasers are commonly found in grocery store checkouts and compact disk players. The time between the invention of a new technology and its common use is getting shorter.

The decade of the 1980s saw more changes in technology than any other decade since the 1780s, when the Industrial Revolution began in Great Britain. At the beginning of the 1990s, experts say the amount of information in our technological world doubles every four to five years. This was not always true. Before the Industrial Revolution, change was very slow, taking hundreds or even thousands of years. Figure 28-2 shows the rate of growth of technology before and after the Industrial Revolution. The increasing rate of growth of technology in this figure is called an exponential rate of growth. **Exponential rate of growth** means an ever-increasing rate of change.

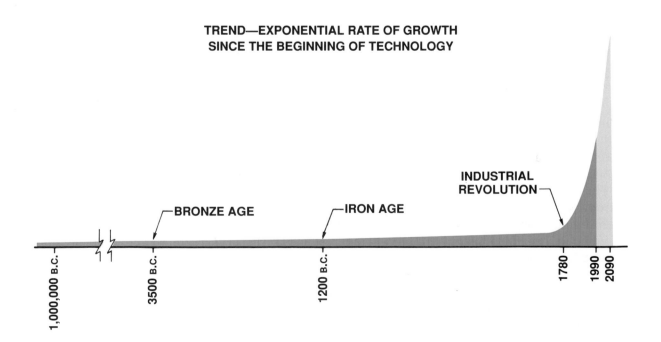

Forecasting the Future

Humans have been trying to predict what might happen in the future since the beginning of time. **Forecasting** is a "tool" used daily by all of us, from planning our day to anticipating a changing traffic light. What has happened in the past is our only guide to what might happen in the future. Therefore, because we have seen a traffic light change on a regular basis from green to yellow to red, we can safely assume that it will continue to do so in the future. If you normally catch the school bus at 7:30 A.M. every day, you can assume this trend will go on. This method of forecasting is called **trend analysis**. Past and present trends are studied and forecasted into the future. However, knowing the past does not automatically tell us what will happen in the future. We often use our knowledge of the past to shape future events. But to create something based only on the past is a difficult process.

Making up stories about the future is another method of forecasting. A "made-up" story about the future is called a **scenario**. Science fiction stories are scenarios. A scenario begins to take shape when we ask the question "what would happen if ...?" We all use a form of scenario while daydreaming when we imagine ourselves in a particular situation. For example, ask yourself, "what would happen if I were the principal in this school?" Writing a scenario is not hard. An imagination and knowledge of the basic facts related to the topic are all you need.

Another way to forecast the future is to ask experts. People who know about a subject are often asked for their opinions about the future of their particular field. Often, a person involved in one job for a number of years begins to see the trends that may give some hint of the future. By asking a group of experts, you can get an average of their opinions on the future.

Mapping is another way to predict a possible future. This method can be as sim-ple as making a list or as complex as a family tree or computer flowchart. By making a map of some event, such as a computer program, the person creating the map is predicting a future event, Figure 1.

Simulation, another forecasting technique, is a broad term for activities like role-playing, modeling, and gaming. When children pretend to be doctors, play with toys, or play board games, they are simulating an activity that they have seen and might do some day. In your production class, you may be asked to participate in a simulation of an enterprise. You may play the role of a manager, production worker, accountant, or some other worker. The production lab for your class is actually a model of the real thing, Figure 2.

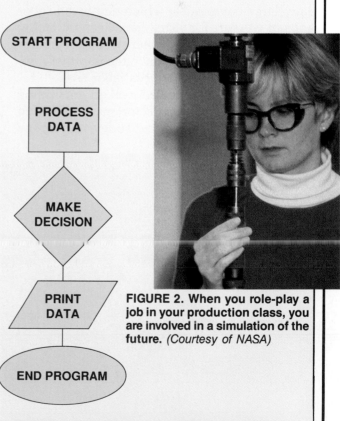

FIGURE 2. When you role-play a job in your production class, you are involved in a simulation of the future. *(Courtesy of NASA)*

START PROGRAM

PROCESS DATA

MAKE DECISION

PRINT DATA

END PROGRAM

FIGURE 1. A computer programmer's flowchart is an example of mapping or predicting the future.

Today's exponential rate of growth of technology is impressive. Almost ninety percent of all our technology has been invented in the last fifty years. The trend of exponential growth will probably continue in the future.

Throughout this text, several trends in production were described. Let's look at a few of these trends and see what impact they might have on the future of production.

Trend #1: Decreased Use of Labor

In the past, production was based on labor. That is, large numbers of people were hired as workers, Figure 28–3. Production was based on human muscle power and physical skills. Labor is one of the largest expenses for a production company. Payroll can cost a company thousands or even millions of dollars each year. Because of this, production companies have been reducing the number of human workers in recent years. Many large companies, which employed thousands of workers for years, have been replacing humans with robots and other automated equipment. Automated equipment often improves product quality while it reduces labor costs.

Most of these unemployed workers are finding new jobs in the service sector; particularly in information-processing jobs. Today, and for the near future, many new jobs will involve collecting, processing, storing, retrieving, and analyzing information. For this reason, we often say that we live in an information age. Production industries also expect more jobs for workers who process information. These workers include engineers, professionals, technicians, and marketing personnel. Eventually information-processing jobs will also be automated. This will lead to a decline in the use of labor for information processing. Remember, change is the only constant, Figure 28–4.

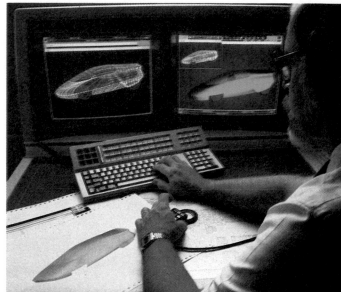

FIGURE 28–4 In the future, many jobs that require physical strength will be replaced by jobs that require mental strength. *(Courtesy of Ford Motor Company)*

Trend #2: Increased Use of Automation

This trend is directly related to the first trend. Increasingly, computers, robots, lasers, and other automated machines are replacing human workers. For the production company, computerized and automated equipment increases productivity, reduces labor costs, and often improves the quality of product. In the long run, automation should make more profits for the shareholders.

A related trend is the need for people who can learn to set up, operate, and maintain automated equipment. These new areas of employment will be in jobs that require mental skills. In the future,

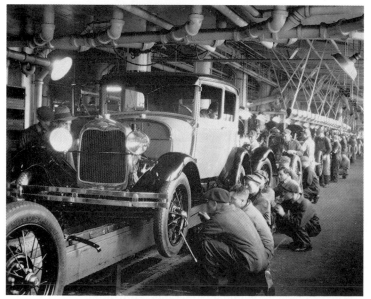

FIGURE 28–3 In the past, large numbers of workers were hired to do physical jobs in production. *(Courtesy of Ford Motor Company)*

Manufactured Construction

One trend that is really growing in production is manufactured construction. **Manufactured construction** involves manufacturing all of the parts for a house or other structure in a factory and assembling them on the construction site. In recent years, the field of manufactured construction has grown rapidly. There are three types of manufactured construction: panelized construction, modular construction, and precut homes.

Panelized Construction

In panelized construction, the various framed parts of a house, such as the floors and walls, are made in a factory as panels, Figure 1. The panels are made of framing members and plywood sheathing. They are delivered to the site on a flatbed truck. Construction workers on the site simply erect the parts and nail them together, Figure 2. The interior and exterior finishing is done in the usual manner.

FIGURE 2. This wall was assembled (panelized) in the factory, then shipped to the site. *(Courtesy of Northern Homes)*

Modular Construction

In modular construction the entire structure, including framing and finishing, is done in a factory. The structure is built in small parts called modules. A module is a boxlike unit that includes several rooms. The modules are combined to make homes, motels, and office buildings, Figure 3. In

FIGURE 1. Carpenters use pneumatic nailers to assemble the units. *(Courtesy of Cardinal Industries, Inc.)*

FIGURE 3. Lifting a manufactued house to its foundation. *(Courtesy of Cardinal Industries, Inc.)*

most modular structures the electrical, plumbing, and climate control systems are built in at the factory. When the structure is built, it is a simple matter to hook up these utilities to the main service lines.

Precut Homes

Precut homes are made for the do-it-yourselfer. A precut home arrives on the building site as a kit. Each part of the house has been measured and cut to size in the factory. The owner must read the working drawings that come with the kit and build the home as if it were a full-size model kit.

Advantages of Manufactured Construction

Manufactured construction is gaining popularity for two main reasons. First, the time required to build a structure is much shorter. Many of the parts arrive on the building site preassembled and ready to install. Second, since manufactured structures are mass produced, they cost less. Also, building the structure faster saves labor costs.

more technologists will be needed. This means more jobs for workers like engineers and technicians. A good technologist should be able to learn to operate new technological tools, process information, and work with other technologists on the job. Also, computer and math skills will be important in the future. Automation will require workers who have knowledge and skills in technology. These new workers must be technologically literate.

If some future predictions come true, we might not need workers in production, or any other industry for that matter. The totally automated factory, where robots and computers do all the work, may become reality. There are already a few totally automated factories. These factories can run twenty-four hours a day with fewer than a dozen workers for maintaining and servicing automated equipment, Figure 28–5.

Trend #3: Increased Global Competition

From the early 1900s until recently, the United States was the world leader in production technologies. Increasing competition from foreign companies has recently undermined this leadership. The production of video equipment is a good example. The video cassette recorder (VCR)

FIGURE 28–5 An important trend in production is the development of the totally automated factory. *(Courtesy of ABB Robotics, Inc.)*

was invented in America. Today, however, not a single VCR is totally manufactured in this country. In less than two decades, Japan has improved its technology to the point where it is a respected world leader in the field of electronic compo-

Biotechnology: Blood Testing

New technology has replaced some of the traditional techniques hospitals use to analyze a patient's blood.

In traditional blood testing, the lab technician gets blood samples and returns to the hospital's laboratory to perform the tests. However, while the lab technician is transporting the blood samples, certain conditions of the blood may change. This is because of the time it takes to get to the lab. This may result in inaccurate test results.

Now, there are chemical analytical tools that can be located in a patient's room. They can also be moved from one room to the next easily. These high-tech "bedside" analyzers can get more accurate information about a patient's blood sample quickly. This may result in a more accurate and faster diagnosis and treatment.

These chemical analytical tools do not replace technicians; however, hospital lab technicians are being faced with learning new technologies.

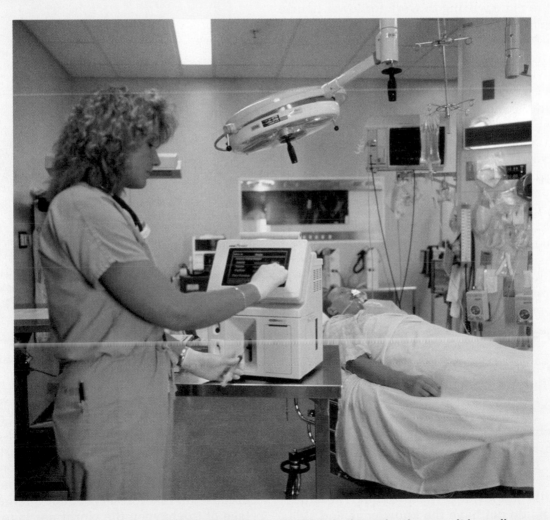

The Gem® Premier is used to analyze a patient's blood sample in the patient's room. It is easily moved from room to room. *(Courtesy of Mallinckrodt Sensor Systems, Inc.)*

nents manufacturing. With the growth and development of Japanese manufacturing technology, American manufacturers have lost part of the market for their products. But not even Japan will be safe in the future. Right behind Japan are developing countries like Korea, Taiwan, and China, who are improving their technology. There will be more and more **global competition** among manufacturing companies.

Another factor related to the increase in global competition is the trend for American manufacturing companies to open plants in developing countries. Many American companies have moved production operations to countries that have lower standards of living. Workers in these countries are more willing to work for lower hourly wages than those wages expected by American workers. In addition, these countries usually do not have labor laws or unions to protect workers. Finally, government regulators, such as OSHA or the Environmental Protection Agency (EPA), are not found in less developed countries. Operating a plant under these conditions can save a company millions of dollars while increasing profits for stockholders. Critics of this system say that it weakens the U.S. economy. Thousands of Americans are put out of work, and foreign countries learn to operate our technology. Often, the decision to move a production plant to another country is made purely on the basis of increasing profits and dividends. Continuing this trend in the future may have an important impact on the American standard of living and economic base.

Trend #4:
Decreased Demand for Unskilled/Semiskilled Labor

As technology grows and develops, it becomes ever more complex. It is harder to learn, understand, operate, analyze, and maintain. As this trend continues, workers will need more mental skills and fewer physical skills. Those people who do not have the knowledge and skills to understand and operate the technology of tomorrow will find themselves unemployed or working at difficult, boring jobs. More and more manufacturers are using complex, computer-

controlled equipment. To run this type of equipment, an employee must be technologically literate. Historically, manufacturing employed large numbers of unskilled and semiskilled workers. With the improvements in technology, these people will find it harder to get a good job, Figure 28–6.

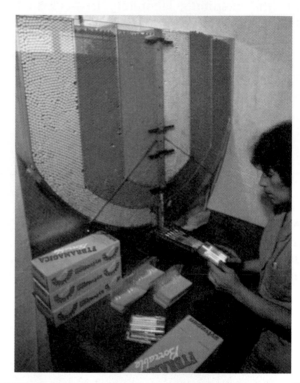

FIGURE 28–6 Unskilled and semiskilled workers will find it harder to get jobs as more automated equipment is introduced. *(Courtesy of Gillette Company)*

Trend #5:
Increased Use of Natural Resources

There is a limited supply of natural resources in our world. As global competition increases, the rate at which the world's natural resources are used will increase. Brazil provides an example of this trend. With its huge forests, Brazil is considered a valuable resource for lumber. Some call Brazil the "lungs of the world" because its vast forests produce a large percentage of the oxygen the world needs to survive. Development in Brazil is ruining thousands of acres of forests every day. Environmentalists are concerned that this loss of forests could harm the ecological balance in our environment. Not only would this decrease the supply of wood available, it would also reduce the amount of oxygen being produced.

Industry is the largest user of energy in our society. As industries expand and grow, so do their energy needs. More production means more of our natural resources will be used to generate energy. Increasingly, production companies will be forced to conserve energy. Technologies like cogeneration will help save natural energy resources. In the area of natural materials, we will need to recycle more. Materials like aluminum, glass, and paper can be recycled at a fraction of the costs of manufacturing from natural materials (in terms of money spent and natural resources consumed), Figure 28–7.

Trend #6: Decreased Employment by Corporate Giants

At one time, many Americans dreamed of working for one of the giant corporations that employed millions of people. Recently, and in the future, people are less likely to work for giant

production firms and more likely to work for small entrepreneurial companies. More than ninety percent of new jobs are found in small businesses with fewer than 100 employees. These businesses are started by entrepreneurs. In the future, entrepreneurs will create the new jobs for the future of production.

These are just a few of the important trends that will help determine the future of production. Two other bright areas are biotechnology and the possibility that someday products may be produced in space.

Deciding Which Futures

Some futurists say technology may grow in two possible directions in the future; hard technology or soft technology. Figure 28–8 compares hard technology and soft technology. **Hard technology** is often equated with high technology that is based on improving productivity and profits without considering the human, environmental, and social impacts. **Soft technology** is referred to as **appropriate technology** (meaning choosing and using technologies that are "best" suited for a particular situation). Soft technology is concerned

FIGURE 28–7 Recycling materials has many advantages. This aluminum ingot was made from recycled beverage cans. *(Courtesy of Reynolds Metals Company)*

HARD TECHNOLOGY	SOFT TECHNOLOGY
Capital intensive	Labor intensive
Environmentally harmful	Environmentally safe
Large energy input	Small energy input
High pollution	Low pollution
Depletes natural resources	Recycles resources
Worldwide trade	Regional trade
Technocrats/ technopeasants	No technocrats/ technopeasants
Quantity production	Quality production
Frequent, serious technological accidents	Few, less serious technological accidents
Operation too complex for lay person to understand	Operation understood by everyone

FIGURE 28–8 Futurists predict two extreme paths for the future of technology—hard technology and soft technology.

Can Technocracy Replace Democracy?

One possible future would find our society and citizens controlled by a small, elite group of people who understand and operate technology. This society would be called a **technocracy**. In our present democracy, we all have an opportunity to decide how technology will be used in the future. In a technocracy, the future would be decided by **technocrats**. Technocrats would use their understanding of technology to control other citizens (technocrats would be technologically literate). Those people who were technologically illiterate would be called **technopeasants**. The technopeasants would do all the work for the technocrats.

Today, our technology is becoming more and more complex and difficult to understand and operate. It seems that only people with scientific and engineering degrees can understand how certain technologies work. If this trend continues, could our democracy be replaced by a technocracy? And if a

FIGURE 1. One possible scenario for the future is a society called a technocracy. (© *Walt Disney Company. Nova City from a scene in the Walt Disney World EPCOT Center Horizons Pavilion presented by General Electric*)

technocracy developed, would you be a technocrat or a technopeasant? Figure 1.

about productivity and profits, but the impacts of technology on people, the environment, and society are always considered. Hard technology would replace human workers with automated equipment. Soft technology would keep people working. Hard technology would cause large amounts of pollution, deplete our natural resources, and consume large amounts of energy. Soft technology would create smaller pollution levels, recycle natural resources, and use less energy.

The future probably won't use only hard or soft technology. It will more likely use some of each, with certain trade-offs. The only way to know for sure is to wait for the future. Or is it? Every individual can help shape the future. In our society, every citizen has the opportunity to participate in the decision-making processes that will determine which path technology will take. Some individuals, particularly entrepreneurs, inventors, production managers, and political leaders can play an even bigger role in determining the future. In order for you to participate in deciding the

future, you must be a technologically literate citizen. Without technological literacy, you won't have the knowledge, skills, and attitudes necessary to make decisions about the future of technology.

Summary

No one can accurately predict the future. The only thing we know for sure is that the future will be different from today. The exponential rate of growth causes technology to double every four to five years. This trend will probably continue to cause technological change in the future. Several trends that may have an impact on the future of production were examined in this chapter. Studying these trends suggests that production in the future will use more automation and fewer human workers. The people who will work in production in the future will probably work for smaller entrepreneurial companies. They will probably be required to have more skills and knowledge than workers of the past. Increasingly, American companies will be forced to compete with smaller countries that are currently developing strong technology bases. The trend of increased use of natural resources will probably continue.

DISCUSSION QUESTIONS

1. Of the six trends that were discussed in this chapter, which do you think will be most important in determining the future? Why?

2. Do you think technology can continue to grow at an exponential rate? What might stop or slow down this growth?

3. What are the advantages and disadvantages of using labor or automated equipment for production?

4. Should American companies that start production facilities in foreign countries be permitted to put thousands of American people out of work to increase company profits? Should American workers or the government try to stop this trend?

5. What will happen to all the workers who cannot find work when total automation is in place? Do companies have a responsibility to keep people working?

CHAPTER ACTIVITIES

 ## Biographies of Important People

OBJECTIVE

One way to learn more about the future is to study the past. In this chapter, you learned that entrepreneurs would create most of the jobs in the future. In this activity, you will write a biography of an important entrepreneur or inventor from the past.

EQUIPMENT AND SUPPLIES

1. Library resources (encyclopedias and reference books)
2. Paper and pencil or computer and word processor

PROCEDURE

1. Pick an important entrepreneur or inventor from the past.
2. Go to the library and ask the librarian to help you find some information on that person.
3. Write a short paper on the person that describes:
   a. the years when he or she lived;
   b. where he or she lived and worked;
   c. his or her invention or idea;
   d. how his or her idea changed our way of life.
4. Make a small poster that describes the person you studied. You might even photocopy a photograph out of a book to paste on your poster.
5. Make a bulletin board with other members of your class. You might use this title: "People Who Made Technology Great."

MATH AND SCIENCE CONNECTIONS

Make a timeline of your biographies. Use drawings of the names of the people your class studied. Make the timeline so the time between years is kept to a certain measurement, such as 1 foot or a certain number of inches. The timeline could be placed around the room on the walls. It will give you an idea of how long ago the people you studied lived.

RELATED QUESTIONS

1. What interesting facts did you learn through your research?
2. What lessons did you learn about being a successful inventor or entrepreneur?
3. Was it easy for the person you studied to get his or her idea accepted?
4. What problems did the person you studied have to overcome?

 # Scenario: The Future of Production

OBJECTIVE

A scenario is a made-up story about the future. In this activity, you will write a scenario about the future of production.

EQUIPMENT AND SUPPLIES

1. Research and reference books and materials about the future of production technologies.
2. Pencil and paper or computer and word processor.

PROCEDURE

1. Divide the class into two groups. Each group will write scenarios about the future of production.
2. Each student in group 1 will write a scenario to answer the question, "What would happen if hard technology developed in the future?"
3. Each student in group 2 will write a scenario to answer the question, "What would happen if soft technology developed in the future?"
3. Each scenario should describe workers needed, working conditions, American standard of living, and environmental and natural resource concerns.
4. For background information, read recent books, magazine articles, and newspaper articles on future developments in production.
5. Consider interviewing experts in various areas of production. Talk to workers in the fields of construction and manufacturing to learn their opinions about the future.
6. When finished, share your paper with the class.

MATH AND SCIENCE CONNECTIONS

Made-up stories about what the possible impact of science and technology on the future will be are called science fiction.

RELATED QUESTIONS

1. What are some other forecasting tools other than a scenario?
2. What are the advantages and disadvantages of both hard and soft technology?
3. What differences were there among the papers in your class? What special or interesting information was found?
4. How accurate do you think the papers were in predicting the future of production?

PRODUCT PLANS

INTRODUCTION

There are two different types of product plans in this section—*fully developed product plans* and *product ideas requiring further development.* The fully developed plans include most of the information your classroom manufacturing enterprise needs to start manufacturing engineering (plant layout, materials handling, and tooling) and quality control engineering (inspection). The fully developed plans include drawings of the product and parts, a parts list, and an operation process chart.

Most of the design engineering and methods engineering processes have been completed with the fully developed plans. However, your teacher might ask you to make prototypes of the products to verify specifications. You will also need to complete certain methods engineering processes,

such as flow process charts and operation sheets. Also, your class might decide to change the product designs, substitute materials, change the combining processes suggested, or improve the efficiency of the operation process chart.

If your class decides to use additional product plans provided by your teacher, the fully developed plans will serve as a guide, giving you some hints for presenting product design ideas.

The product ideas vary in their development. Most of these plans include the basic information you need to complete the necessary design engineering and production engineering processes, such as working drawings, mock-ups and prototypes, operation process charts, plant layout, and tooling. Here is a list of the product plans.

MEMO MINDER

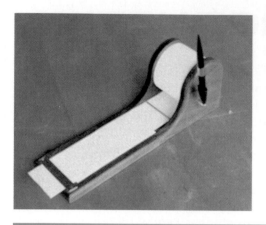

PARTS LIST

Part	Qty.	Part Name	Description
A	1	Base	1/2 T × 3 1/2 W × 13 L Wood
B	2	Sides	1/2 T × 4 1/2 W × 5 5/16 L Wood
C	1	Dowel	1/2 Dia × 3 1/2 L Dowel
D	1	Rod	3/16 Dia × 3 L Welding Rod
E	1	Cutter	.02 T × 1/2 W × 3 3/8 L Brass/Sheet Metal
F	1	Pad	1/16 T × 2 7/8 W × 6 3/8 L Plastic
G	1	Pen	Pen and holder

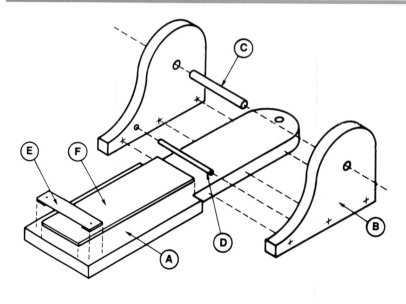

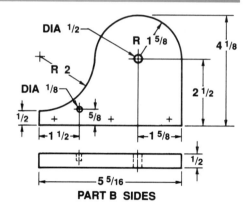

PART B SIDES

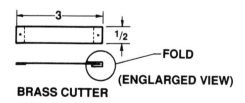

BRASS CUTTER

FOLD

(ENGLARGED VIEW)

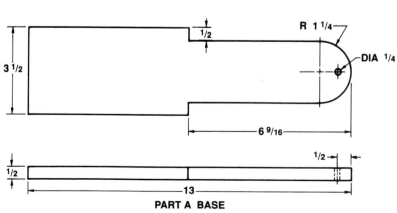

PART A BASE

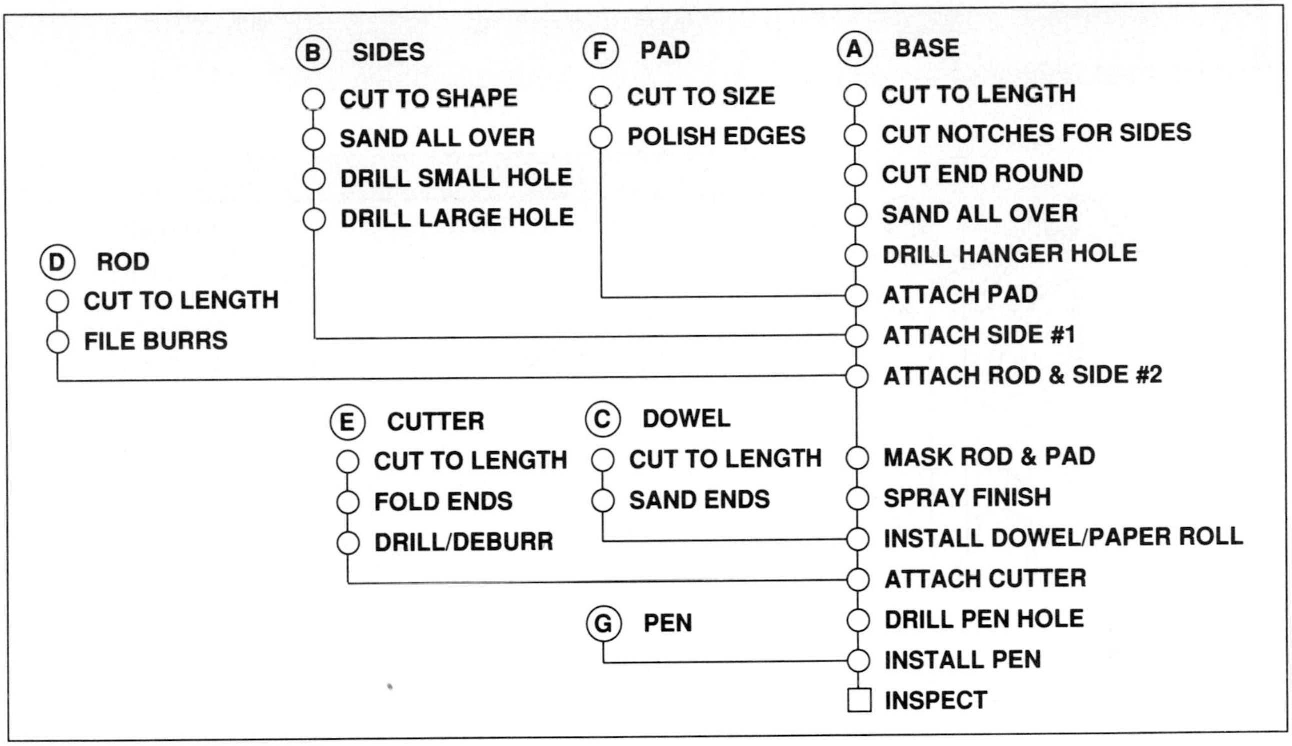

Ⓑ SIDES
○ CUT TO SHAPE
○ SAND ALL OVER
○ DRILL SMALL HOLE
○ DRILL LARGE HOLE

Ⓕ PAD
○ CUT TO SIZE
○ POLISH EDGES

Ⓐ BASE
○ CUT TO LENGTH
○ CUT NOTCHES FOR SIDES
○ CUT END ROUND
○ SAND ALL OVER
○ DRILL HANGER HOLE
○ ATTACH PAD
○ ATTACH SIDE #1
○ ATTACH ROD & SIDE #2

Ⓓ ROD
○ CUT TO LENGTH
○ FILE BURRS

Ⓔ CUTTER
○ CUT TO LENGTH
○ FOLD ENDS
○ DRILL/DEBURR

Ⓒ DOWEL
○ CUT TO LENGTH
○ SAND ENDS

○ MASK ROD & PAD
○ SPRAY FINISH
○ INSTALL DOWEL/PAPER ROLL
○ ATTACH CUTTER

Ⓖ PEN
○ DRILL PEN HOLE
○ INSTALL PEN
☐ INSPECT

NEW ENGLAND CLOCK

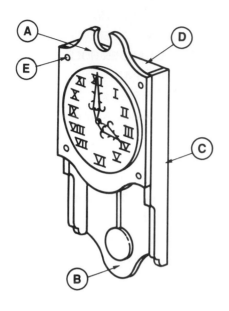

PARTS LIST

Part	Qty.	Part Name	Description
A	1	Face	3/4 T × 10 W × 11 L Wood
B	1	Back	3/4 T × 8 1/2 W × 10 L Wood
C	2	Sides	3/4 T × 2 1/2 W × 13 1/2 L Wood
D	1	Top	3/4 T × 2 1/2 W × 8 1/2 L Wood
E	4	Plugs	Wood Taper or Button
F	1	Movement	Pendulum, Battery Operated
G	1	Bezel/Dial	Glass/Brass

SHELF CLOCK

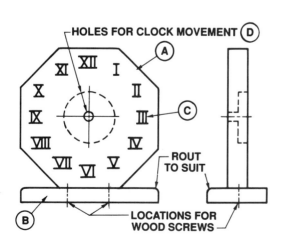

HOLES FOR CLOCK MOVEMENT

ROUT TO SUIT

LOCATIONS FOR WOOD SCREWS

PARTS LIST

Part	Qty.	Part Name	Description
A	1	Face	3/4 T × 8 W × 8 L Wood
B	1	Base	3/4 T × 2 1/4 W × 8 L Wood
C	1 set	Numerals	Plastic Stick-on (Black/Brass)
D	1	Movement	Battery Operated

SPINNING TOP

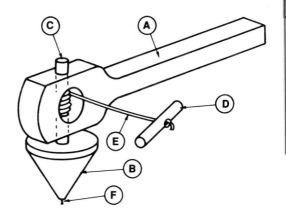

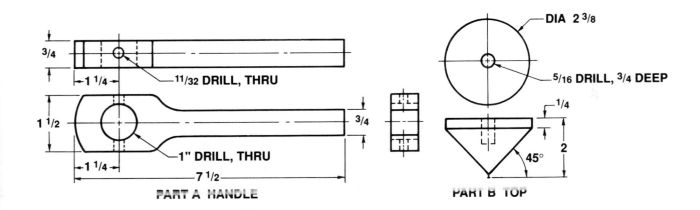

PART A HANDLE

PART B TOP

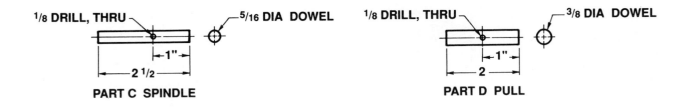

PART C SPINDLE

PART D PULL

PLANTER BOX

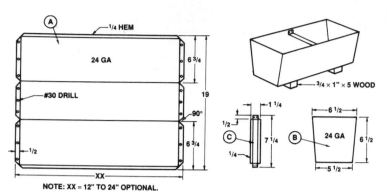

24 GA

¹/₄ HEM

A

6 ³/₄

19

#30 DRILL

90°

6 ³/₄

¹/₂

XX

NOTE: XX = 12" TO 24" OPTIONAL.

³/₄ × 1" × 5 WOOD

1 ¹/₄

¹/₂

C

7 ¹/₄

¹/₄

6 ¹/₂

B

24 GA

6 ¹/₂

5 ¹/₂

PARTS LIST

Part	Qty.	Part Name	Description
A	1	Body	24 Ga. × 19 L sheet metal
B	2	Ends	24 Ga. × 6 ¹/₂ × 6 ¹/₂ sheet metal
C	1	Support	24 Ga. × 1 ¹/₄ × 7 ¹/₄ sheet metal

PLANT STAND

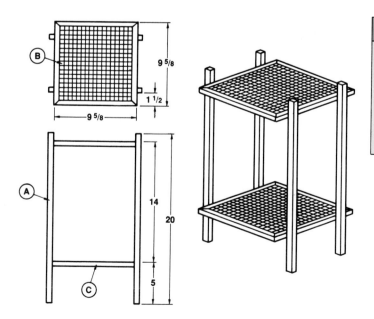

B

9 ⁵/₈

1 ¹/₂

9 ⁵/₈

A

14

20

C

5

PARTS LIST

Part	Qty.	Part Name	Description
A	4	Legs	¹/₂ Sq Steel × 20 L
B	2	Shelf	Perf. Steel—Cut to fit frame
C	8	Frame	¹/₂ × ¹/₂ Angle Steel × 9 ⁵/₈ L

WIND CHIMES

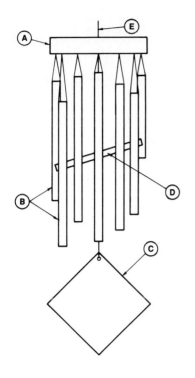

PARTS LIST

Part	Qty.	Part Name	Description
A	1	Base	4" Dia. circle, 3/4" thick Wood
B	12	Rods	Two 8" Long Rods, Two 5" Long Rods
C	1	Wind Catcher	4" Sq. Sheet Metal or Acrylic Sheet
D	1	Center Disk	2 1/2" Dia. C. Al.
E		Twine	

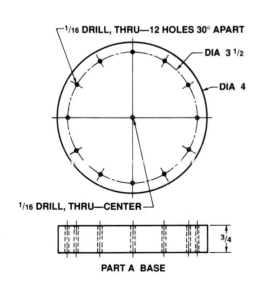

1/16 DRILL, THRU—12 HOLES 30° APART

DIA 3 1/2

DIA 4

1/16 DRILL, THRU—CENTER

3/4

PART A BASE

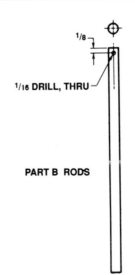

1/8

1/16 DRILL, THRU

PART B RODS

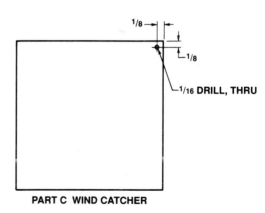

1/8

1/8

1/16 DRILL, THRU

PART C WIND CATCHER

DIA 2 1/2

DIA 1 1/2

1/16 DRILL

3/32

1/4

PART D CENTER DISC

FILING BOX

PARTS LIST

Part	Qty.	Part Name	Description
A	1	Front	1/2 T × 3 1/2 W × 6 1/4 L Wood
B	2	Sides	1/2 T × 4 1/2 W × 3 L Wood
C	1	Bottom	1/2 T × 4 1/2 W × 7 1/4 L Wood
D	1	Back	1/2 T × 7 W × 6 1/4 L Wood
E	1	Lid	1/2 T × 3 5/8 W × 6 1/4 L Wood
F	1	Knob	Ceramic, Brass, or Wood

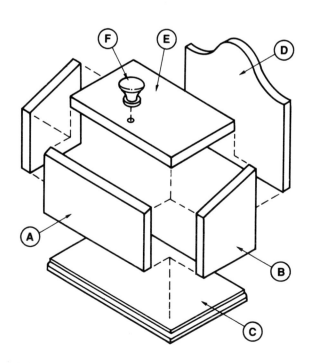

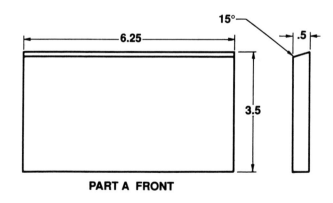

PART A FRONT

15°

6.25

3.5

.5

PARTS LIST

Part	Qty.	Part Name	Description
A	1	Base	¾T x 6 W x 9 L Wood
B	1	Pen Holder	2¾ Square x 3½ High Cast Polyester Plastic or Ceramic or Aluminum
C	1	Metal Clip	1¼ in. Metal Clip
D	1	Note Pad	4 x 5 Approx.

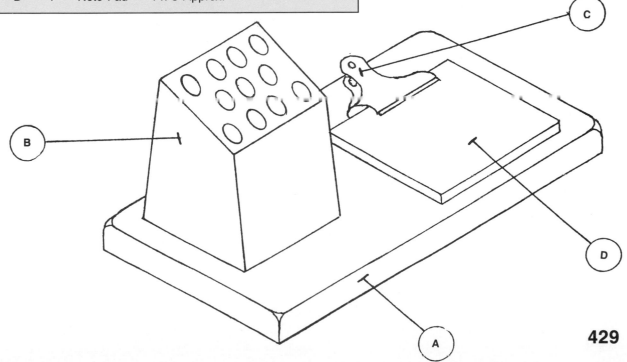

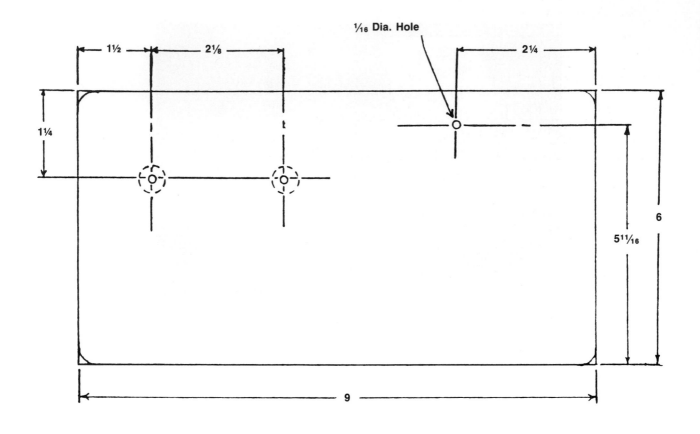

¹⁄₁₆ Dia. Hole

1½

2⅛

2¼

1¼

6

5¹¹⁄₁₆

9

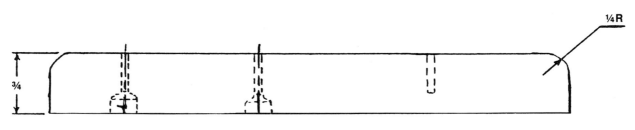

¼R

¾

DRILL AND COUNTERBORE FOR WOOD SCREWS

(A) BASE

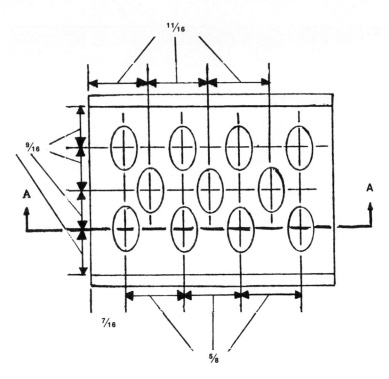

NOTE: ALL HOLES IN PEN HOLDER ARE ⅜ Dia. x 2¼ Deep

SECTION A–A

(B) PEN HOLDER

CASSETTE TAPE CRATE

PARTS LIST

Part	Qty.	Part Name	Description
A	2	Ends	⅝T x 2½ W x 4⅜ L Wood
B	2	Bottoms	5⁄16T x 1½ W x 12½ L Wood
C	4	Sides	5⁄16T x ¾ W x 12½ L Wood
D	2	Handles	½T x 1″ W x 3 L Wood

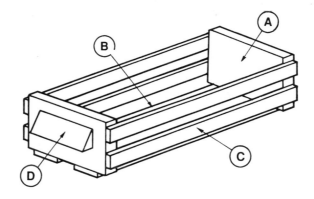

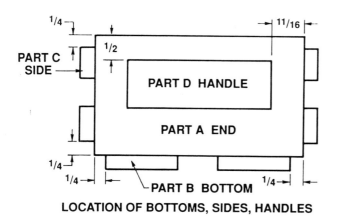

LOCATION OF BOTTOMS, SIDES, HANDLES

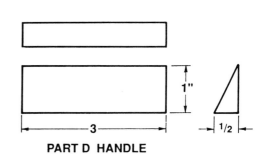

PART D HANDLE

KITCHEN MATE

PARTS LIST

Part	Qty.	Part Name	Description
A	1	Base	1T x 6 W x 5 L Pine
B	1	Holder	⅛T x 5 W x 10 L Plexiglass
C	2	Wood Screws	

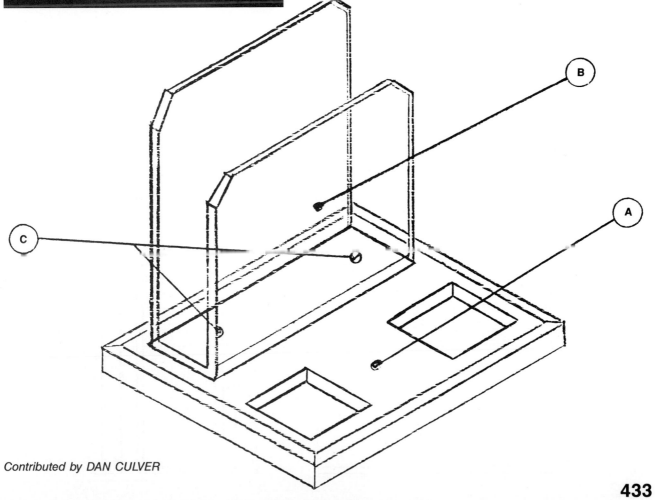

Contributed by DAN CULVER

433

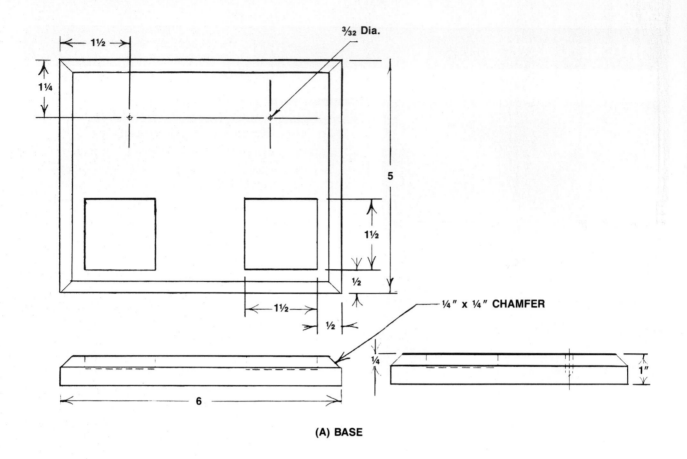

¾₂ Dia.

¼" x ¼" CHAMFER

(A) BASE

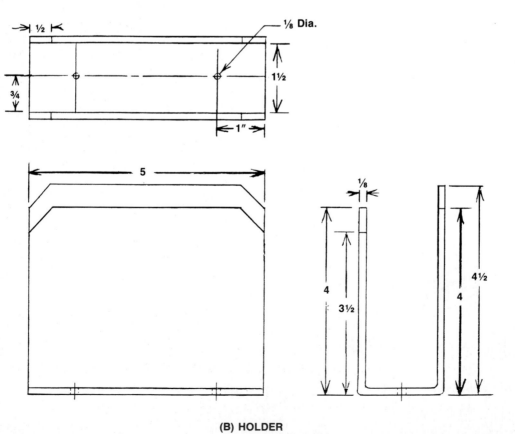

⅛ Dia.

(B) HOLDER

434

SMALL WHAT-NOT BOX

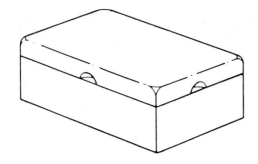

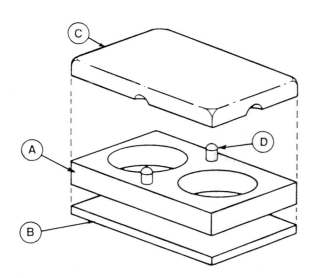

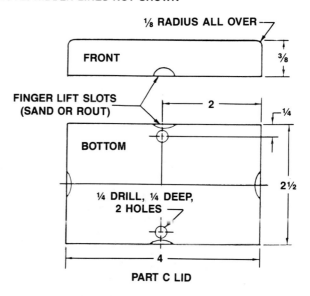

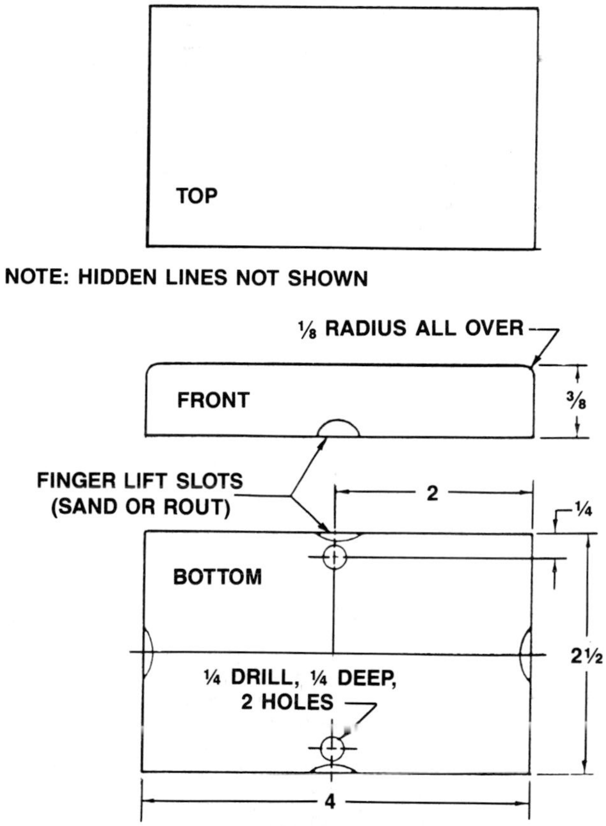

TOP

NOTE: HIDDEN LINES NOT SHOWN

⅛ RADIUS ALL OVER

FRONT ⅜

FINGER LIFT SLOTS (SAND OR ROUT)

2

¼

BOTTOM

2½

¼ DRILL, ¼ DEEP, 2 HOLES

4

PART C LID

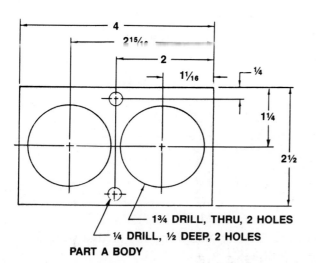

4

2¹⁵⁄₁₆

2

1¹⁄₁₆

¼

1¼

2½

1¾ DRILL, THRU, 2 HOLES

¼ DRILL, ½ DEEP, 2 HOLES

PART A BODY

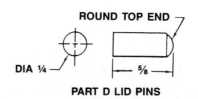

ROUND TOP END

DIA ¼

⅝

PART D LID PINS

435

Toothpick Bridge

INTRODUCTION

Bridges are important structures in our society. They give us a quick and easy way to cross rivers, railroad tracks, and roads. Before a real bridge is built, the designers often make models. The models are tested to make sure the bridge will be strong enough..

PROBLEM

Design, construct, and test a model toothpick bridge (Figure 1).

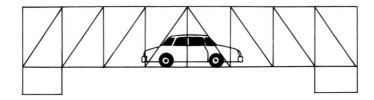

INVESTIGATION

1. Study the different ways bridges are built.
2. Pick a strong bridge design and make a sketch.
3. You are limited to the following materials:
    a. 2 wooden sticks, each measuring $\frac{1}{8}''$ x $\frac{1}{8}''$ x 12″ (for bridge girders)
    b. 100 wooden toothpicks
    c. 1 sheet 8½″ x 11″ paper (for roadbed)
    d. Carpenter's glue
4. Your bridge must span a distance of 10 inches. The girders will overlap one inch on each end (Figure 2).

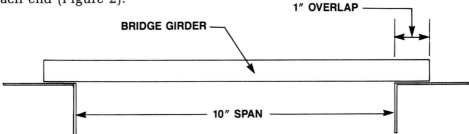

5. The roadbed must be covered by the paper.
6. The top of your bridge must have a flat surface for testing.
7. Each finished bridge will be weighed.
8. Each bridge will be tested by placing weight on top until the bridge fails.
9. A strength rating will be calculated by dividing the test weight by the bridge weight.
10. Your teacher may set other rules for you to follow.

Recycling Center

INTRODUCTION

Recycling is a way to conserve our natural resources. Many communities across the country have started recycling programs. Paper, aluminum cans, and glass can be recycled from our garbage instead of being thrown in the dump.

PROBLEM

Design and produce devices that can be used in a home recycling center.

INVESTIGATION

1. Find out about recycling programs in your community.
2. Ask people what materials they are recycling from their garbage.
3. Decide what type of devices may be needed in a home recycling center.
4. Some ideas for recycling center devices include:
   a. Bag holders (See Figure 1)
   b. Can crushers (See Figure 2)
   c. Paper bundlers
5. Pick a device that you would like to make and draw a rough sketch.
6. Work with your teacher to finish the design and to select materials.
7. Prepare a set of working drawings and a bill of materials.
8. Make a prototype of the device.
9. Share your prototype with your classmates, teachers, and people in the community.
10. Consider mass producing the devices.

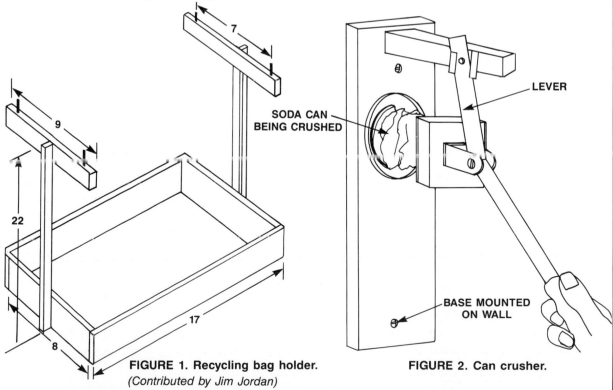

FIGURE 1. Recycling bag holder.
(Contributed by Jim Jordan)

FIGURE 2. Can crusher.

Towering Heights

INTRODUCTION
Construction technology builds many different types of structures. One very important type is the tower. Towers are used to hold lights at sports stadiums, to support electrical power lines, and for communication antennas.

PROBLEM
Design and construct the tallest tower in your class with the given materials (Figure 1).

INVESTIGATION
1. Study the designs of various towers in your community.
2. Make sketches of the type of tower you will build.
3. You are limited to the following materials:
   a. 50 plastic straws
   b. 100 straight pins
   c. 10 feet of kite string
   d. 8 inches of masking tape
4. You may not use the wrappers from the straws.
5. Only the straight pins may be used to connect the straws together.
6. The base of the tower must fit inside a six-inch square drawn on a piece of plywood.
7. The base of the tower may be pinned to the plywood.
8. The kite string may be used for guy wires to support the tower.
9. Use the tape to fasten the guy wires (string) to the floor. The guy wires may be fastened to the floor outside the six-inch square.
10. The tower must be free standing for one minute.

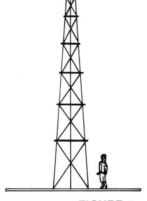

FIGURE 1.

Product Design Brief

INTRODUCTION
Our manager has decided to make and sell a kitchen utensil. The utensil will be used to pull out and push in the rack on an oven.

The figure shows a thumbnail sketch made by our CEO on her ideas for the product. We hope you can find a better idea.

DESIGN PROBLEM
Design an oven rack pusher/puller that matches the design standards that follow.

A contest may be held to decide which team has the best product design ideas.

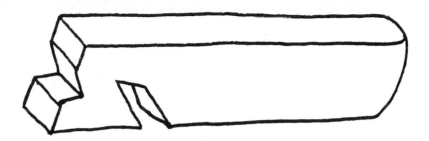

FIGURE 1. Thumbnail sketch of an oven rack pusher/puller.

DESIGN IDEAS

1. Follow all safety rules. Be sure to wear safety glasses when working with tools.
2. Do not test your mock-ups in an oven.
3. If you test your prototype in an oven, be careful.
   - Make sure your teacher or another adult is there.
   - Use a kitchen mitt when testing your prototype in the oven, not a towel. A towel can catch fire.
   - Have the right fire extinguisher on hand in case of fire.

Product Design Standards

1. Product Function
   a. The oven rack pusher/puller will be held in one hand and used to pull out an oven rack six inches. It will then be used to push the rack back in.
   b. The oven rack pusher/puller should not bend or twist.
2. Product Form—the product must not be bigger than:
   .625″ thick x 2.25″ wide x 12″ long
3. Product Appearance—natural wood and colored plastic kitchen utensils are good sellers.
4. Ease of Manufacture—you should be able to make the product using the tools and machines in your lab.
5. Standard Parts—the product should be made from one of the materials available from your teacher.
6. Quality—three product tests will be made:
   a. Heat test: materials used must not burn, catch fire, or melt. They should not let the user's hands get hot.
   b. Drop test: product will be dropped four feet to a hard surface floor. It must not break, crack, or be badly damaged.
   c. Washing test: product must not swell, splinter, or be badly damaged when washed.
7. Costs—the product must sell for less than $2.50.

Tower Power Design Brief

OBJECTIVE

One use of towers is to support electric power lines. There are many other uses for towers, too. Often, tall towers are made of trussed construction. The trussing makes the tower strong and stable. In this activity, you will design and build a tower and test its strength.

PROBLEM

Design and construct a trussed tower structure that will be stronger than any other structure in the class.

GUIDELINES

1. The base of the tower must fit inside an 8″ square.
2. The tower must be a minimum of 6″ in height and a maximum of 8″ in height.
3. The tower must have an opening down through its center that will permit the tower to be placed over a 1″ diameter rod. This is required for testing.
4. Your teacher may limit the amount of glue you can use.
5. Your teacher will set a time limit for completion of the design and construction.
6. A sketch of the tower, showing the front view, is required.
7. You are limited to the materials identified in the Equipment and Supplies list.
8. Your tower *must* have a flat top for testing.

EQUIPMENT AND SUPPLIES

$\frac{3}{32}$″ square balsa wood—26 total linear inches
$\frac{1}{8}$″ square balsa wood—46 total linear inches
$\frac{1}{4}$″square balsa wood—16 total linear inches
Wood glue or hot glue guns and sticks
Utility knife
Straight pins
Grid paper and pencils

PROCEDURE

1. Work in groups of two to design your tower. Study other tower designs for ideas.
2. Construct your tower accurately. Good solid joints are important.
3. Your teacher will conduct the testing. Your tower will be placed in a Tower Testing Device, Figure 1, and crushed.
4. Watch your tower carefully while it is being tested. Try to determine which part fails first. CAUTION: Be sure to follow safety rules.
5. The tower holding the most weight will be the winner.

MATH AND SCIENCE CONNECTIONS

Steel electrical transmission towers are often located in remote areas. It can be difficult to properly maintain the steel by painting to prevent rust. Through scientific advances, a new steel, called COR-TEN™ has replaced the need for painting. COR-TEN™ is not painted. Rust is allowed to form on the outer surface. The rust creates a thin protective covering over the steel that prevents further rusting.

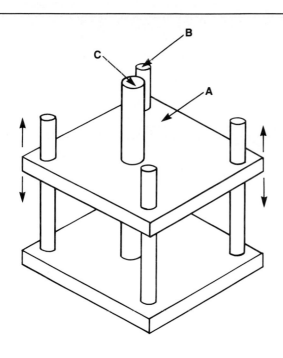

PARTS LIST

A   TOP/BOTTOM ¾T x 12 L x 12 W
B   4 CORNERS ¾ DIA x 12 L
C   CENTER POST 1″ DIA x 12 L

NOTES: 1. All five posts are secured to bottom
2. The top should be free to move up and down
3. Minimum 8-inch square required between 4 corners
4. To test towers:
   a. Remove top
   b. Place tower over center post
   c. Place top on tower
   d. Add weights over center post slowly until tower fails

RELATED QUESTIONS

1. What caused your tower to fail? Which part failed first?
2. What type of design held the most weight? Why do you think it was successful?
3. If your tower was not the strongest, what could you do to improve the design?
4. Why are the tallest electrical transmission towers made of trussed steel while smaller ones are made of wooden poles?

Paper Tower Design Brief #1

Contributed by Neil Eshelman, Williamsport High School, Williamsport, PA

PROBLEM

Design and construct a self-supporting tower as tall as possible.

GUIDELINES

1. You may use no more than two sheets of 8½″ x 11″ paper and 12″ of ¼″ wide masking tape.
2. Work in groups of two. Use several sheets of paper to test out some of your ideas with prototype towers.
3. Sketch your best design idea. Bring it to class on the competition day.
4. On competition day, you will have a time limit of 30 minutes for building.
5. Your tower must be free standing.
6. Your tower must remain standing for 30 seconds during testing.
7. Your teacher will measure the total height of all towers.

MATH AND SCIENCE CONNECTIONS

Gravity is the force that tries to knock over tall towers. Often tall towers will have strong wires attached to them that run to the ground. These wires, called guy wires, help to keep the tower in an upright position and resist gravity.

Paper Tower Design Brief #2

PROBLEM

Design and construct a self-supporting tower that will be lightweight and have a combination of strength and height, Figure 1.

GUIDELINES

1. You may use as many sheets of 8½ ″ x 11 ″ paper and as much ¼ ″ wide masking tape as needed.

2. Work in groups of two. Sketch several designs and discuss which is best.

3. Your tower will be tested using the formula described in the MATH AND SCIENCE CONNECTIONS section.

4. Your tower must support a weight on its top. You should design a flat top. You will decide how much weight the tower will hold.

5. Your tower must support the weight for 20 seconds without collapsing.

6. The base of your tower must be no larger than 20 percent of the total height of the tower. For example, the base of a 30-inch tall tower could be no larger than six inches (30 inches x .20 = 6 inches).

7. Your teacher will limit the amount of time for this activity.

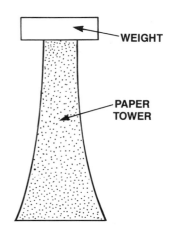

FIGURE 1.
Design a lightweight, tall, strong tower.

MATH AND SCIENCE CONNECTIONS

Tower testing following formula:

$$\frac{\text{WEIGHT HELD X HEIGHT}}{\text{TOWER WEIGHT}}$$

The tower that produces the highest number using this formula will be the best tower. You must consider a combination of the three factors in the formula.

TECHNOLOGY STUDENT ASSOCIATION

The Technology Student Association (TSA) is a national organization for students in technology education (industrial arts, industrial technology, industrial education). Technology students in middle schools, junior high schools, and senior high schools can join TSA. There is even an association similar to TSA available for college students.

TSA begins in the classroom and lab. Students can organize a TSA chapter, elect officers, hold meetings, and plan school, community, and social activities. Students can meet with other TSA chapters at state and national conferences. Most states in the United States are affiliated with the national TSA organization. The national conference is held each summer in a different city. State conferences are usually held in the spring. At both the state and national conferences, students can meet and compete with other TSA students.

These students are observing a computer-controlled lathe used in manufacturing. (Photo by Arvid Van Dyke)

Purpose and Goals of TSA

The main purpose of TSA is to prepare students to live and work in our technological society. The TSA motto is "Learning to live in a technical world." The specific goals of TSA include the following:

Leadership. TSA promotes leadership by asking students to plan and run club meetings, competitive events, and school, community, and club activities.

Technology. TSA helps students understand the impacts of technology on their future and prepares them to function in a society dominated by technology.

Knowledge. TSA promotes the development of consumer and career knowledge. Also, many of the competitive events in TSA give students opportunities for in-depth study in technology, the development of craftsmanship and creativity, and the development of interests in leisure and recreational activities related to technology.

Respect. TSA promotes respect for the dignity of work, for quality craftsmanship, and for the worth of individuals as they contribute to group efforts.

Motivation. TSA motivates students to excel in their scholastic efforts, to be the best they can be, and to prepare for the challenges of living in our technological world.

Choices. TSA gives students the information they need to help them make meaningful and informed decisions about their future educational and occupational choices.

Transition. TSA helps students make the transition from the classroom to the community as workers, consumers, and involved citizens.

A student's entry in a TSA Technology Processes Contest. *(Photo by Arvid Van Dyke)*

These students are using a robot to simulate a manufacturing procedure. *(Photo by Arvid Van Dyke)*

Production-Related TSA Activities

Many TSA chapters are involved in activities in the school and community. These activities can be related to production. TSA chapters organize tours of local manufacturing plants or invite people from industry to speak at local or regional meetings. Your TSA chapter could provide a service to the school and raise money for club activities by mass-producing school products, such as bookcases, desk sets, or products that display the school mascot. During Technology Education Week or a school's open house, TSA students can demonstrate a mass-production line for other students, teachers, administrators, and parents.

Some TSA chapters get involved in civic and community activities. In November and December, TSA students mass-produce toys for children. In the spring, TSA students build benches or garbage cans for a local community park. Some TSA students even join in community clean-up activities and take aluminum cans or newspapers to recycling centers.

Of course, many of the competitive events held at regional, state, and national levels can be related to production. Speeches, research papers, technology process display, and projects can focus on high-interest topics in production. Many states even have contests where students compete for the best mass-production run or enterprise activity.

If your school does not have a TSA chapter, organize one. If there is a TSA chapter in your school, join and become an active participant in an organization dedicated to preparing students to live and work in our technological society. A good way to start is by competing in TSA contests. Try the bridge design contest described below.

Design A Bridge Contest

Statement of the Problem

Design a bridge to withstand a certain load in accordance with the following engineering specifications.

Engineering Specifications

1. Bridges are to be constructed from no more than thirty (30) feet of 3-32″ square balsa wood and model airplane type of cement. While glue may be used to laminate the wood, it cannot be used to coat the entire piece. Paint may be used if desired. Contestants must supply their own materials. (Use of any other than the type of specified material will disqualify the contestant.)

2. Bridge must be free standing and be of the following dimensions:
   a. total length — 16″
   b. total height — 10″
   c. total width — 2″
   d. maximum height above water — 5″
   e. the stream is 10″ wide.

3. Bridges cannot be fastened to any common base. They will be tested by placing a standard 3½″ x 7⅞″ brick (approx. 3 lb) on the center of the span. The bridge should be able to support the total weight as specified below.

4. The event will be judged in accordance with the following parameters:
   a. Function — bridge must withstand the full load for at least 15 seconds.
   b. Form — bridge must be aesthetically attractive.
   c. Specifications — bridges must conform to all engineering specifications outlined above.

The Winner

The winning bridges will be placed on display in a suitable location in the school.

Abutments — Foundations on the ends of a bridge.

Acid rain — Rainwater with a pH lower than 5.6.

Acrylic latex — Common name for water-based paints.

Adhesive — A type of bonding process that uses glue, cement, or some other tacky substance to cause materials to stick together at a joint area.

Advertising — Making consumers aware of the products they can buy.

Advertising campaign — Methods used to advertise a product.

AGV — Automated Guidance Vehicle; a robotic materials handling system.

Air pollution — Occurs whenever fuel or other materials are burned.

Alkyd-resin — Another name for oil-based paints.

Alloy — A combination of two or more materials.

Annealing — A thermal conditioning process that makes materials softer.

Anthropomorphic — Having human characteristics.

Applied research — The search for practical uses for the new ideas found through pure research.

Apprenticeship program — Training new workers for construction jobs.

Appropriate technology — Choosing and using technologies that are ''best'' suited for a particular application.

Arc welders — Use an electric arc to weld metals.

Architect — A person who designs the appearance of a structure.

Articles of incorporation — Application for incorporation filed with the state.

Assembly drawing — A drawing that shows what the final product will look like when it is assembled.

Assets — The things a company owns.

Automated beam builder — Machine that will make beams for space structures.

Automation — When most or all of the machines and processes run with little or no human control. To perform work without the aid of people.

Balance sheet — Reports the assets and liabilities of the company.

Bar chart — A technique used to schedule materials and equipment for a construction project.

Bartering — Another word for trading.

Batch manufacturing — Another term for intermittent manufacturing.

Batter boards — Erected at each corner of a construction site several feet outside the building lines to help workers find the corners during excavation.

Bearing-wall construction — When solid walls support the structure.

Bedrock — The earth's hard rock crust.

Bid — An offer to build a structure for a certain amount of money.

Bill of materials — A list of all the parts, components, and hardware needed to make the product.

Biopesticides — Biogenetically engineered pesticides that kill pests, but do not harm people.

Blooming mill — Where metal ingots are rolled into blooms.

Board foot — A board measuring one inch thick, twelve inches wide, and one foot long.

Board of directors — A group of people who set policies and goals for a corporation.

Bond — In construction, an arrangement of bricks, such as a stacked bond or a running bond.

Bonding — A process used to combine solid materials.

Bonds — Money loaned to a company by a group of people instead of a bank.

Bottlenecks — A point in a production line where parts back up and are delayed.

Bottom chords — Bottom members of a roof truss.

Bottom line — Net income shown on an income statement.

Box beam — A girder made up of a lumber frame covered with plywood.

Box nails — Flat-headed, thin-shanked nails.

Brainstorming — When groups of people talk about their ideas. One method of creating a list of possible solutions to a problem.

Break-even analysis — A report used to calculate the break-even point.

Budget — A plan or guess of the amount of money that will be needed.

Builder's level — Used to check levelness between two points.

Building codes — Laws that control construction.

Building permit — A permit giving the builder permission to construct a structure.

Bylaws — Specific rules used to guide management in operating a corporation.

CAD — Computer-aided drawing.

CADD — Computer-aided drawing and design.

CAD/CAM — Computer-aided drawing/computer-aided manufacturing.

CAM — Computer-aided manufacturing; controlling several processes at one time with computers.

Casing — Interior and exterior molding used around window and door frames.

Casting — A forming process where liquid materials are cast (poured) into a cavity.

Catalyst — A material that causes chemical reactions to occur.

Ceramics — Compounds of oxygen and silicates. Pottery, china, porcelain, and glass are examples of ceramic materials.

CEO — Chief Executive Officer of a company.

Chalk line — A layout tool. Uses chalk to mark a long straight line.

CIM — Computer-integrated manufacturing.

Civil construction — Any field of construction that is closely related to earthwork.

Clamps — Tools used to hold parts during processing.

Claw hammer — Used to force nails into wood.

Coating — A process that places one or more layers of one material on top of another material.

Cohesive — A type of bonding process where heat, pressure, chemicals, or combinations of these are used to cause molecules of materials to cross a joint area and permanently mix or cohere.

Collar beam — Beams placed between roof rafters for extra support.

Combining — Process used to add one part to another.

Combining tools — Add one part to another.

Common nails — Flat-headed, smooth-shanked nails.

Communication technology — The study of sending and receiving information.

Composites — Combinations of two or more materials.

Compressing — A process that squeezes materials into desired shapes.

Compression strength — Allows a material to resist crushing.

Computer — A machine that can be programmed to accept, access, and display information.

Computer simulation — Designing and testing product ideas using computers.

Conditioning — Processes that change the inside structure of materials.

Conduit — Metal or plastic tubing with wires running inside of it.

Conservation — Making wise use of available materials.

Construction specifications — Written documents that give detailed information not shown on the working drawings.

Construction technology — The building of structures that cannot be moved.

Continuous manufacturing — Making a large number of one product using mass production and unskilled workers.

Contour lines — Lines on topographical drawings that show the elevation above sea level.

Contract — A legal agreement between the contractor and the people paying for the structure.

Contractor — Plans the building of a project according to a written agreement.

Cornice — Trim at the edge of a roof.

Cottage industry — A small production plant set up in the home (cottage).

Countersink — Used to taper the pilot hole for flat-head screws.

CSI Format — Construction Specification Institute Format for specification writing.

Custom manufacturing — One person making one product by hand, or making one-of-a-kind products using highly skilled workers.

Cut — When soil is removed to create the desired grade.

Dead load — The weight of all the parts of the structure.

Debugging — The process of tuning up a production line so that it will run smoothly.

Deck — The flat surface made by a floor in a floor frame.

Degrees of freedom — Number of ways a robot arm can move.

Department — A small part of a company.

Deployables — Structures that will practically build themselves, like umbrellas.

Design engineering — The process of creating product ideas.

Design standards — Rules used to guide product designers.

Designer — A person who creates new ideas for products.

Detail drawings — Multiview drawings of all the parts of a product or structure. Provide specific details on how certain parts of a product or structure should be built.

Development — A type of applied research that tries to make new products.

Digester — A furnace used to burn materials that cannot be recycled.

Directing — Guiding workers.

Dividends — The money paid to stockholders out of the company's profits.

Dividers — Tools used to lay out circles and arcs.

Division of Labor — When each worker performs a different material process.

Drafters — People who make working drawings and plans for structures.

Drafting — Using tools to do technical drawings.

Drying — A thermal conditioning process that removes moisture from materials.

Ductility — Allows a material to be drawn into a wire shape.

Ducts — Sheet metal or plastic pipes that supply cool or warm air to rooms.

EBIT — Earnings Before Interest and Taxes.

Effluent — Fouled water and solid waste.

Elasticity — Allows a material to return to its original shape after being changed.

Elevations — Drawings of a structure as it is seen from its different sides.

Embankment — Ramp to an elevated structure such as a bridge.

Eminent domain — The right of the government to buy land for public uses and to pay a fair price for that land.

Energy and power technology — Provides the muscle needed to communicate information, transport things, construct structures, and manufacture products.

Enterprise — A business or company.

Environmental impact statement — A report of all important aspects of the environment where construction will take place.

Equal Employment Opportunity Act — Laws that protect the rights of workers and people looking for work.

Estimate — A carefully made calculation of costs.

Estimating — Making a prediction.

Excavation — Digging the soil with heavy earthworking equipment.

Exhaustible — Raw material resources found in the earth that are not living and cannot be renewed.

Experimental studies — Market research that experiments to determine consumer reaction to new products.

Exponential rate of growth — An ever-increasing rate of change.

Fascia — Cornice trim nailed to the ends of the rafters.

Fascia header — Lumber nailed to the overhang end of roof rafters.

Fatigue strength — Allows a material to resist failure after repeated forces in opposite directions.

Feedback — The final step in the system model.

Ferrous — Metals which contain iron.

Fiber-reinforced composite — Material composed of fibers used to reinforce a base of plastic.

Fill — When soil is added to fill low spots.

Finance — Another word for money.

Finish plumbing — All plumbing fixtures such as sinks, lavatories, and toilets.

Finishing nails — Nails with very small heads and thin shanks.

Firing — A thermal conditioning process. Involves heating clay-based ceramic materials to make them hard.

Fixed costs — Costs that stay the same, or fixed, for a company for a certain time period.

Fixed-position layout — A plant layout where the product remains in a fixed position and parts are brought to the product.

Fixtures — Attach to a machine and position parts being processed.

Flexible manufacturing — Making many versions of the same product.

Float — A tool used to smooth out the unevenness in the surface of freshly placed concrete.

Flow process chart — A chart that lists all the processes needed to make one part of a product.

Flux — Material used to make the solder flow all around the pieces of pipe that are being connected.

Footing — A strip of concrete placed around the perimeter of the structure.

Forecasting — Planning the future.

Form — The size and shape of a product or structure.

Forming — Changing the outside shape or inside structure of materials.

Forming tools — Used to change the outside shape or inside structure of materials.

Frieze — A horizontal piece on a cornice that is placed against the wall of a building.

Frost line — The depth to which the earth freezes.

Function — How a structure will be used.

Futurists — People who study the future.

Gable roof — A type of roof with two sloping sides that meet at the ridge board. The gable is the triangle formed between the wall and the roof at the ends of the house.

Gambrel roof — A roof type similar to the gable roof, but with a more gradual top slope.

General contractor — A contractor who uses subcontractors.

Geodesic dome — Structure made of a series of triangles placed on the surface of a sphere.

Girders — Beams that run the length of the building and support joists.

Global competition — Competition from countries from around the world.

Go–no-go gage — Used during quality control inspections in manufacturing.

Greenhouse effect — Warming of the earth caused by excess carbon dioxide in the atmosphere. Similar to solar heating by sunlight entering through glass.

Gross pay — Hourly wage rate multiplied by the number of hours worked.

Grounded — Electrical wiring connected to a metal rod that is driven into the earth.

Grout — Mortarlike material used to fill joints between ceramic tiles.

Gussets — Plywood or metal plates nailed over the joints on a roof truss.

Gypsum — A natural rock taken from mines. Used to make plaster of paris.

Hand tool — The simplest form of tool. Powered only by human muscle.

Hard technology — High technology based on improving productivity and profits without considering the human, environmental and social impacts.

Hardening — A thermal conditioning process that makes materials harder.

Hardwoods — Woods that come from trees with broad leaves.

Hazardous waste — Includes materials that are explosive or highly reactive, flammable, corrosive, toxic, or poisonous.

Header — A band of lumber that runs around the top of the sill and ties together the joists, or a piece that transfers loads to the sides of a door or window structure.

Heap leach mining — Mining for gold, silver, and other minerals using percolating chemicals.

Heavy construction — The building of large structures such as roads, highways, bridges, tunnels, dams, power plants, and manufacturing factories.

Herbicides — Chemicals that kill weeds, but allow crops to grow.

Hierarchical order — A form of management structure.

Hinge gain — A recess made in a door edge and a door jamb to accept a hinge.

Hip roof — A roof that slopes on all four sides.

Hire — To give a person a job.

Historical studies — Market research of how well a previously-made product sold.

Human resources — Manufacturing department that finds and hires workers.

HVAC — Heating, ventilating, and air conditioning.

Impact — An effect on something or the result of something.

Impact strength — Allows a material to absorb energy during impacts.

Income statement — Reports the income and expenses for the company.

Industrial construction — Building facilities for processing or controlling materials, energy, or other resources.

Industrial Revolution — Refers to the changes in production that happened between the years 1750 and 1850.

Ingots — Slabs of metal.

Innovations — Improved technology.

Inputs — Everything that goes into starting a system.

Interest — Extra money paid above the amount of a loan.

Intermittent manufacturing — Making limited quantities of a product using skilled and semiskilled workers.

Inventions — New technology.

Investors — People who give money to an enterprise.

Jack studs — Extra studs added below a windowsill for added strength.

Jambs — Top and side trim on a door.

Jigs — Guide the path of a tool on a part being processed.

JIT — Just-in-time manufacturing.

Job application — A form filled out by people interested in a job.

Job description — A written statement of the experience, skills, education, and training needed for a job.

Job lot — A quantity of products made in one job using intermittent manufacturing.

Joint compound — Plaster material used to smooth joints between wallboard.

Joists — Closely spaced floor beams that rest on the foundation walls or girders.

Journeyman — A skilled worker who has finished an apprenticeship.

Just-in-time manufacturing — Making and delivering products *just in time* to be sold. Abbreviated JIT.

Kanban — Japanese inventory system developed by Toyota. Used in JIT.

Laborers — Unskilled construction workers.

Landscape plan — Working drawing of the design for a landscape.

Laser — Light Amplification by Stimulated Emissions of Radiation.

Lath — A wall covering similar to wallboard. Also, wire netting used with stucco.

Layered composite — Two or more materials sandwiched together by an adhesive or other binder.

Levitators — A system for processing materials without containers.

Liabilities — What a company owes other companies and businesses.

Life cycle — The number of months or years the product is expected to be wanted by consumers.

Light construction — The building of houses, smaller apartment buildings, and smaller commercial and business buildings.

Live load — The weight of people, furniture, and rain and snow.

Load-bearing walls — Walls that support the weight of the structure.

Loan — Money given to a company by a bank on a temporary basis.

Lot — A certain number of products made with intermittent manufacturing.

Machines — Powered tools that stay still during manufacturing.

Maintenance — Keeping a structure in good working condition.

Malleability — Allows a material to be pounded, rolled, and formed into sheets.

Management — Reaching goals through the work of others.

Management department — Department in manufacturing that makes sure all the other departments work together.

Mansard roof — A roof similar to the hip roof.

Manufacturing engineer — An engineer who plans the layout of tools, machines, and people on a production line.

Manufacturing technology — The making of products in a factory.

Mapping — Forecasting method that draws maps or diagrams.

Market research — Studying consumers to determine their product needs and wants.

Marketing — Department that advertises and sells the product, or the process of actually selling the product.

Mass media — An advertising method using media that reach large numbers of people, like television, radio, and newspapers.

Mass production — Making a large number (a mass) of the same product at one time.

Materials flowchart — A drawing of the plant layout.

Materials handling — The transportation system used to move materials and parts through the plant.

Measuring line — An imaginary line along the center of the rafter. Used to determine rafter length.

Mechanical advantage — An increase in a force.

Mechanical fastening — A process that uses threaded and nonthreaded fasteners to combine materials.

Methods engineer — A person who plans the sequence of processes needed to make parts and assemble a finished product.

Microgravity — The almost nonexistent gravity of space.

Micrometer — Used for accurate measurements to a thousandth of an inch on precision parts.

Microspheres — Tiny plastic beads manufactured in space.

Mock-up — A scale model of the finished product.

Modeling — Making a replica of the finished object to scale.

Mold — A cavity used for forming liquid materials.

Molding — In manufacturing, a forming process in which a liquid or semiliquid material is forced into a cavity. In construction, refers to wood trim for walls, doors, and windows.

Mortar — Material made of portland cement, hydrated lime, and sand mixed with water. Used to bond bricks and blocks.

Mortgage — Guarantees a loan by permitting the lender to assume ownership of the real estate if the payments are not made.

National Electrical Code — Specifies the design of safe electrical systems.

Net pay — Gross pay minus deductions.

Neutral buoyancy tank — Underwater building tank used to simulate microgravity.

Nonferrous — Metals which contain little or no iron.

Operation process chart — A chart listing all the processes needed to make one product.

Operation sheet — A list that details the work to be done at each workstation.

Organizing — Trying to get teamwork from employees.

Outputs — The results of the processes in a system.

Particle composites — Material made of small bits of reinforcement in a base.

Partitions — Walls that support only their own weight.

Partnership — A business with two or more owners who share responsibilities.

Penny size — Used to specify nail size.

Percolation test — Used to check drainage of soil.

Pesticide — Chemical designed to kill pests.

Photovoltaic — Panels that convert the light energy of the sun into electrical current.

Pictorial drawings — Drawings that represent how an object looks to the eye.

Piers — Columns of concrete used as a foundation.

Pigments — The coloring materials in paint.

Pile — A column of steel or wood used as a foundation.

Pilot run — A practice session when the production line is tested to make sure all systems are working.

Pitch — The steepness of the roof.

Plan views — Drawings that look down from above a structure.

Planning — Deciding what to do and how to do it.

Plant layout — The way tools, machines, and other workstations are arranged.

Plaster — A gypsum product used for finishing interior walls.

Plastic laminate — A thin sheet of plastic that protects the wood used in making the cabinet.

Plasticity — Allows a material to be easily formed into various shapes and to retain the shape it is formed into.

Pliers — Hand tools used to hold small parts.

Plot plan — A drawing that shows the location of a structure on the site.

Plumb bob — A pointed weight attached to a string. Used to check verticalness.

Pneumatic — Air-powered.

Pneumatic structures — Structures with a plastic sheet shell that is supported by air pressure.

Pollution — Any harmful changes in air, water, or land caused by scrap or waste.

Polymerization — Chemical reaction used to make plastics.

Polymers — Wood and plastic materials.

Pop rivet guns — Tools used to set pop rivets.

Portland cement — An ingredient in concrete. Made mainly of clay and limestone or chalk.

Power hand tools — Improved hand tools with an external power source.

Prestressed concrete beams — Beams cast with tightened steel cables inside them.

Primary manufacturing — The process of converting raw materials into materials for production.

Primer — A special paint that sticks to the surface better than regular paint.

Problem solving — Finding solutions to problems or finding better ways of doing something.

Process layout — A plant layout where tools and machines used for similar processes are grouped together.

Processes — The actions that occur to make the system work safely and properly.

Product layout — A plant layout with a smooth flow of materials from start to finish.

Product planning — Making sure new products are added, old ones are removed, and questionable ones are changed.

Product service — Making sure the product works correctly and that the consumer knows how to use the product.

Production engineering — Planning which manufacturing system will be used to make the product.

Production technology — The study of how products and structures are made.

Productivity — The number of products made per worker hour.

Profit — The money left over after a product is sold and all expenses have been paid.

Prototype — The first full-size working model of the finished product.

Purchase order — An order giving a vendor permission to send materials to the company.

Purchasing — Obtaining something in exchange for money.

Pure research — The search for new ideas without thinking of any real uses for the ideas; also called basic research.

Pylons — Foundation supports under the center of a bridge.

QC circle — A group of workers responsible for controlling product quality.

Quality control — Controlling product quality.

Quality control engineer — An engineer who makes sure the finished product matches the working drawings.

R value — The resistance of a material to the flow of heat.

Rafters — Roof framing members that extend from the wall plates to the ridge board.

Rebar — Steel rods used to reinforce concrete.

Reclamation — Process used to make contaminated industrial materials (such as waste water) safe to put back into the environment.

Recycling — The process of reusing scrap materials.

Recycling programs — Aid consumers in the recycling and reusing of materials.

Remodeling — Changing the inside or outside of a structure to improve its usefulness or appearance.

Rendering — A sketch that shows the final product with all its details.

Renewable — Raw material resources that are living and can be renewed.

Resources — Raw materials, energy sources, and water.

Ridge board — A board nailed between the rafters at the top of a roof.

Rise — The height from the top of the wall plates to the top of the roof.

Robot — A reprogrammable multifunctional manipulator designed to move material, parts, tools, or other specialized devices through variable programmed motions for the performance of a variety of tasks.

Rolls — Cylindrical tools that form metal into shapes.

Rough plumbing — The installation of main supply lines, main sewer lines, and all branch plumbing.

Rough sketches — Small sketches of product ideas; also called thumbnail sketches.

Run — The width covered by one rafter.

Sales forecasts — Guesses or predictions about how many products will be sold in the future.

Salespeople — workers who find ways to get people to buy the product.

Sash — Glass in a window and the frame that holds it.

Scenario — A make-up story about the future.

Scrap — Output of a process that cannot easily be reused by the company that makes it.

Screeding — Using a straightedge to push excess concrete into low spots.

Secondary manufacturing — Using manufacturing processes to convert standard industrial materials into finished products.

Section views — Show the inside elevation of a structure.

Semiskilled — Workers who run machines and use the special tools made by skilled workers.

Separating tools — Used to remove excess material.

Service panel — A box in the house which is connected to the electrical service line outside.

Setback — The distance from the street to the front of the building.

Sheathing — The first layer of wall covering on the outside of the walls.

Shed roof — A simple sloped roof with no ridge.

Shellac — Clear finish that produces a very fine finish on wood.

Silicates — One of the most common materials on earth. Found in sands and clays.

Sill — A floor wood member that rests directly on top of the foundation, or a horizontal bottom member of a rough window opening.

Simulation — Forecasting technique, such as role-playing, modeling, and gaming.

Skeleton-frame construction — When walls are made of individual members with space between them.

Sketching — Drawing objects by hand.

Skilled — Workers who design and make the special tools for manufacturing.

Slab-on-grade — Using the concrete-slab first floor as the foundation of a building.

Slump test — A test of the quality of a concrete mixture.

Sod — A blanket of existing grass that is grown on a farm and transplanted.

Soffit — The covering on the underside of the rafters on a cornice.

Soft technology — Also called appropriate technology.

Software — A set of coded instructions written to control the operations of a computer.

Softwoods — Woods that come from cone-bearing trees with needles.

Solar orientation — The position of a building in relation to the sun, wind, trees, and other structures.

Solder — A soft metal made by combining tin and lead; used to join piping.

Soldering gun — A tool used for electric soldering.

Sole proprietorship — A business whose owner and operator are the same person.

Span — The total width covered by the rafters.

Spirit level — Uses an air bubble in a fluid to measure level and plumb.

Spot welders — Used to weld two pieces of metal together on one spot.

Squares — Tools used to measure 90-degree (square) angles.

Standard stock — Materials in standard forms or shapes.

Stock — Certificates people buy to gain partial ownership of a company.

Stockholders — People who buy shares of stock in a corporation.

Stretching — A process that pulls materials into desired shapes.

Structures — Buildings, roads, bridges, and pipelines are examples of structures built with construction technology.

Subcontract — To hire another company or group to do part of the work.

Subcontractors — Hire workers who specialize in certain construction trades.

Subfloor — The first layer of material applied over the floor joists.

Substructure — The foundation of a structure.

Subsystems — Smaller systems within a system.

Superalloy — Superior alloys that can be made in space, but not on earth.

Superstructure — The part of the structure above the foundation (substructure).

Supervisors — Managers who make sure workers follow rules made by the upper-level managers.

Survey — A set of questions that asks consumers what they think of a product.

Synthetic — Human-made.

System — A group of parts that work together to do some task or job.

System model — Breaking down complex systems into inputs, processes, outputs, and feedback.

Tail — The portion of the rafter that extends from the wall outward to create overhang.

Tape measures — Rules with flexible steel blades.

Technician — A skilled worker.

Technocracy — A society controlled by technocrats.

Technocrats — People who control other citizens with their knowledge of technology.

Technologist — A person who studies technology.

Technology — The use of tools, materials, and processes to meet human needs and wants.

Technopeasants — People who are technologically illiterate.

Tempered glass — Glass that is three to five times stronger than regular glass.

Tensile strength — Allows a material to resist being pulled apart.

Thermoplastics — Plastics that can be reheated and reshaped.

Thermoset — Material that cannot be reheated and reshaped.

Thermostat — An automatic switch used to sense the temperature in the house.

Thumbnail sketches — Small sketches of product ideas; also called rough sketches.

Tilt-up construction — Reinforced concrete panels cast on site and tilted up for walls.

Tooling — Special tools and devices used by production workers.

Tooling up — Putting together the workers, plant layout, tooling, and materials handling systems.

Tools — Extend human abilities in doing the work of processing (changing) materials or information.

Top chords — Top members of a roof truss.

Topographical drawings — Drawings of the rise and fall of the land.

Transfer station — Place where garbage is separated from reusable materials.

Transportation technology — Moving people and things using vehicles.

Trap — A curved area of plumbing that fills with water to prevent sewer gases from entering the house.

Trend — A pattern in recent events in society.

Trend analysis — Studying recent patterns.

Trial and error — A method of problem solving; involves trying something and sometimes making mistakes.

Trimmer — An extra stud that supports a header and strengthens the full stud.

Trowel — A tool used to produce a smooth surface on concrete.

Underlayment — A second layer of plywood laid over the subfloor.

Unemployment — When workers are displaced from their jobs.

Unit costs — The costs to manufacture each unit (product).

Unskilled — Workers who do hard, routine jobs.

Utilities — Service required in homes, such as water, sewage, and electricity.

Variable costs — Costs that vary — that is, that go up or down — with the number of products made.

Vehicle — The liquid in which the pigment and other ingredients are mixed to make paint.

Vendor — A supplier or seller of materials.

Veneer — Thin layers of wood. Used to make plywood.

Vent — An opening that allows air pressure to enter the plumbing system and break the suction of the trap.

Wallboard — A fireproof wall covering made from gypsum plaster.

Waste — Any material that is thrown away and never reused or recycled.

Webs — Braces between top and bottom chords on a roof truss.

Wind bracing — Diagonal members added to skeleton-frame construction.

Work envelope — The area in which the robot can reach and work.

Work hardening — Another name for mechanical conditioning processes.

Working drawings — A set of drawings and plans for a product.

Workstation — The place in a production line, such as machines and benches, where operations occur.

Zoning boards — Groups that study the needs of the community for new structures.

INDEX